T0180199

COVID-19

Moones Rahmandoust · Seyed-Omid Ranaei-Siadat
Editors

COVID-19

Science to Social Impact

 Springer

Editors
Moones Rahmandoust
Protein Research Center
Shahid Beheshti University
Tehran, Iran

Seyed-Omid Ranaei-Siadat
Protein Research Center
Shahid Beheshti University
Tehran, Iran

ISBN 978-981-16-3110-8   ISBN 978-981-16-3108-5   (eBook)
https://doi.org/10.1007/978-981-16-3108-5

This Springer imprint is published by the registered company Springer Nature Singapore Pte Ltd.
The registered company address is: 152 Beach Road, #21-01/04 Gateway East, Singapore 189721,
Singapore

Preface

Public health and world economy have always been put to challenges by the emergence of various pathogen outbreaks, including viral infections. Coronavirus disease 2019 (COVID-19), as the most recent infection, which originated from severe acute respiratory syndrome Coronavirus 2 (SARS-CoV-2) is affecting numerous people in many countries. Ever since its initial discovery in late 2019, there has been a monumental increase in morbidity and mortality rate around the world. It has caused various disorders and complications, propelling scientists and pharmaceutical companies to develop practical vaccines and employ the drug repurposing approach to inhibit the disaster. Thus far, since there was no well-known treatment to cure COVID-19 patients, another challenge has also raised to rapidly and accurately detect the virus, to provide efficient isolation of those who are infected as a highly critical emergency.

"CoViD-19: Science to Social Impact", as the name implies, takes a comprehensive look at the different aspects of the science involved in the field of the emerging global crisis. It consists of seven chapters written by professional scholars in each field and will introduce and cover the state-of-the-art achievements on the subject worldwide.

In Chap. 1, a brief introduction is provided on the history of the virus and the various pathogenesis stages of the SARS-CoV-2 infection, taking the necessary biology of the virus into account. The ongoing clinical trials as well as their pertaining phase of approval are discussed in the Chap. 2, in detail, considering the potent therapeutics and medications with respect to their target, as to be the pathogenesis involved with both the direct harm inflicted by the virus and the excessive inflammatory and immune response from the host.

Further details on the structure of the virus and its various structural, nonstructural and accessory proteins are presented in Chap. 3 of the book. On account of the decisive roles of the proteases in polyprotein processing and viral replication, the main protease enzyme and its inhibitors are discussed more extensively in Chap. 4. Vaccinating against the coronavirus disease, as the most promising perspective to control the global crisis, is available through various scientific strategies which are introduced and discussed in Chap. 5 of this book.

As mentioned earlier, along with all the approaches taken to cure the emerging viral infection, the need for early, accurate, low-cost and simple diagnosis of COVID-19 is still a very important challenge to face. Therefore, Chaps. 6 and 7 extensively introduce common and novel strategies for identifying and monitoring the disease, as a necessity of isolating infected patients at an appropriate time, so as to break the viral transmission chain effectively.

In the last three chapters, the influence of the outbreak is studied from the social point of view. The effects and consequences of the pandemic on the development of some countries are investigated in Chap. 8. When the ninth chapter looks at the international relations, as they are affected by the world's pandemic crisis, Chap. 10 studies the influence of the emerging situation on Iran's diplomacy.

The editors of the book are very much grateful to all the scholars who have contributed scientifically and technologically in the war against the emerging worldwide crisis of COVID-19, with special thanks to the young scientists who have invested their knowledge and time to gather the presented manuscript most accurately, addressing the state-of-the-art achievements in the field.

Tehran, Iran   Moones Rahmandoust
                                                          Seyed-Omid Ranaei-Siadat

Contents

Contributors

Hossein Abolhassani Biotechnology Group, Faculty of Chemical Engineering, Tarbiat Modares University, Tehran, Iran

Sareh Arjmand Protein Research Center, Shahid Beheshti University, Tehran, Iran

Ghazal Bashiri Biotechnology Group, Faculty of Chemical Engineering, Tarbiat Modares University, Tehran, Iran

Fataneh Fatemi Protein Research Center, Shahid Beheshti University, Tehran, Iran

Abdulrahman Ghassemlou Protein Research Center, Shahid Beheshti University, Tehran, Iran

Behrad Ghiasi Protein Research Center, Shahid Beheshti University, Tehran, Iran

Samin Haghighi Poodeh Protein Research Center, Shahid Beheshti University, Tehran, Iran

Amir Mohammad Haji-Yousefi Faculty of Economics and Political Science, Shahid Beheshti University, Tehran, Iran

Zahra Hassani Nejad Institute of Biochemistry and Biophysics, University of Tehran, Tehran, Iran

Hasan Kouchakzadeh Protein Research Center, Shahid Beheshti University, Tehran, Iran

Heidarali Masoudi Faculty of Economics and Political Science, Shahid Beheshti University, Tehran, Iran

Mahdi Montazeri Protein Research Center, Shahid Beheshti University, Tehran, Iran

Moones Rahmandoust Protein Research Center, Shahid Beheshti University, Tehran, Iran

Shokouh Rezaei Protein Research Center, Shahid Beheshti University, Tehran, Iran

Mahmood Sariolghalam School of Economics and Political Science, Shahid Beheshti University, Tehran, Iran

Yahya Sefidbakht Protein Research Center, Shahid Beheshti University, Tehran, Iran

Seyed Abbas Shojaosadati Biotechnology Group, Faculty of Chemical Engineering, Tarbiat Modares University, Tehran, Iran

Seyed Ehsan Ranaei Siadat Rahpouyan Fanavar Sadegh Company, Pardis Technology Park, Tehran, Iran;
Sobhan Recombinant Protein, Tehran, Iran

Abbreviations

-1 PRF	Programmed -1 ribosomal frameshifting
3CLpro	3-chymotrypsin-like protease
ACE2	Angiotensin-Converting Enzyme II
ALI	Acute lung injury
AMPs	Antibody mimic proteins
ARDS	Acute respiratory distress syndrome
AuNIs	Gold nano-islands
BIP	Backward inner primer
CDC	Centers for disease control and prevention
cDNA	Complementary DNA
CLIA	Chemiluminescence immunoassay
CoV	Coronavirus
COVID-19	Coronavirus disease 2019
COX	Cyclooxygenase
CP	Convalescent plasma
CQ	Chloroquine
CRISPR	Clustered regularly interspaced short palindromic repeats
CT	Computed tomography
CTD	C-terminal domain
CXR	Chest radiograph
ddPCR	Droplet digital PCR
DETECTR	DNA endonuclease-targeted CRISPR trans reporter
DMV	Double-membrane vesicle
dsRNA	Double-strand RNA
E protein	Envelope protein
EC	Extracellular
EC50	Half-maximal effective concentration
ELISA	Enzyme-linked immunosorbent assay
ERGIC	Endoplasmic reticulum-Golgi intermediate compartment
ExoN	Exoribonuclease
FAM	Fluorescein amidite
FDA	Food and drug administration

FET	Field-effect transistor
FIP	Forward inner primer
FOP	Forward outer primer
FP	Fusion peptide
GIS	Gold nanoparticle immunochromatographic strip
GRAVY	Grand average of hydropathicity
HBV	Hepatitis B virus
HC	Heavy chains
HCoV	Human coronavirus
HCQ	Hydroxychloroquine
HCV	Hepatitis C virus
HE	Hemagglutinin esterase
Hel	Helicase
HIV	Human immunodeficiency virus
HMOX1	Heme oxygenase-1
HR	Heptad repeat
IFN	Interferon
Ig	Immunoglobins
IL	Interleukin
ISG15	Interferon-stimulated gene 15
IVIG	Intravenous immunoglobulin
JAK-STAT	Janus kinase signal transducer and activator of transcription
LAMP	Loop-mediated isothermal amplification
LBDD	Ligand-based drug discovery
LC	Light chains
LFIA	Lateral flow immunoassay
LMWH	Low molecular weight heparin
LSPR	Localized surface plasmon resonance
LUS	Lung ultrasound
M protein	Membrane protein
mAb	Monoclonal antibody
MD	Molecular dynamics
MERS-CoV	Middle East respiratory syndrome coronavirus
Mpro	Main protease
MTase	Methyltransferase
mTORC1	Mammalian target of rapamycin complex 1
MW	Molecular weight
N	Nucleocapsid
N protein	Nucleocapsid protein
NCBI	National center for biotechnology information
nCoV-2019	Novel Coronavirus 2019
NGS	Next-generation sequencing
NK	Natural killer
NO	Nitric oxide
NSAID	Non-steroidal anti-inflammatory drug

NSP	Non-structural protein
NTD	N-terminal domain
NTP	Nucleoside triphosphate
NTPase	Nucleoside triphosphate hydrolase
ORF	Open reading frames
PCR	Polymerase chain reaction
PLpro	Papain like protease
PPT	Plasmonic photothermal
RBD	Receptor binding domain
RBM	Receptor binding motif
RdRp	RNA-dependent RNA polymerase
RDT	Rapid diagnostic test
rhACE2	Recombinant human angiotensin-converting enzyme 2
rhIFN	Recombinant human interferon
rhmAb	Recombinant humanized monoclonal antibody
RPA	Recombinase polymerase amplification
RTC	Replicase-transcriptase complex
RT-RAA	Reverse transcript recombinase-aided amplification
S protein	Spike protein
SAM	S-adenosylmethionine
SARS	Severe acute respiratory syndrome
SARS-CoV	Severe acute respiratory syndrome coronavirus
SBDD	Structure-based drug discovery
SELEX	Systematic evolution of ligands by exponential enrichment
SHERLOCK	Specific high-sensitivity enzymatic reporter unlocking
SSBP	Single-stranded DNA-binding protein
ssRNA	Single-strand RNA
TM	Transmembrane
TMD	Transmembrane domain
TMPRSS2	Transmembrane protease
TNF-α	Tumor necrosis factor-α
UBL	Ubiquitin-like modifier
USP	Ubiquitin specific protease
UTR	Untranslated region
VEGF	Vascular endothelial growth factor
VLP	Virus-like particle
WHO	World Health Organization
ZBD	Zinc binding domain

List of Figures

List of Tables

Chapter 1
Introduction to the Virus and Its Infection Stages

**Hossein Abolhassani, Ghazal Bashiri, Mahdi Montazeri,
Hasan Kouchakzadeh, Seyed Abbas Shojaosadati,
and Seyed Ehsan Ranaei Siadat**

1.1 Introduction

The public health and world economy have always been put to challenges by the emergence of a pathogen outbreak. Viral infection has been considered one of the major causes of morbidity and death in the world [1, 2]. Since the beginning of this century, three zoonotic outbreaks caused by coronaviruses (CoVs) have arisen. The diversity of CoVs originates from their capacity for mutation and recombination during replication [3, 4]. In 2003, the first identified severe disease caused by the severe acute respiratory syndrome coronavirus 1 (SARS-CoV-1) in China, infected 8422 people mostly in China and Hong Kong, and caused 916 deaths (with a mortality rate of approximately 11%) before being restrained. Almost a decade later, the second outbreak of severe infection arose in 2012 in Saudi Arabia with Middle East respiratory syndrome coronavirus (MERS-CoV), which affected 2494 people and caused 858 deaths (with a mortality rate of approximately 34%) [5, 6]. The third outbreak of severe infection caused by severe acute respiratory syndrome coronavirus 2 (SARS-CoV-2) was first documented in December 2019 in Wuhan city, China [7]. SARS-CoV-2, a novel class of the *Coronaviridae* family closely related to SARS coronavirus 1 (SARS-CoV-1) (with approximately 80% sequence identity), causes many different disorders in patients and can generate complications

H. Abolhassani · G. Bashiri · S. A. Shojaosadati (✉)
Biotechnology Group, Faculty of Chemical Engineering, Tarbiat Modares University, Tehran, Iran
e-mail: shoja_sa@modares.ac.ir

M. Montazeri · H. Kouchakzadeh (✉)
Protein Research Center, Shahid Beheshti University, Tehran, Iran
e-mail: h_kouchakzadeh@sbu.ac.ir

S. E. R. Siadat
Rahpouyan Fanavar Sadegh Company, Pardis Technology Park, 20th Km of Damavand Road, Tehran, Iran

© The Author(s), under exclusive license to Springer Nature Singapore Pte Ltd. 2021
M. Rahmandoust and S.-O. Ranaei-Siadat (eds.), *COVID-19*,
https://doi.org/10.1007/978-981-16-3108-5_1

like acute respiratory distress syndrome (ARDS) as well as acute lung injury (ALI) or myocarditis, in some cases leading to death [8–10].

From a phylogenetic perspective, analyzed with the available full genome sequences and by the comparison with other members of the CoVs family, bats are more likely to be the reservoirs of SARS-CoV-2; nevertheless, the identity of the intermediate host(s) has yet been mysterious considering the fact that direct transmission from bats is implausible [6, 11].

Coronavirus disease 2019 (COVID-19) caused by SARS-CoV-2, with a higher outbreak than SARS-CoV-1, was distinguished as a pandemic by the World Health Organization (WHO) on March 11, 2020, while the epidemic has been incessantly growing and affecting roughly every country by high human-to-human transmission around the world [12, 13]. Up to date (January 3, 2021), 83,326,479 confirmed cases with the number of 1,831,703 deaths are recorded worldwide during the time that the numbers are constantly increasing [14]. The symptoms of COVID-19 are nonspecific, and the infection is mainly characterized by influenza-like symptoms including cough, fever, fatigue, dyspnea, and other symptoms such as headache, hemoptysis, and diarrhea in a few patients manifesting after few days (within 14 days) while fever is the most prevalent symptom [6, 15]. However, the severe cases are involved with pneumonia, ARDS, acute heart, liver, and gastrointestinal injury, renal failure, immune failure, compromised coagulation, and even death with approximately 2%–3% mortality [16–18]. Pneumonia and Lymphopenia often occur in severe patients during the second or third week of symptomatic COVID-19 infection [16].

The widespread disaster caused by COVID-19 besides the lack of pragmatic solutions and with no signs of abating have created an urgent, unfulfilled demand for successful therapies, and propelled the scientists, researchers, and pharmaceutical companies to respond to this situation by developing practical vaccines and antivirals to combat the disease [19, 20]. The countries most affected by the pandemic have attempted various treatment strategies to fight the disease including employing the existing antivirals and immunomodulatory drugs, convalescent plasma, different modes of oxygen therapy, or mechanical ventilation [16, 21]. Most patients have received multiple potential remedies and therapeutics, as employed so far, based on the experience in facing with SARS-CoV-1, MERS-CoV, human immunodeficiency virus (HIV), influenza, hepatitis B virus (HBV), and hepatitis C virus (HCV) [22]; yet, there are no specific available drugs against COVID-19. However, in order to manage the pandemic, patient isolation at early stages, social distancing, self-quarantine, hand hygiene, wearing face masks, and provision of supportive medical care alongside some restrictions have been proposed [23, 24].

It is believed that new drug discovery is a time-consuming, costly, and arduous scientific process [25, 26]. A more effective attitude toward COVID-19 therapy could be the drug repurposing approach and employing existing drug databases. Drug repurposing is a process in which new applications of approved or investigational drugs are discovered. Hence, exploring the existing antivirals and other drugs against the emerging health problem is a more feasible strategy [26, 27]. Meanwhile, many in-progress studies seeking potential novel and repurposed drugs and vaccines are currently under development and in clinical testing phases [19, 28–30].

In a sequential structure, by taking voluminously the biology and pathophysiology of different CoVs and particularly the structure of SARS-CoV-2 into account, in this chapter, we aim to comprehensively describe the potential treatment mechanisms for COVID-19 and highlight the therapeutic interventions, medications, biological, and natural products that may have a promising role in suppressing COVID-19 based on current evidence.

1.2 Coronaviruses History and SARS-CoV-2 Biology

The CoVs family are a class of enveloped single-stranded positive-sense RNA (+ssRNA) viruses that belongs to the order *Nidovirales*, suborder *Coronavirineae*, family *Coronaviridae*, and Subfamily *Orthocoronavirinae*. The morphology of CoVs appears round or oval with an average diameter of 60–140 nm [6, 11, 31]. The Subfamily Orthocoronavirinae comprises four genera of Aalphacoronavirus, Betacoronavirus, Gammacoronavirus, and Deltacoronavirus ($\alpha-/\beta-/\gamma-/\delta$-CoV) [11]. The different CoVs affect diverse host range. The α- and β-CoVs are only able to infect mammals while gamma and delta species have a wider host range consisting of avian species [32, 33].

So far, seven CoVs have been identified as human-CoVs (HCoVs) capable of infecting humans (HCoV-229E, HCoV-NL63, HCoV-OC43, HCoV-HKU1, SARS-CoV-1, MERS-CoV, and SARS-CoV-2), while three of them (SARS-CoV-1, MERS-CoV, and SARS-CoV-2) induce severe diseases in individuals [11, 13, 34]. Prior to the outbreak of SARS-CoV-1, only HCoV-OC43 and HCoV-229E were recognized (in the 1960s). However, after the emergence of SARS-CoV-1 in 2003, two common HCoVs (HCoV-NL63 in 2004 and HCoV-HKU1 in 2005) were known with low pathogenicity, and they generally cause mild upper respiratory tract infections associated with symptoms similar to a common cold in humans with immunodeficiency [1, 35] (Table 1.1).

The SARS-CoV-2 is an enveloped RNA of β-CoV genera that belongs to the bat-derived *Coronaviridae* family and shares approximately 50% similarity with MERS-CoV and about 80% genomic homology with SARS-CoV-1. Hence, SARS-CoV-2 and SARS-CoV-1 could be considered of the same species with a high virological resemblance [10, 36]. The SARS-CoV-2 as a majorly airborne infected pathogen transmit through close contact and droplets with a fatality rate of approximately 2–3% which is lower than SARS-CoV-1 (fatality rate of 10%) and MERS-CoV (fatality rate of 35%). Nonetheless, SARS-CoV-2 is considerably more epizootic that brings its epidemiological dynamics [31, 37].

SARS-CoV-2 has a +ssRNA with a genome size of 29.9 kilobases (kb) [13]. CoVs genome is considered as one of the largest RNA viral genomes with a size ranging from 27 to 32 kb. SARS-CoV-2 genome is bounded by 5′-caps and 3′-poly-adenine tails (poly(A) tail) and contains multiple large open reading frames (ORFs) that encode structural, nonstructural, and a variety of accessory proteins of the virus (Fig. 1.1) [10, 11, 31]. Regular CoVs entail at least six ORFs in their

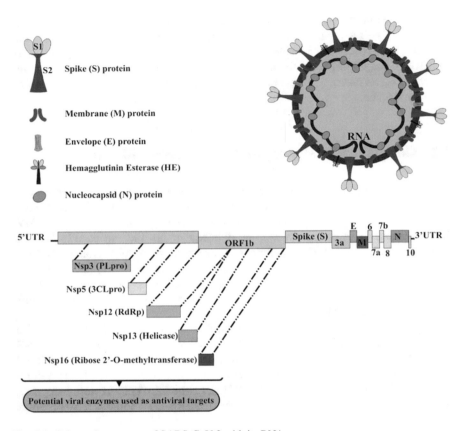

Fig. 1.1 Schematic structure of SARS-CoV-2 with its RNA genome

genome [38]. At the 5′ end, the RNA genome represents two large ORFs (ORF1a and ORF1b) that entail two-thirds of the whole genome flanked by 5′ and 3′ untranslated regions (UTRs). ORF1a and ORF1b are responsible for encoding 15–16 nonstructural proteins (NSPs). On the other hand, the structural and accessory proteins are encoded by the ORFs transcribed from the 3′ one-third of the RNA genome; a nested set of subgenomic mRNAs (sgmRNAs) [35, 38, 39]. Studies have indicated that there is a conserved gene order of 5′UTR-replicase-S-E-M-N-3′UTR in all members of the CoVs family [31, 40].

1.3 Structural Proteins

The 3′ one-third of the CoVs genome has a ranging from 270 to 500 nucleotides and is followed by a poly(A) tail. It consists of several ORFs that can encode four structural proteins, including spike (S), envelope (E), membrane (M), and nucleocapsid (N) proteins crucial for viral infusion and replication, non-structural proteins, and hemagglutinin esterase (HE) for some β-CoVs [35, 40]. The schematic of SARS-CoV-2 virion with its structural proteins is represented in Fig. 1.1.

1.3.1 Spike (S) Protein

The S protein with a crown-like appearance is a homotrimeric glycosylated transmembrane protein that is the most outward envelope of the CoVs. The S protein has critical roles in the infection process [31, 41]. It facilitates the attachment of the virus to the host cell receptors (i.e., angiotensin-converting enzyme 2), induces fusion between viral and host cell membranes, and adjusts the CoVs tropism. Thus, S protein is the main target of the neutralizing antibodies [11, 42]. The S protein is able to be divided into two functional subunits of the S1 and S2 [41]. The S1 subunit has been reported to be involved in the virus attaching to the host cell and serves as the receptor-binding domain (RBD) [43]. The S2 subunit is composed of a fusion peptide (FP) and two heptad repeat regions (HR1, HR2) which have decisive roles in stimulating virus–cell fusion [44].

1.3.2 Envelope (E) Protein

The E protein is the smallest structural protein consisting of 76–109 amino acids which has a molecular weight of 8.4–12 kDa [31, 45]. The E protein is an integral membrane protein with no charge that plays significant functions in ion channel activity, the host cell stress response inhibition, envelope formation, implication in pathogenesis, and virion trafficking [11, 40]. The E protein structure comprises a free charged transmembrane domain (TMD) surrounded by a negatively charged amino-terminal (N-terminal) and a variable charged carboxyl-terminal (C-terminal) [40, 46, 47].

The TMD amino acid sequence of E protein is composed of 25 amino acid residues [48, 49] mostly including non-polar leucine and valine amino acids which bring a strong hydrophobicity to the protein. The hydrophobic region consists of at least one α-helix associated with the formation of membrane ion conductive pore (viropurine) [50, 51]. The N-terminal of E protein is a short hydrophobic domain located at the outside of the virion membrane and contains 7–12 amino acid residues [46]. The E protein has a long hydrophilic C-terminal placed inside the virion and is embodied by conserved cysteine and proline residues [46]. The conserved cysteine residue in the C-terminal is a target site for palmitoylation. The conserved proline residue, besides, is located in the center of a β-coil-β motif in β- and γ-CoVs and acts as a signal for the transfer of protein to the Golgi apparatus [48, 52].

1.3.3 Membrane (M) Protein

The M glycoprotein has been identified as the most abundant structural protein in CoVs [31, 53] that plays an important role in the virion assembly and envelope formation [11]. Structurally, this protein contains 217–230 amino acids and has three TMDs surrounded by a long intravirion C-terminal and a short extravirion glycosylated N-terminal [54]. The CoVs M protein is known as the major regulator of the virion assembly by interacting with other structural proteins (Table 1.2) [55, 56].

1.3.4 Nucleocapsid (N) Protein

The N phosphoprotein with 422 amino acids has a molecular weight of 43–46 kDa [31, 57]. The N protein has multiple tasks and is associated with both the virion and the host cell. The main function of this protein is to form the ribonucleoprotein (RNP) in the ribonucleocapsid known as the packaging mechanism of CoVs by binding to the viral RNA [11, 58]. This protein also contributes to the virus assembly, translation, replication, and transcription. In the host cells, N protein is able to stimulate deregulation of the cell cycle [59–61], restrict the interferon production [62], up-regulate the cyclooxygenase-2 (COX2) expression and the AP1 activity [59, 63], and induce apoptotic cell death by serum-deprived cells [64].

Structurally, N protein consists of three highly conserved domains of NTD, CTD, and linker region (LKR) that binds to the RNA of the CoV [65, 66]. NTD has substantial roles in the synthesis of subgenomic RNA and other processes necessary for RNA remodeling [67]. LKR has a serine/arginine-rich (SR) motif including several phosphorylation sites which are involved in N protein interactions with M protein, heterogeneous nuclear RNPA1 (hnRNP-A1), and RNA with high binding affinity [66, 67]. CTD, also called the dimerization domain, has residues responsible for self-association to form homodimers. It, also, assists the formation of homo-oligomers through a domain-swapping mechanism [67]. A summary list of structural proteins and their functions is presented in Table 1.3.

Table 1.1 A summary list of the genome structure of pathogenic HCoVs [55, 56]

Virus	Genome structure	Natural host	Discovery year	Symptoms	Cellular entry receptors
HCoV-229E	5′UTR-ORF 1a-ORF 1b-[S]-3a-3b-[E]-[M]-[N]-3′UTR	Bats	1966	Mild respiratory tract infections	Human aminopeptidase N (APN)
HCoV-OC43	5′UTR-ORF 1a-ORF 1b-2(HE)-[S]-5-[E]-[M]-[N]-3′UTR	Rodents	1967	Mild respiratory tract infections	9-O-Acetylated sialic acid (9-O-Ac-Sia)
SARS-CoV-1	5′UTR-ORF 1a-ORF 1b-[S]-3a,3b-[E]-[M]-6–7a,7b–8a,8b–[N],9b(I)-3′UTR	Bats	2003	Severe acute respiratory syndrome (10% mortality rate)	Angiotensin-converting enzyme 2 (ACE2)
HCoV-NL63	5′UTR-ORF 1a-ORF 1b-[S]-3-[E]-[M]-[N]-3′UTR	Bats	2004	Mild respiratory tract infections	ACE2
HCoV-HKU1	5′UTR-ORF 1a-ORF 1b-(HE)-[S]-4-[E]-[M]-[N]-3′UTR	Rodents	2005	Pneumonia	(9-O-Ac-Sia)
MERS-CoV	5′UTR-ORF 1a-ORF 1b -[S]-3-4a,4b-5-[E]-[M]-8b-[N]-3′UTR	Bats	2012	Severe acute respiratory syndrome (35% mortality rate)	Dipeptidyl peptidase 4 (DPP4)
SARS-CoV-2	5′UTR-ORF 1a-ORF 1b-[S]-3a-[E]-[M]-6–7a,7b-8-[N],10-3′UTR	Bats	2019	Severe acute respiratory syndrome (2%–3% mortality rate)	ACE2

Table 1.2 Interactions of M protein with other structural proteins of SARS-CoV [55, 56]

Type of interaction	Function
M-M	Viral envelope formation
M-S	Retention of the S protein in the endoplasmic reticulum–Golgi intermediate compartment (ERGIC) and its incorporation into new virions
M-N	Providing stability for the N proteins
M-E	Viral envelope production and release of virus-like particles (VLPs)

1.4 Nonstructural or Functional Proteins

Two large overlapping ORF1a and ORF1b are located at the 5'UTR CoV genome while there is a frameshift between ORF1a and ORF1b which encodes two polypeptides of pp1a and pp1ab. The polypeptides are processed by viral proteases into 16 NSPs; among them, nsp1–nsp11 are encoded in ORF1a, and nsp12–16 are encoded in ORF1b [68, 69]. The NSPs have a fundamental role in CoVs replication [35]. In fact, these proteins have a wide variety of enzymatic functions including viral papain-like protease (PLpro), 3-chymotrypsin-like protease (3CLpro) also known as the main protease, RNA-dependent RNA polymerase (RdRp), and helicase that participate in the viral RNA replication, transcription, and translation [35]. Accordingly, the medications that suppress COVID-19 functional enzymes could inhibit the replication and assembly of new viral RNAs [70].

1.5 Accessory Proteins

Accessory genes distributed among structural genes (S, E, M, N) vary in number and sequence, and encode various accessory proteins in different types of CoVs [71, 72]. Most of the accessory proteins are not essential for virus replication. On the other hand, they associate with the virus–host interactions and viral pathogenesis. Moreover, some of these proteins can modulate the interferon signaling pathways and the production of pro-inflammatory cytokines in the human body [72, 73]. Yet, the molecular functions of many accessory proteins are to be discovered [31, 74]. The major accessory gene sequence of SARS-CoV-2 is similar to SARS-CoV-1, while the exact genome sequence of SARS-CoV-2 is still unclear. The accessory protein 10 has been hypothesized as unique to SARS-CoV-2 [75]. The accessory gene sequence of different HCoVs are listed in Table 1.1 [31], and the SARS-CoV-2 accessory proteins and their proposed functions are presented in Table 1.3.

Table 1.3 Structural, nonstructural, and accessory proteins of SARS-CoV-2 with their proposed functions

Viral protein	Function
Structural proteins	
S	S1: Receptor recognition S2: Membrane fusion and anchoring the S protein into the viral membrane
E	Virion assembly, budding, envelope formation, and pathogenesis
M	Virion assembly and envelope formation
N	Packaging the viral genome into a ribonucleoprotein (RNP)
Nonstructural proteins	
Nsp1	Targeting the host cell translation machinery, host mRNA degradation, translation inhibition
Nsp2	Unknown
Nsp3	Papain-like protease (PLpro)
Nsp4	Double-membrane vesicle (DMV) formation
Nsp5	3-chymotrypsin-like protease (3CLpro)
Nsp6	DMV formation
Nsp7	RNA-dependent RNA polymerase (RdRp) cofactor
Nsp8	RdRp cofactor, primase
Nsp9	Binding of single-stranded RNA
Nsp10	nsp14 and nsp16 cofactor
Nsp11	Unknown
Nsp12	RdRp, nucleotidyltransferase
Nsp13	Helicase, RNA5′ triphosphatase
Nsp14	3′ to 5′ exoribonuclease, proofreading, RNA cap formation, guanosineN7 methyltransferase
Nsp15	Endoribonuclease, evasion of immune response
Nsp16	RNA cap formation, ribose 2′-O-methyltransferase
Accessory proteins	
3a	Facilitation of virion assembly and evading the host immune system, participation in the ion channel
6	β-interferon antagonist, cellular DNA synthesis stimulation
7a	Facilitation of virion assembly, inducing inflammatory responses
7b	Attenuation activity
8	Interfering with the host immune response (feasible)
10	Unknown functions

1.6 Stages of Infection and Potential Treatment Mechanisms

Principally, the pathogenicity steps of COVID-19 can be divided into the four dominant following stages including the viral entry by binding the S protein of the virus to angiotensin-converting enzyme 2 (ACE2) receptor of the host cell, fusion of the viral and host cell membranes, viral RNA injection into the host cell cytoplasm, and viral RNA replication and the releases of newly generated SARS-CoV-2 viral particles [68]. Each of the steps in the disease progression introduces an opportunity for a potential treatment to prevent COVID-19 while disrupting the normal SARS-CoV-2 life cycle (Fig. 1.2) [19, 71].

1.6.1 Inhibiting Viral Entry and Fusion

Binding the S1 subunit of SARS-CoV-2 to surface cellular receptors (ACE2) is the first step of the SARS-CoV-2 life cycle and its infection. This binding causes the viral entry followed by fusion between viral and host cell membranes [76]. Following the receptor binding, there are two main pathways for viral fusion which the virus employs depending on the type of host cell and access to the membrane proteases:

(1) plasma membrane fusion also termed early pathway is exploited by the most CoVs to fuse into the host cell plasma membrane in the presence of membrane proteases such as transmembrane protease, serine 2 (TMPRSS2) [31, 77].
(2) pH-dependent endocytosis pathway is used by CoVs in the absence of plasma membrane proteases to enter their target cells employing clathrin- and non-clathrin-mediated endocytosis pathways. The endosomal pH decreases following the entry of the virus into the host cell. The increasing acidity can trigger endosomal proteases cathepsin L leading to the S2-dependent lysosomal membrane fusion [31, 78].

1.6.1.1 Targeting S Protein

The functional subunits of SARS-CoV-2 S protein (S1, S2) are made up of different domains [41]. The S1 subunit (amino acids 12–680) consists of NTD and CTD. RBD in the SARS-CoV-2 is located at the S1-CTD and has a critical role in attachment to the host cell, while S1-NTD has an important role in binding sugar receptor molecule [35, 42]. Thus, the S1 subunit is an essential target for monoclonal neutralizing antibodies (such as m396, CR3014) and vaccine design [4, 79]. The S2 subunit consists of several conserved domains including fusion peptides, the transmembrane domain, cytoplasmic domain, and heptad repeats (HR1 and HR2) [31, 80, 81]. The heptad repeats and fusion peptides are involved in virus–cell fusion. The S2 subunit, moreover, contributes to anchoring the S protein into the viral membrane [31, 44].

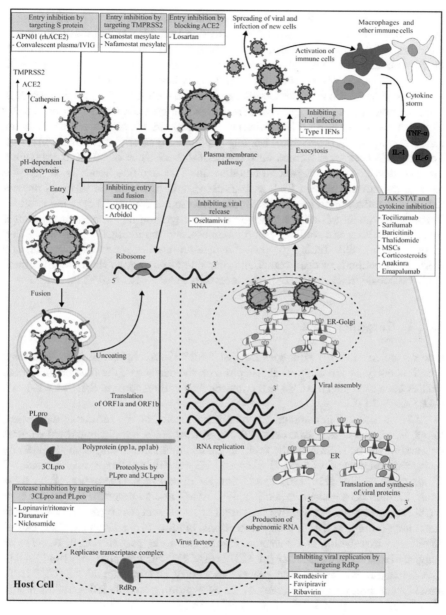

Fig. 1.2 Schematic representation of SARS-CoV-2 life cycle and stages of the Infection in the host cells with the potential therapeutic targets for COVID-19 treatment

The HR2 specific antibodies as well as the synthetic HR2 peptides have demonstrated the potential of blocking SARS-CoV-1 infection [35, 82, 83]. Normally, there is a high rate of mutation in the RBD structure in the S1 subunit. On the other hand, the S2 subunit structure is well conserved. Consequently, antibodies targeting the S2 region could render better protection against COVID-19 [35, 84–87].

The post-translational modifications appear in SARS-CoV-2 S protein including disulfide bond formation, glycosylation, and palmitoylation. Glycosylation usually occurs on the ectodomain (N-terminal) of the S protein, the modification which has a crucial function in the receptor binding and the antigenicity of S protein [31, 40]. The disulfide bonds are formed on account of the existence of 30–50 conservative cysteine residues in the S protein ectodomain. The disulfide bonds have decisive roles in the correct protein folding, trafficking, and trimerization of the S protein [31, 40]. Palmitoylation applies its modification to the conserved cysteine residues in the endodomain tail of the S proteins. Palmitoylation has important tasks in S protein trafficking and folding, virion assembly, and the interaction between S and M proteins [31, 88]. Taking the post-translational modifications into account, any therapeutic methods or drug compounds capable of preventing these modifications could be considered as possible procedures to prevent SARS-CoV-2 infection.

1.6.1.2 Targeting ACE2

One of the most direct strategies against SARS-CoV-2 infection is to neutralize the virus entry by targeting viral receptors on the surface of the host cells. Recent studies have recognized ACE2 as a common host cell receptor in SARS-CoV-1 and SARS-CoV-2 [79].

ACE2 is a zinc-dependent carboxypeptidase located on the surface of the kidney, lungs, heart, and other tissue cells [30, 40, 89]. ACE2 is a multifunctional enzyme countering the activation of the renin-angiotensin-aldosterone system (RAAS) by converting angiotensin (Ang) II (AspArgValTyrIleHisProPhe) to angiotensin 1-7 (AspArgValTyrIleHisPro) [90]. In addition, ACE2 assists the regulation of intestinal neutral amino acid transportation [91, 92]. ACE2 and angiotensin II receptor type 2 (AT2R) have a lung-protective function but AngII receptor type 1(AT1R) causes acute lung failure. In SARS-CoV-1 infections, binding of S protein to ACE2 causes ACE2 downregulation, and consequently increases the level of AngII. Kuba et al. suggested that recombinant human ACE2 (rhACE2) can competitively bind SARS-CoV-2 in a mimicking direction, prevent viral entry, and regulate RAAS [93, 94]. To some extent, blocking or changing ACE2 by exploiting soluble RBD, ACE2 monoclonal antibodies, rhACE2 (such as APN01), and AT1R inhibitors (such as losartan, valsartan, and ramipril) could be an interesting therapeutic method for blocking the SARS-CoV-2 entry, owing to the fact that the host ACE2 protein lack changes and the virus is incapable of escaping from this therapeutic strategy [95, 96]. In addition, some recent studies have shown that chloroquine (CQ) and hydroxychloroquine (HCQ) have the potential to be utilized as drugs to interfere in the glycosylation process of ACE2 and block the viral entry and fusion [96].

1.6.1.3 Inhibiting TMPRSS2

TMPRSS2, an enzyme belonging to the serine protease family, has four domains: type II transmembrane domain, class A receptor, cysteine-rich scavenger receptor, and the protease domain. TMPRSS2 has been found active in many tissues in the human body like the prostate, lungs, intestine, liver, kidneys, pancreas, upper airways, and bronchi [97, 98]. TMPRSS2 inhibition has been recently employed as a potential method for prostate cancer and several inflammatory pathologies treatment [99]. The S protein is cleaved at two domains of S1/S2 and S2 subunit by TMPRSS2. Priming or cleavage at the S1/S2 domain induces the release of the S1 Subunit RBD for binding to ACE2, while the cleavage of the S2 domain facilitates the release of the fusion S2 domains like HR1, HR2, and the fusion peptide [7, 98, 100–102]. TMPRSS2 plays a significant role in SARS-CoV-2 entry and its dispersion. Therefore, inhibition of this protease by some therapeutic approaches using antiandrogens, serine protease inhibitors like camostat, mesylate, nafamostat, and bromhexine could be possibly an interesting strategy for antiviral intervention [17, 79, 96, 103, 104].

1.6.1.4 Inhibition of Cathepsin L

Cathepsin L, the lysosomal cysteine protease, found in mammalian lysosomes is highly expressed in malignant tumors, and has been reported to be involved in the entry of the Ebola virus into its host cells. Cathepsin L cleaves the S protein into the S1 and S2 subunits in lysosomal acidic pH. Subsequently, the S1 subunit is attached to ACE2 with high affinity while the S2 subunit is a trigger directly associated with the membrane fusion [31, 44, 105, 106].

Some studies have indicated that pH neutralization of endolysosomes by CQ, HCQ, and ammonium chloride (NH_4Cl) could be effective against viral entry and fusion [107–109]. Besides, cathepsin protease inhibition by aloxistatin (E64d), flavonoids, bafilomycin A1, quercetin [79, 110, 111], and also incorporation into cell membranes and intervention into the hydrogen bonding network of phospholipids by umifenovir (arbidol) could be potential treatment options to prevent receptor binding and to block SARS-CoV-2 fusion [107, 112]. In consideration of the fact that MERS-CoV and SARS-CoVs can enter the host cells employing both plasma membrane fusion and pH-dependent endocytosis pathway, the inhibition of both proteases (cathepsin and TMPRSS2) should be considered for a vigorous block of antiviral entry [98, 113]. This approach could block the viral entry and fusion, while not interfering with other main protease activities such as T-cell activation and antiviral antibody production in normal immune responses [114].

1.6.2   Inhibiting Viral Replication, Translation, and Release

The genomic replication cycle generally starts after receptor binding, membrane fusion, and uncoating (the release of viral RNA into the host cell cytoplasm) [31]. As mentioned in the previous section, several ORFs encoding structural proteins (N, S, E, M), 16 NSPs, and a variety of accessory proteins are nested in the SARS-CoV-2 genome. 5'UTR constitutes approximately two-thirds of the CoV genome including ORF1a and ORF1b which are responsible for translating polyproteins pp1a and pp1ab, respectively [31, 115]. By the conduct of two cysteine proteases (PLpro and 3CLpro), upon the proteolytic cleavage, 16 NSPs release from pp1a (nsp1–11) and pp1ab (nsp12–16), any of which has specific functions presented in Table 1.3 [35].

NSPs participate in the viral RNA translation, transcription, and replication. Particularly, NSPs with PLpro, 3CLpro, helicase (Hel), RdRp functions play crucial roles in the forming of the replicase-transcriptase complex (RTC). By the mediating ER and Golgi, RTC replicates and synthesizes nested set of subgenomic mRNAs, responsible for encoding structural and accessory proteins [35, 116, 117]. Viruses generally use host cell proteins in their replication cycle for several purposes like cell attachment and entry, initiation and regulation of RNA replication and transcription, protein synthesis, and progeny virion assembly. The newly synthesized RNAs are either employed for generating more NSPs and RTCs by translation or packed into new virions for infecting other host cells [35]. The SARS-CoVs virion assembly established at the ERGIC requires the structural proteins (S, E, M, N) to become mature [118]. It has been reported that after translation, four structural proteins enter the secretory pathway in the ERGIC where they are added to the structure and assembled into virions [118, 119].

1.6.2.1   Inhibiting 3CLpro and PLpro

PLpro (nested in nsp3) and 3CLpro (nested in nsp5) are two classes of cysteine proteases encoded by the SARS-CoV-2 genome [120]. These proteases are involved in polyproteins (pp1a and pp1ab) proteolytic cleavage and consequently in the viral maturation. 3CLpro, also known as the main protease (Mpro), consists of three domains: the N-terminal finger domain, the catalytic domain, and the C-terminal domain [35, 121]. 3CLpro is a multifunctional protease that cleaves a significant number of 11 conserved sites in the pp1ab C-proximal and seven conserved sites in the pp1a C-proximal causing itself and other replicative machinery releases such as RdRp, helicase, and three RNA processing domains. 3CLpro also plays an important role in the interferon signaling inhibition. On the other hand, PLpro at nsp3, cleaves polyproteins at three sites releasing nsp1, nsp2, nsp3, and nsp4 amino terminus [31]. This cysteine protease also has critical functions in double-membrane vesicles (DMVs) formation [122], antagonism of interferons, DeISGylation, and deubiquitination which generally help CoVs escape from the host immune system [123, 124].

On account of the decisive roles of these proteases in polyproteins processing and viral replication, 3CLpro and PLpro inhibition by utilizing therapeutics such as darunavir, ASC09 [79, 125], bromhexine [79, 125], quercetin [79], thiopurine analogs (6-thioguanine, 6-mercaptopurine) [123], and acetaldehyde dehydrogenase inhibitors (disulfiram) [123, 126], could be effective treatment mechanisms for treating SARS-CoV-2 infected patients. In this regard, some recent studies have also shown that HIV protease inhibitors like lopinavir/ritonavir possess a modest antiviral activity against COVID-19 [41, 127, 128].

1.6.2.2 Inhibiting RdRP/nsp12

SARS-CoV-2 is incapable of accessing its host cell nucleus and subsequently, the replication action, thus, it encodes the RdRP to transcript and replicate itself [31, 130]. RdRp is located in the nsp12 C-terminal and operates as the transcription and replication core. To conduct this activity, RdRp requires its cofactors (nsp7–nsp8 heterodimer and the second nsp8 subunit) [130]. Structurally, the RdRp domain in nsp12 C-terminal is composed of three subdomains, a fingers domain, a palm domain, and a thumb domain. Besides, a β-hairpin and a nidovirus-specific extension domain constitute nsp12 N-terminal [31, 131–133]. The RdRp has a determining role in the SARS-CoV-2 replication and transcription, a function that is inactive for the virus in the host cell nucleus. Hence, inhibition of this polymerase activity by medications such as remdesivir [134–136], sofosbuvir, and favipiravir could be a practical treatment option for interrupting the RNA transcription leading to the prevention of COVID-19 [137–139].

1.6.2.3 Inhibiting NTPase/Helicase (nsp13)

CoVs nsp13 is an NTPase/helicase dual-function enzyme belonging to the helicase superfamily-1 (SF1). Generally, as a motor protein, it employs the energy obtained from the hydrolysis of natural nucleotides and deoxynucleotides triphosphates (such as ATP, dATP, and GTP) to unwind the RNA helix in the direction of 5′-3′ providing the situation for the RdRP and RTC to perform their tasks. nsp13 has several domains. Zinc binding domain (ZBD) is a high conservative domain located in the nsp13 N-terminal while there is a stem region next to it. These domains render structural stability to the enzyme. Previous studies on MERS-CoV have demonstrated that nsp13 N-terminal consists of 15 Cys/His (CH) rich domains with three zinc atoms. The CH domains involve in the coordination of the three zinc ions. The stem region is connected to the two catalytic RecA1 and RecA2 helicase domains by a beta domain in the nsp13 structure. It has also been suggested that RecA1, RecA2, and the beta domain are associated with regulating oligonucleotide binding. It should be noted that SARS-CoV-1 helicase can unwind DNA helix as well [31, 140–142]. To some extent, the helicase cooperates with RdRP and RTC toward the viral replication. Hence, the inhibition of this enzyme by a range of therapeutic drugs such as bismuth salts, flavonoids, scutellarein, and myricetin should be taken into consideration to interrupt viral metabolism without affecting normal cells [143–146].

1.6.3 Enhancing the Innate Immune System

Various approaches have been identified to be utilized to enhance the innate immune system responses [147]. The innate immune system comprises phagocytic cells (macrophages and neutrophils), natural killer cells (NKs), mast cells, basophils, eosinophils, dendritic cells (DCs), and lymphoid cells [148]. The innate immunity can induce its effects through inflammatory cytokines and the innate immune system [149]. NKs, mesenchymal stem cells (MSCs), and recombinant interferons (IFNs) have been employed as potential treatment options to enhance the innate immune system in the battle with COVID-19 [4, 28, 147].

1.6.4 Immunomodulation, Inflammatory Response Attenuating

The inflammatory reaction mostly occurs in severe COVID-19 patients leading to complications like ARDS [150]. ARDS is a medical condition involving uncontrolled inflammation in the lungs which could hinder oxygen to the lungs. The fluid accumulation in the distal parts of the lung might be by ARDS, and consequently, the disturbance in the surfactant and pulmonary capillary endothelial cells [9, 10]. Severe COVID-19 infection is mainly associated with the cytokine storm, the secretion of cytokines including IL-1β, IL-1RA, IL-6, IL-7, IL-8, IL-2R, TNF-α [150, 151]. This might suggest that the pathogenesis could be partly originating from inflammatory responses. Taking this concept into account, targeting SARS-CoV-2 alone with antiviral medications might not be enough to fight the pathogenic disease [150]. In this regard, various medications such as convalescent plasma, interleukin inhibitors, anticytokines, immunosuppressants, non-steroidal anti-inflammatory drugs (NSAIDs), anticoagulation agents, corticosteroids, antibiotics, and many recombinant proteins and monoclonal antibodies have been utilized as immunomodulators and anti-inflammatory therapeutics [4, 147, 150, 152].

It is worth noting that severe and critical COVID-19 patients with complications and secondary infections repeatedly face symptoms and distress with the possible risk of death urging a call to action for providing proactive symptom control and palliative care in response to the pandemic [153–155. Elevated vascular endothelial growth factor (VEGF) level is also indicated in patients with ARDS which could cause endothelial injury and microvascular permeability increase [156]. The most common symptoms observed in patients with severe COVID-19 infection are breathlessness, agitation, drowsiness, pain, and delirium where employing medications such as opioids benzodiazepine and monoclonal antibodies could be partially effective. Oxygen supplements can also be provided to hypoxic patients to help them through reducing the subjective work of breathing [4, 154, 157].

1.6.5 Vaccine

Vaccine development is considered one of the most rapid and economical strategies against COVID-19 [158]. The purpose of developing vaccines is to induce long-standing immunity without exposure to the disease's maximum brunt to control the COVID-19 pandemic, prevent its higher outbreak, and eventually inhibit its future recurrence. For vaccine preparation, it is crucial to induce protective T- and B-cell immune responses. After assessing the SARS-CoV-2 genome, many vaccines have been developing in different phases while some of them have indicated promising results for obtaining the United States (US) Food and Drug Administration (FDA) approval [159–161].

In general, antiviral vaccines fall into one of the following categories: inactive or live-attenuated viruses, virus-like particles, viral vectors, DNA-based, mRNA-based, and protein-based vaccines [4, 162]. Vaccines majorly are directed against the S glycoprotein [16]. Roughly, one-third of the COVID-19 vaccine candidates and patents have concentrated on the type of protein-based vaccines including the S protein subunit vaccine and by using S1 RBD of the S protein [162, 163]. Moreover, the S protein could be expressed by employing viral vector, DNA-based, and mRNA-based vaccines for antibody production in the human body [16].

S-trimer vaccines are made to resemble the S protein of CoVs, the protein that links to the ACE2 receptor on alveolar cells and facilitates the viral entry into cells. The S protein has been an attractive target for developing vaccines on account of the fact that it elicits the immune response during the disease progression and neutralizing antibodies are directed against it [21, 161].

DNA vaccines perform by infusion of DNA plasmids encoding antigens in host cells. This function induces cellular and humoral antigen-specific immunity bringing an immune response toward the target disease [19, 162]. In most cases, DNA vaccines have demonstrated safety to the human body without obvious off-target effects or peculiar toxicities [164]. In a like manner, viral vectors, like adenoviruses, could be employed as vectors to deliver genetic code for antigens, majorly S protein in the case of COVID-19, into cells instructing them to make a great amount of antigen leading to triggering the immune response [161, 165]. mRNA vaccines, mostly encapsulated in lipid nanoparticles, on the other hand, cause the production of antigens while in the cytosol without incorporating into the host cell genome and bringing the risk of mutation [166–168].

On the other hand, in live-attenuated or inactive virus vaccines, viruses with diminished pathogenesis or viruses inactivated by physical and chemical techniques like UV light, β-propiolactone, or formaldehyde are employed to stimulate the immune system [16]. Alternatively, VLPs as multimeric structures can mimic the three-dimensional conformation of native viruses and could be utilized as vaccine delivery agents capable of directly stimulating immune cells. VLPs are inherently safer than attenuated or inactivated virus vaccines for individuals due to lack of any infectious genetic materials. Furthermore, they possess privileged adjuvant properties and can induce both innate and adaptive immune responses [161, 169].

1.7   Conclusion

By the abrupt emergence of SARS-CoV-2 in December 2019, numerous people in many countries have been involved with the virus; among them, many patients have encountered various disorders and serious complications like ARDS, ALI, and cytokine storm, leading to mortality in lots of cases. As the world is still in limbo from COVID-19, many types of research and studies are being performed in order to dampen the pandemic by employing practical strategies like developing vaccines and drug repurposing approach. Yet, there are no particular available drugs against COVID-19; meanwhile, some therapeutics have demonstrated high potency against the infection, and several classes of medications are currently under development and in clinical testing phases.

Taking the biology of SARS-CoV-2 and pathogenesis stages of the infection into account, the virus exploits the inherent functions of both its components (e.g., structural, non-structural, and accessory proteins) and the host cell compartments to consecutively enter and fuse with the host cells, inject its RNA into the cytoplasm, and get replicated to infect other cells and individuals. Consequently, each of the mentioned steps represents potential targets to inhibit the infection toward diminishing the pandemic. In particular, the S Protein, 3CLpro, PLpro, RdRP, and NTPase/helicase of the virus alongside ACE2, TMPRSS2, and Cathepsin L of the host cells are associated with the infection and are potential candidates to be inhibited for COVID-19 treatment. However, prior to attacking each of the targets, proper diagnosis and identification of the pathogenesis steps are necessitated. Hence, any intervention into the infection by potential medications and therapeutics need to be accompanied by precise recognition of the viral progression stages toward practical management and treatment of COVID-19 while avoiding any probable side effects and deterioration. It should be noted that delayed treatment with the medications may limit the effectiveness of the agents.

References

1. Zhu N, Zhang D, Wang W, Li X, Yang B, Song J, Zhao X, Huang B, Shi W, Lu R (2020) A novel coronavirus from patients with pneumonia in China, 2019. New Engl J Med

2. M. Nicola, Z. Alsafi, C. Sohrabi, A. Kerwan, A. Al-Jabir, C. Iosifidis, M. Agha, R. Agha, The socio-economic implications of the coronavirus and COVID-19 pandemic: a review. Int J Surg (2020)

3. Maurya VK, Kumar S, Bhatt ML, Saxena SK (2020) Therapeutic development and drugs for the treatment of COVID-19, coronavirus disease 2019 (COVID-19). Springer, pp 109–126

4. Tu Y-F, Chien C-S, Yarmishyn AA, Lin Y-Y, Luo Y-H, Lin Y-T, Lai W-Y, Yang D-M, Chou S-J, Yang Y-P (2020) A review of SARS-CoV-2 and the ongoing clinical trials. Int J Mol Sci 21(7):2657

5. Lu R, Zhao X, Li J, Niu P, Yang B, Wu H, Wang W, Song H, Huang B, Zhu N (2020) Genomic characterisation and epidemiology of 2019 novel coronavirus: implications for virus origins and receptor binding. The Lancet 395(10224):565–574

6. Singhal T (2020) A review of coronavirus disease-2019 (COVID-19). Indian J Pediatr 1–6

7. Kumar GV, Jeyanthi V, Ramakrishnan S (2020) A short review on antibody therapy for COVID-19. New Microbes New Infect 100682

8. Harapan H, Itoh N, Yufika A, Winardi W, Keam S, Te H, Megawati D, Hayati Z, Wagner AL, Mudatsir M (2020) Coronavirus disease 2019 (COVID-19): a literature review. J Infect Public Health

9. Zhou M, Zhang X, Qu J (2020) Coronavirus disease 2019 (COVID-19): a clinical update. Front Med 1–10

10. Jin Y, Yang H, Ji W, Wu W, Chen S, Zhang W, Duan G (2020) Virology, epidemiology, pathogenesis, and control of COVID-19. Viruses 12(4):372

11. Chen Y, Liu Q, Guo D (2020) Emerging coronaviruses: genome structure, replication, and pathogenesis. J Med Virol 92(4):418–423

12. Lai C-C, Shih T-P, Ko W-C, Tang H-J, Hsueh P-R (2020) Severe acute respiratory syndrome coronavirus 2 (SARS-CoV-2) and corona virus disease-2019 (COVID-19): the epidemic and the challenges. Int J Antimicrob Agents 105924

13. Guo Y-R, Cao Q-D, Hong Z-S, Tan Y-Y, Chen S-D, Jin H-J, Tan K-S, Wang D-Y, Yan Y (2020) The origin, transmission and clinical therapies on coronavirus disease 2019 (COVID-19) outbreak–an update on the status. Military Med Res 7(1):1–10

14. WHO (2021) COVID-19 weekly epidemiological update. https://www.who.int/emergencies/diseases/novel-coronavirus-2019/situation-reports/. Accessed 6 Jan 2021

15. Sun P, Lu X, Xu C, Sun W, Pan B (2020) Understanding of COVID-19 based on current evidence. J Med Virol 92(6):548–551

16. Ita K (2020) Coronavirus DIsease (COVID-19): current status and prospects for drug and vaccine development. Arch Med Res

17. Giovane RA, Rezai S, Cleland E, Henderson CE (2020) Current pharmacological modalities for management of novel coronavirus disease 2019 (COVID-19) and the rationale for their utilization: a review. Rev Med Virol 30(5):

18. Jamwal S, Gautam A, Elsworth J, Kumar M, Chawla R, Kumar P (2020) An updated insight into the molecular pathogenesis, secondary complications and potential therapeutics of COVID-19 pandemic. Life Sci 118105

19. Chary MA, Barbuto AF, Izadmehr S, Hayes BD, Burns MM (2020) COVID-19: therapeutics and their toxicities. J Med Toxicol 16(3):101007

20. Wu R, Wang L, Kuo H-CD, Shannar A, Peter R, Chou PJ, Li S, Hudlikar R, Liu X, Liu Z (2020) An update on current therapeutic drugs treating COVID-19. Curr Pharmacol Rep 1

21. Salvi R, Patankar P (2020) Emerging pharmacotherapies for COVID-19. Biomed Pharmacotherapy 110267

22. Zhang Y, Xu Q, Sun Z, Zhou L (2020) Current targeted therapeutics against COVID-19: based on first-line experience in china. Pharmacol Res 104854

23. Vellingiri B, Jayaramayya K, Iyer M, Narayanasamy A, Govindasamy V, Giridharan B, Ganesan S, Venugopal A, Venkatesan D, Ganesan H (2020) COVID-19: a promising cure for the global panic. Sci Total Environ 138277

24. Sohrabi C, Alsafi Z, O'Neill N, Khan M, Kerwan A, Al-Jabir A, Iosifidis C, Agha R (2020) World Health Organization declares global emergency: a review of the 2019 novel coronavirus (COVID-19). Int J Surg

25. Taylor D (2015) The pharmaceutical industry and the future of drug development
26. Singh TU, Parida S, Lingaraju MC, Kesavan M, Kumar D, Singh RK (2020) Drug repurposing approach to fight COVID-19. Pharmacol Rep 1–30
27. Shah B, Modi P, Sagar SR (2020) In silico studies on therapeutic agents for COVID-19: drug repurposing approach. Life Sci 117652
28. Alnefaie A, Albogami S (2020) Current approaches used in treating COVID-19 from a molecular mechanisms and immune response perspective. Saudi Pharm J
29. Sanders JM, Monogue ML, Jodlowski TZ, Cutrell JB (2020) Pharmacologic treatments for coronavirus disease 2019 (COVID-19): a review. JAMA 323(18):1824–1836
30. Xian Y, Zhang J, Bian Z, Zhou H, Zhang Z, Lin Z, Xu H (2020) Bioactive natural compounds against human coronaviruses: a review and perspective. Acta Pharmaceutica Sinica B
31. Artika IM, Dewantari AK, Wiyatno A (2020) Molecular biology of coronaviruses: current knowledge. Heliyon e04743
32. C.S.G. of the International (2020) The species Severe acute respiratory syndrome-related coronavirus: classifying 2019-nCoV and naming it SARS-CoV-2. Nat Microbiol 5(4):536
33. Paules CI, Marston HD, Fauci AS (2020) Coronavirus infections—more than just the common cold. JAMA 323(8):707–708
34. Rabaan AA, Al-Ahmed SH, Haque S, Sah R, Tiwari R, Malik YS, Dhama K, Yatoo MI, Bonilla-Aldana DK, Rodriguez-Morales AJ (2020) SARS-CoV-2, SARS-CoV, and MERS-CoV: a comparative overview. Infez Med 28(2):174–184
35. V'kovski P, Kratzel A, Steiner S, Stalder H, Thiel V (2020) Coronavirus biology and replication: implications for SARS-CoV-2. Nat Rev Microbiol 1–16
36. Wang L-S, Wang Y-R, Ye D-W, Liu Q-Q (2020) A review of the 2019 Novel Coronavirus (COVID-19) based on current evidence. Int J Antimicrob Agents 105948
37. Petrosillo N, Viceconte G, Ergonul O, Ippolito G, Petersen E (2020) COVID-19, SARS and MERS: are they closely related? Clin Microbiol Infect (2020)
38. L. Mousavizadeh, S. Ghasemi, Genotype and phenotype of COVID-19: Their roles in pathogenesis, Journal of Microbiology, Immunology and Infection (2020)
39. van Boheemen S, de Graaf M, Lauber C, Bestebroer TM, Raj VS, Zaki AM, Osterhaus AD, Haagmans BL, Gorbalenya AE, Snijder EJ (2012) Genomic characterization of a newly discovered coronavirus associated with acute respiratory distress syndrome in humans. MBio 3(6)
40. Masters PS (2006) The molecular biology of coronaviruses. Adv Virus Res 66:193–292
41. Ou X, Liu Y, Lei X, Li P, Mi D, Ren L, Guo L, Guo R, Chen T, Hu J (2020) Characterization of spike glycoprotein of SARS-CoV-2 on virus entry and its immune cross-reactivity with SARS-CoV. Nat Commun 11(1):1–12
42. Tortorici MA, Walls AC, Lang Y, Wang C, Li Z, Koerhuis D, Boons G-J, Bosch B-J, Rey FA, de Groot RJ (2019) Structural basis for human coronavirus attachment to sialic acid receptors. Nat Struct Mol Biol 26(6):481–489
43. Shang J, Ye G, Shi K, Wan Y, Luo C, Aihara H, Geng Q, Auerbach A, Li F (2020) Structural basis of receptor recognition by SARS-CoV-2. Nature 581(7807):221–224
44. Hulswit R, De Haan C, Bosch B-J (2016) Coronavirus spike protein and tropism changes. Adv Virus Res 29–57
45. Kuo L, Hurst KR, Masters PS (2007) Exceptional flexibility in the sequence requirements for coronavirus small envelope protein function. J Virol 81(5):2249–2262
46. Schoeman D, Fielding BC (2019) Coronavirus envelope protein: current knowledge. Virol J 16(1):1–22
47. Du Y, Zuckermann FA, Yoo D (2010) Myristoylation of the small envelope protein of porcine reproductive and respiratory syndrome virus is non-essential for virus infectivity but promotes its growth. Virus Res 147(2):294–299
48. Ruch TR, Machamer CE (2012) The coronavirus E protein: assembly and beyond. Viruses 4(3):363–382
49. Heinz F, Collett M, Purcell R, Gould E, Howard C, Van Regenmortel MHV, Fauquet CM, Bishop DHL, Carstens EB, Estes MK et al (2000) Virus taxonomy. In: Seventh Report of the International Committee on Taxonomy of Viruses. Academic Press, San Diego, pp 859–878

50. Wu Q, Zhang Y, Lü H, Wang J, He X, Liu Y, Ye C, Lin W, Hu J, Ji J (2003) The E protein is a multifunctional membrane protein of SARS-CoV. Genom Proteomics Bioinformatics 1(2):131–144
51. McClenaghan C, Hanson A, Lee S-J, Nichols CG (2020) Coronavirus proteins as Ion channels: current and potential research. Front Immunol 11:2651
52. Sarkar M, Saha S (2020) Structural insight into the role of novel SARS-CoV-2 E protein: a potential target for vaccine development and other therapeutic strategies. PLoS ONE 15(8):
53. Ujike M, Taguchi F (2015) Incorporation of spike and membrane glycoproteins into coronavirus virions. Viruses 7(4):1700–1725
54. De Haan CA, Smeets M, Vernooij F, Vennema H, Rottier P (1999) Mapping of the coronavirus membrane protein domains involved in interaction with the spike protein. J Virol 73(9):7441–7452
55. Arndt AL, Larson BJ, Hogue BG (2010) A conserved domain in the coronavirus membrane protein tail is important for virus assembly. J Virol 84(21):11418–11428
56. Thomas S (2020) The structure of the membrane protein of SARS-CoV-2 resembles the sugar transporter semiSWEET
57. Chang C-K, Hou M-H, Chang C-F, Hsiao C-D, Huang T-H (2014) The SARS coronavirus nucleocapsid protein–forms and functions. Antiviral Res 103:39–50
58. Dutta NK, Mazumdar K, Gordy JT (2020) The nucleocapsid protein of SARS–CoV-2: a target for vaccine development. J Virol 94(13)
59. Wurm T, Chen H, Hodgson T, Britton P, Brooks G, Hiscox JA (2001) Localization to the nucleolus is a common feature of coronavirus nucleoproteins, and the protein may disrupt host cell division. J Virol 75(19):9345–9356
60. Surjit M, Liu B, Chow VT, Lal SK (2006) The nucleocapsid protein of severe acute respiratory syndrome-coronavirus inhibits the activity of cyclin-cyclin-dependent kinase complex and blocks S phase progression in mammalian cells. J Biol Chem 281(16):10669–10681
61. Mu J, Fang Y, Yang Q, Shu T, Wang A, Huang M, Jin L, Deng F, Qiu Y, Zhou X (2020) SARS-CoV-2 N protein antagonizes type I interferon signaling by suppressing phosphorylation and nuclear translocation of STAT1 and STAT2. Cell discovery 6(1):1–4
62. Kopecky-Bromberg SA, Martínez-Sobrido L, Frieman M, Baric RA, Palese P (2007) Severe acute respiratory syndrome coronavirus open reading frame (ORF) 3b, ORF 6, and nucleocapsid proteins function as interferon antagonists. J Virol 81(2):548–557
63. Yan X, Hao Q, Mu Y, Timani KA, Ye L, Zhu Y, Wu J (2006) Nucleocapsid protein of SARS-CoV activates the expression of cyclooxygenase-2 by binding directly to regulatory elements for nuclear factor-kappa B and CCAAT/enhancer binding protein. Int J Biochem Cell Biol 38(8):1417–1428
64. Surjit M, Liu B, Jameel S, Chow VT, Lal SK (2004) The SARS coronavirus nucleocapsid protein induces actin reorganization and apoptosis in COS-1 cells in the absence of growth factors. Biochem J 383(1):13–18
65. Huang Q, Yu L, Petros AM, Gunasekera A, Liu Z, Xu N, Hajduk P, Mack J, Fesik SW, Olejniczak ET (2004) Structure of the N-terminal RNA-binding domain of the SARS CoV nucleocapsid protein. Biochemistry 43(20):6059–6063
66. Zeng W, Liu G, Ma H, Zhao D, Yang Y, Liu M, Mohammed A, Zhao C, Yang Y, Xie J (2020) Biochemical characterization of SARS-CoV-2 nucleocapsid protein. Biochem Biophys Res Commun
67. McBride R, Van Zyl M, Fielding BC (2014) The coronavirus nucleocapsid is a multifunctional protein. Viruses 6(8):2991–3018
68. Astuti I (2020) Severe acute respiratory syndrome coronavirus 2 (SARS-CoV-2): an overview of viral structure and host response. Diab Metabolic Synd Clin Res Rev
69. Subissi L, Imbert I, Ferron F, Collet A, Coutard B, Decroly E, Canard B (2014) SARS-CoV ORF1b-encoded nonstructural proteins 12–16: replicative enzymes as antiviral targets. Antiviral Res 101:122–130
70. Zhang W, Zhang P, Wang G, Cheng W, Chen J, Zhang X (2020) Recent advances of therapeutic targets and potential drugs of COVID-19. Die Pharmazie Int J Pharm Sci 75(5):160–162

71. Zhou P, Yang X-L, Wang X-G, Hu B, Zhang L, Zhang W, Si H-R, Zhu Y, Li B, Huang C-L (2020) A pneumonia outbreak associated with a new coronavirus of probable bat origin. Nature 579(7798):270–273
72. Michel CJ, Mayer C, Poch O, Thompson JD (2020) Characterization of accessory genes in coronavirus genomes
73. Liu DX, Fung TS, Chong KK-L, Shukla A, Hilgenfeld R (2014) Accessory proteins of SARS-CoV and other coronaviruses. Antiviral Res 109:97–109
74. Kim D, Lee J-Y, Yang J-S, Kim JW, Kim VN, Chang H (2020) The architecture of SARS-CoV-2 transcriptome. Cell
75. Tang X, Wu C, Li X, Song Y, Yao X, Wu X, Duan Y, Zhang H, Wang Y, Qian Z (2020) On the origin and continuing evolution of SARS-CoV-2. Natl Sci Rev
76. Walls AC, Park Y-J, Tortorici MA, Wall A, McGuire AT, Veesler D (2020) Structure, function, and antigenicity of the SARS-CoV-2 spike glycoprotein. Cell
77. Matsuyama S, Nao N, Shirato K, Kawase M, Saito S, Takayama I, Nagata N, Sekizuka T, Katoh H, Kato F (2020) Enhanced isolation of SARS-CoV-2 by TMPRSS2-expressing cells. Proc Natl Acad Sci 117(13):7001–7003
78. Santos IDA, Grosche VR, Bergamini FRG, Sabino-Silva R, Jardim AC (2020) Antivirals against coronaviruses: candidate drugs for SARS-coV-2 treatment? Front Microbiol 11:1818
79. Depfenhart M, de Villiers D, Lemperle G, Meyer M, Di Somma S (2020) Potential new treatment strategies for COVID-19: is there a role for bromhexine as add-on therapy? Internal Emerg Med 1
80. Tripet B, Howard MW, Jobling M, Holmes RK, Holmes KV, Hodges RS (2004) Structural characterization of the SARS-coronavirus spike S fusion protein core. J Biol Chem 279(20):20836–20849
81. Li F, Li W, Farzan M, Harrison SC (2005) Structure of SARS coronavirus spike receptor-binding domain complexed with receptor. Science 309(5742):1864–1868
82. Lip K-M, Shen S, Yang X, Keng C-T, Zhang A, Oh H-LJ, Li Z-H, Hwang L-A, Chou C-F, Fielding BC (2006) Monoclonal antibodies targeting the HR2 domain and the region immediately upstream of the HR2 of the S protein neutralize in vitro infection of severe acute respiratory syndrome coronavirus. J Virol 80(2):941–950
83. Lai S-C, Chong PC-S, Yeh C-T, Liu LS-J, Jan J-T, Chi H-Y, Liu H-W, Chen A, Wang Y-C (2005) Characterization of neutralizing monoclonal antibodies recognizing a 15-residues epitope on the spike protein HR2 region of severe acute respiratory syndrome coronavirus (SARS-CoV). J Biomed Sci 12(5):711–727
84. Rockx B, Donaldson E, Frieman M, Sheahan T, Corti D, Lanzavecchia A, Baric RS (2010) Escape from human monoclonal antibody neutralization affects in vitro and in vivo fitness of severe acute respiratory syndrome coronavirus. J Infect Dis 201(6):946–955
85. Zhang H, Wang G, Li J, Nie Y, Shi X, Lian G, Wang W, Yin X, Zhao Y, Qu X (2004) Identification of an antigenic determinant on the S2 domain of the severe acute respiratory syndrome coronavirus spike glycoprotein capable of inducing neutralizing antibodies. J Virol 78(13):6938–6945
86. Keng C-T, Zhang A, Shen S, Lip K-M, Fielding BC, Tan TH, Chou C-F, Loh CB, Wang S, Fu J (2005) Amino acids 1055 to 1192 in the S2 region of severe acute respiratory syndrome coronavirus S protein induce neutralizing antibodies: implications for the development of vaccines and antiviral agents. J Virol 79(6):3289–3296
87. Elshabrawy HA, Coughlin MM, Baker SC, Prabhakar BS (2012) Human monoclonal antibodies against highly conserved HR1 and HR2 domains of the SARS-CoV spike protein are more broadly neutralizing. PLoS ONE 7(11):
88. Fung TS, Liu DX (2018) Post-translational modifications of coronavirus proteins: roles and function. Future Virol 13(6):405–430
89. Tai W, He L, Zhang X, Pu J, Voronin D, Jiang S, Zhou Y, Du L (2020) Characterization of the receptor-binding domain (RBD) of 2019 novel coronavirus: implication for development of RBD protein as a viral attachment inhibitor and vaccine. Cell Mol Immunol 17(6):613–620

90. Vaduganathan M, Vardeny O, Michel T, McMurray JJ, Pfeffer MA, Solomon SD (2020) Renin–angiotensin–aldosterone system inhibitors in patients with Covid-19. N Engl J Med 382(17):1653–1659

91. Turner AJ (2015) ACE2 cell biology, regulation, and physiological functions. The protective arm of the renin angiotensin system (RAS), p 185

92. Camargo SM, Vuille-dit-Bille RN, Meier CF, Verrey F (2020) ACE2 and gut amino acid transport. Clin Sci 134(21):2823–2833

93. Kuba K, Imai Y, Rao S, Gao H, Guo F, Guan B, Huan Y, Yang P, Zhang Y, Deng W (2005) A crucial role of angiotensin converting enzyme 2 (ACE2) in SARS coronavirus–induced lung injury. Nat Med 11(8):875–879

94. Imai Y, Kuba K, Penninger JM (2008) The discovery of angiotensin-converting enzyme 2 and its role in acute lung injury in mice. Exp Physiol 93(5):543–548

95. Sarkar C, Mondal M, Torequl Islam M, Martorell M, Docea AO, Maroyi A, Sharifi-Rad J, Calina D (2020) Potential therapeutic options for COVID-19: current status, challenges, and future perspectives. Front. Pharmacol 11:1428

96. Scavone C, Brusco S, Bertini M, Sportiello L, Rafaniello C, Zoccoli A, Berrino L, Racagni G, Rossi F, Capuano A (2020) Current pharmacological treatments for COVID-19: What's next? Br J Pharmacol

97. Luan B, Huynh T, Cheng X, Lan G, Wang H-R (2020) Targeting proteases for treating COVID-19. J Proteome Res 19(11):4316–4326

98. Hoffmann M, Kleine-Weber H, Schroeder S, Krüger N, Herrler T, Erichsen S, Schiergens TS, Herrler G, Wu N-H, Nitsche A (2020) SARS-CoV-2 cell entry depends on ACE2 and TMPRSS2 and is blocked by a clinically proven protease inhibitor. Cell

99. Ko C-J, Hsu T-W, Wu S-R, Lan S-W, Hsiao T-F, Lin H-Y, Lin H-H, Tu H-F, Lee C-F, Huang C-C (2020) Inhibition of TMPRSS2 by HAI-2 reduces prostate cancer cell invasion and metastasis. Oncogene 39(37):5950–5963

100. Luan B, Huynh T, Cheng X, Lan G, Wang H-R (2020) Targeting proteases for treating COVID-19. J Proteome Res

101. Belouzard S, Chu VC, Whittaker GR (2009) Activation of the SARS coronavirus spike protein via sequential proteolytic cleavage at two distinct sites. Proc Natl Acad Sci 106(14):5871–5876

102. Tomlins SA, Rhodes DR, Perner S, Dhanasekaran SM, Mehra R, Sun X-W, Varambally S, Cao X, Tchinda J, Kuefer R (2005) Recurrent fusion of TMPRSS2 and ETS transcription factor genes in prostate cancer. Science 310(5748):644–648

103. Hoffmann M, Schroeder S, Kleine-Weber H, Müller MA, Drosten C, Pöhlmann S (2020) Nafamostat mesylate blocks activation of SARS-CoV-2: new treatment option for COVID-19. Antimicrob Agents Chemother

104. Mikkonen L, Pihlajamaa P, Sahu B, Zhang F-P, Jänne OA (2010) Androgen receptor and androgen-dependent gene expression in lung. Mol Cell Endocrinol 317(1–2):14–24

105. Fujimoto T, Tsunedomi R, Matsukuma S, Yoshimura K, Oga A, Fujiwara N, Fujiwara Y, Matsui H, Shindo Y, Tokumitsu Y (2020) Cathepsin B is highly expressed in pancreatic cancer stem-like cells and is associated with patients' surgical outcomes. Oncol Lett 21(1):1–1

106. Roshy S, Sloane BF, Moin K (2003) Pericellular cathepsin B and malignant progression. Cancer Metastasis Rev 22(2–3):271–286

107. Yang N, Shen H-M (2020) Targeting the endocytic pathway and autophagy process as a novel therapeutic strategy in COVID-19. Int J Biol Sci 16(10):1724

108. Aguiar AC, Murce E, Cortopassi WA, Pimentel AS, Almeida MM, Barros DC, Guedes JS, Meneghetti MR, Krettli AU (2018) Chloroquine analogs as antimalarial candidates with potent in vitro and in vivo activity. Int J Parasitol Drugs Drug Resistance 8(3):459–464

109. Liu J, Cao R, Xu M, Wang X, Zhang H, Hu H, Li Y, Hu Z, Zhong W, Wang M (2020) Hydroxychloroquine, a less toxic derivative of chloroquine, is effective in inhibiting SARS-CoV-2 infection in vitro. Cell Discov 6(1):1–4

110. Schrezenmeier E, Dörner T (2020) Mechanisms of action of hydroxychloroquine and chloroquine: implications for rheumatology. Nat Rev Rheumatol 1–12

111. Yang J-K, Zhao M-M, Yang W-L, Yang F-Y, Zhang L, Huang W, Fan C, Hou W, Jin R, Feng Y (2020) Cathepsin L plays a key role in SARS-CoV-2 infection in humans and humanized mice and is a promising target for new drug development. medRxiv

112. Zhu Z, Lu Z, Xu T, Chen C, Yang G, Zha T, Lu J, Xue Y (2020) Arbidol monotherapy is superior to lopinavir/ritonavir in treating COVID-19. J Infect 81(1):e21–e23

113. Cannalire R, Stefanelli I, Cerchia C, Beccari AR, Pelliccia S, Summa V (2020) SARS-CoV-2 entry inhibitors: small molecules and peptides targeting virus or host cells. Int J Mol Sci 21(16):5707

114. Liu T, Luo S, Libby P, Shi G-P (2020) Cathepsin L-selective inhibitors: a potentially promising treatment for COVID-19 patients. Pharmacol Therap 107587

115. Fung TS, Liu DX (2019) Human coronavirus: host-pathogen interaction. Annu Rev Microbiol 73:529–557

116. Van Hemert MJ, Van Den Worm SH, Knoops K, Mommaas AM, Gorbalenya AE, Snijder EJ (2008) SARS-coronavirus replication/transcription complexes are membrane-protected and need a host factor for activity in vitro. PLoS Pathog 4(5):

117. Qiu Y, Xu K (2020) Functional studies of the coronavirus nonstructural proteins. STEMedicine 1(2):e39–e39

118. Woo J, Lee EY, Lee M, Kim T, Cho Y-E (2019) An in vivo cell-based assay for investigating the specific interaction between the SARS-CoV N-protein and its viral RNA packaging sequence. Biochem Biophys Res Commun 520(3):499–506

119. Tang T, Bidon M, Jaimes JA, Whittaker GR, Daniel S (2020) Coronavirus membrane fusion mechanism offers as a potential target for antiviral development. Antiviral Res 104792

120. Mitra K, Ghanta P, Acharya S, Chakrapani G, Ramaiah B, Doble M (2020) Dual inhibitors of SARS-CoV-2 proteases: pharmacophore and molecular dynamics based drug repositioning and phytochemical leads. J Biomol Struct Dyn 1–14

121. Novak J, Rimac H, Kandagalla S, Pathak P, Grishina M, Potemkin V (2020) Proposition of a new allosteric binding site for potential SARS-CoV-2 3CL protease inhibitors by utilizing molecular dynamics simulations and ensemble docking

122. Oudshoorn D, Rijs K, Limpens RW, Groen K, Koster AJ, Snijder EJ, Kikkert M, Bárcena M (2017) Expression and cleavage of middle east respiratory syndrome coronavirus nsp3-4 polyprotein induce the formation of double-membrane vesicles that mimic those associated with coronaviral RNA replication. MBio 8(6)

123. Clemente V, D'Arcy P, Bazzaro M (2020) Deubiquitinating enzymes in coronaviruses and possible therapeutic opportunities for COVID-19. Int J Mol Sci 21(10):3492

124. Ruzicka JA (2020) Identification of the antithrombotic protein S as a potential target of the SARS-CoV-2 papain-like protease. Thromb Res 196:257–259

125. Huynh T, Wang H, Luan B (2020) In silico exploration of molecular mechanism of clinically oriented drugs for possibly inhibiting SARS-CoV-2's main protease. J Phys Chem Lett

126. Lin M-H, Moses DC, Hsieh C-H, Cheng S-C, Chen Y-H, Sun C-Y, Chou C-Y (2018) Disulfiram can inhibit MERS and SARS coronavirus papain-like proteases via different modes. Antiviral Res 150:155–163

127. Baden LR, Rubin EJ (2020) Covid-19—the search for effective therapy. Mass Medical Soc

128. Sheahan TP, Sims AC, Leist SR, Schäfer A, Won J, Brown AJ, Montgomery SA, Hogg A, Babusis D, Clarke MO (2020) Comparative therapeutic efficacy of remdesivir and combination lopinavir, ritonavir, and interferon beta against MERS-CoV. Nature communications 11(1):1–14

129. Qu J-M, Cao B, Chen R-C (2020) Treatment of COVID-19. COVID-19 55

130. Gaurav A, Al-Nema M (2019) Polymerases of coronaviruses: structure, function, and inhibitors, viral polymerases. Elsevier, pp 271–300

131. Gao Y, Yan L, Huang Y, Liu F, Zhao Y, Cao L, Wang T, Sun Q, Ming Z, Zhang L (2020) Structure of the RNA-dependent RNA polymerase from COVID-19 virus. Science 368(6492):779–782

132. Jiang Y, Yin W, Xu HE (2020) RNA-dependent RNA polymerase: Structure, mechanism, and drug discovery for COVID-19. Biochem Biophys Res Commun

133. Kirchdoerfer RN, Ward AB (2019) Structure of the SARS-CoV nsp12 polymerase bound to nsp7 and nsp8 co-factors. Nat Commun 10(1):1–9
134. Wang M, Cao R, Zhang L, Yang X, Liu J, Xu M, Shi Z, Hu Z, Zhong W, Xiao G (2020) Remdesivir and chloroquine effectively inhibit the recently emerged novel coronavirus (2019-nCoV) in vitro. Cell Res 30(3):269–271
135. Martinez MA (2020) Compounds with therapeutic potential against novel respiratory 2019 coronavirus. Antimicrob Agents Chemotherapy 64(5)
136. Tao YY, Tang LV, Hu Y (2020) Treatments in the COVID-19 pandemic: an update on clinical trials. Taylor & Francis
137. Dong L, Hu S, Gao J (2020) Discovering drugs to treat coronavirus disease 2019 (COVID-19). Drug Discov Therap 14(1):58–60
138. Ju J, Li X, Kumar S, Jockusch S, Chien M, Tao C, Morozova I, Kalachikov S, Kirchdoerfer R, Russo JJ (2020) Nucleotide analogues as inhibitors of SARS-CoV polymerase. BioRxiv
139. Ramezankhani R, Solhi R, Memarnejadian A, Nami F, Hashemian SM, Tricot T, Vosough M, Verfaillie C (2020) Therapeutic modalities and novel approaches in regenerative medicine for COVID-19. Int J Antimicrob Agents 106208
140. Mickolajczyk KJ, Shelton PM, Grasso M, Cao X, Warrington SR, Aher A, Liu S, Kapoor TM (2020) Force-dependent stimulation of RNA unwinding by SARS-CoV-2 nsp13 helicase. BioRxiv
141. Jia Z, Yan L, Ren Z, Wu L, Wang J, Guo J, Zheng L, Ming Z, Zhang L, Lou Z (2019) Delicate structural coordination of the Severe Acute Respiratory Syndrome coronavirus Nsp13 upon ATP hydrolysis. Nucl Acids Res 47(12):6538–6550
142. Hao W, Wojdyla JA, Zhao R, Han R, Das R, Zlatev I, Manoharan M, Wang M, Cui S (2017) Crystal structure of Middle East respiratory syndrome coronavirus helicase. PLoS Pathog 13(6):
143. Russo M, Moccia S, Spagnuolo C, Tedesco I, Russo GL (2020) Roles of flavonoids against coronavirus infection. Chem-Biol Interact 109211
144. Shu T, Huang M, Di Wu YR, Zhang X, Han Y, Mu J, Wang R, Qiu Y, Zhang D-Y, Zhou X (2020) SARS-coronavirus-2 Nsp13 possesses NTPase and RNA helicase activities that can be inhibited by bismuth salts. Virologica Sinica 1
145. Yu M-S, Lee J, Lee JM, Kim Y, Chin Y-W, Jee J-G, Keum Y-S, Jeong Y-J (2012) Identification of myricetin and scutellarein as novel chemical inhibitors of the SARS coronavirus helicase, nsP13. Bioorg Med Chem Lett 22(12):4049–4054
146. Keum Y-S, Jeong Y-J (2012) Development of chemical inhibitors of the SARS coronavirus: viral helicase as a potential target. Biochem Pharmacol 84(10):1351–1358
147. Nittari G, Pallotta G, Amenta F, Tayebati SK (2020) Current pharmacological treatments for SARS-COV-2: a narrative review. Eur J Pharmacol 173328
148. Warrington R, Watson W, Kim HL, Antonetti FR (2011) An introduction to immunology and immunopathology. Allergy Asthma Clin Immunol 7(S1):S1
149. Christiaansen A, Varga SM, Spencer JV (2015) Viral manipulation of the host immune response. Curr Opin Immunol 36:54–60
150. Magro G (2020) COVID-19: review on latest available drugs and therapies against SARS-CoV-2. Coagulation inflammation cross-talking. Virus Res 198070
151. Qin C, Zhou L, Hu Z, Zhang S, Yang S, Tao Y, Xie C, Ma K, Shang K, Wang W (2020) Dysregulation of immune response in patients with COVID-19 in Wuhan, China. Clin Infect Dis
152. Shen C, Wang Z, Zhao F, Yang Y, Li J, Yuan J, Wang F, Li D, Yang M, Xing L (2020) Treatment of 5 critically ill patients with COVID-19 with convalescent plasma. JAMA 323(16):1582–1589
153. Khosravani H, Steinberg L, Incardona N, Quail P, Perri G-A (2020) Symptom management and end-of-life care of residents with COVID-19 in long-term care homes. Can Fam Physician 66(6):404–406
154. Lovell N, Maddocks M, Etkind SN, Taylor K, Carey I, Vora V, Marsh L, Higginson IJ, Prentice W, Edmonds P (2020) Characteristics, symptom management and outcomes of 101 patients with COVID-19 referred for hospital palliative care. J Pain Symptom Manage

155. Yang X, Liu Y, Liu Y, Yang Q, Wu X, Huang X, Liu H, Cai W, Ma G (2020) Medication therapy strategies for the coronavirus disease 2019 (COVID-19): recent progress and challenges. Expert Rev Clin Pharmacol 13(9):957–975

156. Thickett DR, Armstrong L, Christie SJ, Millar AB (2001) Vascular endothelial growth factor may contribute to increased vascular permeability in acute respiratory distress syndrome. Am J Respir Crit Care Med 164(9):1601–1605

157. Ekström MP, Bornefalk-Hermansson A, Abernethy AP, Currow DC (2014) Safety of benzo-diazepines and opioids in very severe respiratory disease: national prospective study. BMJ 348

158. Speiser DE, Bachmann MF (2020) COVID-19: mechanisms of vaccination and immunity. Vaccines 8(3):404

159. Lurie N, Saville M, Hatchett R, Halton J (2020) Developing Covid-19 vaccines at pandemic speed. N Engl J Med 382(21):1969–1973

160. Haque A, Pant AB (2020) Efforts at COVID-19 vaccine development: challenges and successes. Vaccines 8(4):739

161. Dong Y, Dai T, Wei Y, Zhang L, Zheng M, Zhou F (2020) A systematic review of SARS-CoV-2 vaccine candidates. Sig Transduction Targeted Therapy 5(1):1–14

162. Liu C, Zhou Q, Li Y, Garner LV, Watkins SP, Carter LJ, Smoot J, Gregg AC, Daniels AD, Jervey S (2020) Research and development on therapeutic agents and vaccines for COVID-19 and related human coronavirus diseases. ACS Publications

163. WHO (2021) Draft landscape of COVID-19 candidate vaccines. https://www.who.int/public ations/m/item/draft-landscape-of-covid-19-candidate-vaccines

164. Saade F, Petrovsky N (2012) Technologies for enhanced efficacy of DNA vaccines. Expert Rev Vaccines 11(2):189–209

165. Oroojalian F, Haghbin A, Baradaran B, Hemat N, Shahbazi M-A, Baghi HB, Mokhtarzadeh A, Hamblin MR (2020) Novel insights into the treatment of SARS-CoV-2 infection: an overview of current clinical trials. Int J Biol Macromol

166. Pardi N, Hogan MJ, Porter FW, Weissman D (2018) mRNA vaccines—a new era in vaccinology. Nat Rev Drug Discovery 17(4):261

167. NNanomedicine and the COVID-19 vaccines. Nat Nanotechnol 15(12):963–963

168. McKay PF, Hu K, Blakney AK, Samnuan K, Brown JC, Penn R, Zhou J, Bouton CR, Rogers P, Polra K (2020) Self-amplifying RNA SARS-CoV-2 lipid nanoparticle vaccine candidate induces high neutralizing antibody titers in mice. Nature Commun 11(1):1–7

169. Ghorbani A, Zare F, Sazegari S, Afsharifar A, Eskandari MH, Pormohammad A (2020) Development of a novel platform of virus-like particle (VLP)-based vaccine against COVID-19 by exposing epitopes: an immunoinformatics approach. New Microbes New Infect 38:

Chapter 2
Ongoing Clinical Trials and the Potential Therapeutics for COVID-19 Treatment

Hossein Abolhassani, Ghazal Bashiri, Mahdi Montazeri, Hasan Kouchakzadeh, Seyed Abbas Shojaosadati, and Seyed Ehsan Ranaei Siadat

Keywords SARS-CoV-2 · COVID-19 · Combination therapy · Clinical trials · Drug repurposing · Antivirals

2.1   Introduction

It is demonstrated that Coronavirus disease 2019 (COVID-19) pathogenesis is involved with both the direct harm inflicted by severe acute respiratory syndrome coronavirus 2 (SARS-CoV-2) and on the other hand, excessive inflammatory and immune response from the host [1]. Taking the biology and viral pathogenesis of SARS-CoV-2, and the potential treatment mechanisms of the virus into account thoroughly in previos chapter, many therapeutics and medications have been proposed to be efficacious against the COVID-19 pandemic [1–3]. On account of the fact that there are no particular treatment options available for COVID-19, the drug repurposing approach has been taken into consideration as a promising strategy for the treatment of SARS-CoV-2 infection [4, 5]. Among them, antivirals have demonstrated satisfactory inhibitory effects against COVID-19 in vitro, in vivo, and in clinical

Hossein Abolhassani, Ghazal Bashiri, and Mahdi Montazeri contributed equally to this work.

H. Abolhassani · G. Bashiri · S. A. Shojaosadati (✉)
Biotechnology Group, Faculty of Chemical Engineering, Tarbiat Modares University, Tehran, Iran
e-mail: shoja_sa@modares.ac.ir

M. Montazeri · H. Kouchakzadeh (✉)
Protein Research Center, Shahid Beheshti University, Tehran, Iran
e-mail: h_kouchakzadeh@sbu.ac.ir

S. E. R. Siadat
Rahpouyan Fanavar Sadegh Company, Pardis Technology Park, 20th Km of Damavand Road, Tehran, Iran

© The Author(s), under exclusive license to Springer Nature Singapore Pte Ltd. 2021
M. Rahmandoust and S.-O. Ranaei-Siadat (eds.), *COVID-19*,
https://doi.org/10.1007/978-981-16-3108-5_2

conditions as well [5, 6]. On the other hand, as the severe patients are generally associated with the acute respiratory distress syndrome (ARDS), acute lung injury (ALI), and cytokine storm, immunomodulators and anti-inflammatory drugs, as well as biological products, have been employed aiming to enhance the innate immune system and alleviate the damage caused by the deregulated inflammatory responses to manage the infection and control the symptoms leading to surviving the severe patients [7, 8]. Many therapeutic strategies and medications acting on targets of the virus or on the targets of the host have been proposed and are being developed in several clinical studies to be evaluated regarding their safety and efficacy against COVID-19 (Table 2.1) [1, 4]. The therapeutic interventions, medications, biological, and natural products alongside combination therapy approach that may have a promising role in suppressing COVID-19 is highlighted based on current evidence in this chapter./Para>

2.2   Antivirals/Anti-HIV and Antimalarials

Viruses are obligate, intracellular parasites containing either RNA or DNA that utilize host cells for their reproduction [107]. Viruses such as HIV, herpes simplex, varicella-zoster, respiratory syncytial, cytomegalovirus, HBV, HCV, or influenza virus are known to be associated with the development of a wide range of infections [108]. In the early 1950s, with research on anticancer drugs, advancements in developing antiviral chemotherapy, particularly in compounds preventing viral replication, are commenced [109]. Having expertise in the mechanisms of viral replication has assisted scientists in comprehending the viral life cycle, thereby finding potential antiviral agents for each step of replication [109]. The efficacy of antiviral agents heavily depends on their potency and therapeutic index. That is, besides their damaging effects on viruses, they should remain non-toxic to the host cells. In this regard, target sites special to viruses, without any human homolog, can aid in achieving a high therapeutic index [110]. Since the outbreak of a new infection, COVID-19, health professionals have been trying to find proper drugs for the treatment of infected patients [110].

However, repurposing available antiviral/anti-HIV and antimalarial drugs, with known safety, dosages, and pharmacokinetic properties, is recently gaining attention, given the limited time and high cost required for discovering new drugs [5, 110]. In this regard, many antivirals have been employed to test their efficiency and safety against COVID-19 (Table 2.1). As a result, several therapeutics such as remdesivir, favipiravir, arbidol as well as the combination of lopinavir and ritonavir are identified as potent agents against COVID-19 by WHO [5].

Table 2.1 Potential therapeutics with their clinical trials for the treatment of COVID-19

Therapeutic	Target	Effect	Clinical trial phase (NCT number) (https://clinicaltrials.gov)	Refs.
Antivirals				
Remdesivir	RdRP (virus)	Inhibiting viral replication	Phase 3 (NCT04501952), Phase 3 (NCT04292899)	[9–14]
CQ/HCQ	Endosoamal pH, ACE2 (host)	Inhibiting viral entry and post entry	Phase 4 (NCT04331600), Phase 4 (NCT04382625)	[5, 15–18]
Kaletra (lopinavir/ritonavir)	3CL$_{pro}$ (virus)	Inhibiting protease activity in the replication cycle	Phase 4 (NCT04252885), Phase 3 (NCT04328012), Phase 3 (NCT04321174)	[19, 20]
Umifenovir (arbidol)	S protein/ACE2 interaction (virus)	Inhibiting viral entry	Phase 4 (NCT04260594)	[21–23]
Favipiravir	RdRP (virus)	Inhibiting viral replication	Phase 3 (NCT04600895), Phase2 (NCT04358549)	[24–27]
Oseltamivir (tamiflu)	Neuraminidase (virus)	Inhibiting release of viral particles from host cells	Phase 4 (NCT04255017), Phase 3 (NCT04558463)	[28, 29]
Ribavirin	Inosine monophosphate dehydrogenase, RdRP (virus)	Inhibiting viral replication	Phase 3 (NCT04392427), Phase 2 (NCT04605588)	[5, 30, 31]
Darunavir/cobicistat	3CL$_{pro}$ (virus)	Inhibiting protease activity in the replication cycle	Phase 3 (NCT04252274)	[32–34]
Tenofovir/emtricitabine	Nucleoside reverse transcriptase (virus)	Inhibiting viral replication	Phase 2/3 (NCT04359095), Phase 2/3 (NCT04519125)	[35, 36]
Camostat mesylate	TMPRSS2 (host)	Inhibiting viral entry and fusion	Phase 3 (NCT04608266), Phase 2/3 (NCT04608266)	[37–39]
Nafamostat mesylate	TMPRSS2 (host)	Inhibiting viral entry and fusion	Phase 2/3 (NCT04473053)	[16, 40]

(continued)

Table 2.1 (continued)

Therapeutic	Target	Effect	Clinical trial phase (NCT number) (https://clinicaltria ls.gov)	Refs.
APN01	S protein (virus)	Inhibiting viral entry	Phase 2 (NCT04335136)	[3, 41, 42]
Molnupiravir	RdRP (virus)	Inhibiting viral replication	Phase 2/3 (NCT04575584), Phase 2/3 (NCT04575597)	[43–45]
Sofosbuvir	RdRp (virus)	Inhibiting viral replication	Phase 2/3 (NCT04497649)	[46, 47]
Ivermectin	Nuclear transport process (virus)	Inhibiting viral replication	Phase 3 (NCT04530474), Phase 2/3 (NCT04422561)	[4, 5, 48]
Losartan	ACE2 (host)	Inhibiting viral entry	Phase 3 (NCT04606563), Phase 2 (NCT04312009)	[38, 49]
Valsartan	ACE2 (host)	Inhibiting viral entry	Phase 4 (NCT04335786)	[7, 49]
Famotidine	3CLpro (virus) (feasible)	Inhibiting protease activity in the replication cycle	Phase 3 (NCT04370262), Phase 3 (NCT04389567)	[50, 51]
Novaferon	Replication system (virus (	Inhibiting viral replication	Phase 3 (NCT04669015)	[52, 53]
Bromhexine	TMPRSS2 (host)	Inhibiting viral entry and fusion	Phase 4 (NCT04405999)	[7, 49]
Nitazoxanide	Immune interferon response (host) (feasible)	reducing cytokine storm (feasible)	Phase 4 (NCT04406246) Phase 3 (NCT04359680)	[5, 6, 54]
Nelfinavir	3CLpro (virus)	Inhibiting protease activity in the replication cycle	–	[55–57]
Auranofin	Redox enzymes, endoplasmic reticulum (ER) (host)	Inhibiting viral replication, reducing cytokine storm	–	[58–62]
Carmofur	3CLpro (virus)	Inhibiting protease activity in the replication cycle	–	[63, 64]

(continued)

Table 2.1 (continued)

Therapeutic	Target	Effect	Clinical trial phase (NCT number) (https://clinicaltrials.gov)	Refs.
Niclosamide	3CLpro (virus)	Inhibiting protease activity in the replication cycle	Phase 3 (NCT04558021), Phase 2/3 (NCT04603924)	[6, 65]
Galidesivir	RdRp (virus)	Inhibiting viral replication	Phase 1 (NCT03891420)	[38, 66, 67]
Azvudine	Reverse transcriptase (virus)	Inhibiting viral replication	Phase 3 (NCT04668235)	[38, 68]
Quercetin	S protein/ACE2 interaction (virus) (feasible)	Inhibiting viral entry and pathogenesis, reducing cytokine storm	N/A phase (NCT04377789), Phase 2 (NCT04536090)	[69]
Immunomodulators and Anti-inflammatory Drugs				
NK cells	Immune system (host)	Boosting innate and adaptive immunity	Phase 1/2 (NCT04365101), Phase 1 (NCT04280224)	[1, 7, 62]
MSCs	Immune system (host)	Immunomodulation, reducing cytokine storm, regenerating tissues	Phase 3 (NCT04371393), phase 2 (NCT04466098)	[70, 71]
type I IFNs (IFN-α, IFN-β)	IFNAR signaling (host)	Boosting immunity against viral infection and replication	Phase 3 (NCT04320238), Phase 2 (NCT04385095)	[4, 8, 49, 72–74]
CP/IVIG	S protein (virus)	Boosting immunity against viral entry and pathogenesis	Phase 3 (NCT04418518), Phase 2/3 (NCT04374526)	[75, 76]
Tocilizumab	IL-6 (host)	Reducing cytokine storm	Phase 4 (NCT04377750), Phase 3 (NCT04320615)	[38, 77]
Sarilumab	IL-6 (host)	Reducing cytokine storm	Phase 4 (NCT02735707), Phase 3 (NCT04327388)	[38, 78]

(continued)

Table 2.1 (continued)

Therapeutic	Target	Effect	Clinical trial phase (NCT number) (https://clinicaltrials.gov)	Refs.
Eculizumab	C5-activated complement (host)	Reducing cytokine storm and tissue damage induced by inflammation	Phase 2 (NCT04346797)	[79, 80]
Bevacizumab	VEGF (host)	Reducing endothelial injury and microvascular permeability induced by ARDS and ALI (feasible)	Phase 2 (NCT04275414), Phase 2 (NCT04344782)	[1, 81]
Infliximab	TNF-α (host)	Reducing cytokine storm	Phase 3 (NCT04593940), Phase 2 (NCT04425538)	[82–84]
Anakinra	IL-1 (host)	Reducing cytokine storm	Phase 3 (NCT04680949), Phase 3 (NCT04362111)	[85, 86]
Emapalumab (in combination with anakinra)	INF-γ and IL-1 (host)	Reducing cytokine storm	Phase2/3 (NCT04324021)	[87, 88]
Meplazumab	CD147 (host)	Inhibiting viral entry, reducing cytokine storm	Phase 2/3 (NCT04586153), Phase 1/2 (NCT04275245)	[89, 90]
Sirolimus	mTORC1 (host)	Inhibiting viral replication, reducing cytokine storm (feasible)	Phase 2 (NCT04341675), Phase 1/2 (NCT04482712)	[91, 92]
Baricitinib	JAK (host)	Reducing cytokine storm	Phase 3 (NCT04421027), Phase 2/3 (NCT04340232)	[3, 5]
Ruxolitinib	JAK (host)	Reducing cytokine storm	Phase 3 (NCT04377620), Phase 3 (NCT04362137)	[49, 93]

(continued)

Table 2.1 (continued)

Therapeutic	Target	Effect	Clinical trial phase (NCT number) (https://clinicaltrials.gov)	Refs.
Imatinib	Tyrosine kinase (host)	Inhibiting viral entry, reducing cytokine storm (feasible)	Phase 3 (NCT04394416), Phase 3 (NCT04422678)	[62, 94]
Tofacitinib	JAK (host)	Reducing cytokine storm	Phase 2 (NCT04415151), Phase 2 (NCT04469114)	[38, 62]
Cyclosporine	IL-2 (host)	Reducing cytokine storm	Phase 4 (NCT04392531), Phase 2 (NCT04492891)	[4]
Fingolimod	Sphingosine-1-phosphate receptor (host)	Inhibiting inflammation and autoimmune reaction by sequestering T cells in lymph nodes	Phase 2 (NCT04280588/withdrawn)	[1, 95]
Thalidomide	TNF-α (host)	Reducing cytokine storm	Phase 2 (NCT04273529), Phase 2 (NCT04273581)	[1, 96]
NSAIDs	Cyclooxygenase enzymes (host)	Inhibiting the production of prostaglandins and inflammation (feasible)	Phase 3 (NCT04325633), Phase 4 (NCT04334629)	[38, 97, 98]
Low molecular weight heparins	Factor Xa (host)	Reducing coagulation	Phase 4 (NCT04584580) Phase 3 (NCT04401293)	[62, 99]
Azithromycin	S protein/ACE2 interaction (virus) (feasible)	Inhibiting viral entry and translation (feasible)	Phase 3 (NCT04381962), Phase 3 (NCT04332107)	[4, 100]
Methylprednisolone	Steroid receptors in inflammatory cells (host)	Immunosuppressing, reducing cytokine storm	Phase 4 (NCT04263402), Phase 3 (NCT04438980)	[85, 101]
Dexamethasone	Steroid receptors in inflammatory cells (host)	Immunosuppressing, reducing cytokine storm	Phase 4 (NCT04663555), phase 3 (NCT04327401)	[102, 103]

(continued)

Table 2.1 (continued)

Therapeutic	Target	Effect	Clinical trial phase (NCT number) (https://clinicaltrials.gov)	Refs.
Vitamin D	Immune system (host)	Boosting immunity, reducing cytokine storm	Phase 4 (NCT04411446), Phase 4 (NCT04552951)	[38, 104]
Vitamin C	Immune system (host)	Boosting immunity, reducing cytokine storm	Phase 3 (NCT04401150), Phase 2 (NCT04363216)	[85, 105]
Zinc	Immune system (host)	Inhibiting viral entry and pathogenesis, reducing cytokine storm	Phase 4 (NCT04621461)	[69, 106]

2.2.1 *Remdesivir*

By the arisen of the COVID-19 pandemic caused by SARS-CoV-2, remdesivir is being considered as one of the highly potential therapeutic agents for the treatment of COVID-19 [111]. The researches on developing remdesivir commenced by the cooperation between the United State Army Medical Research Institute of Infectious Diseases (USAMRIID) and the Gilead—the U.S. Centers for Disease Control and Prevention (CDC) to provide potential anti-viral therapeutic agents against RNA-based viruses, namely, Ebola virus and the Coronaviridae family viruses [9]. This led to the compilation of a library for nucleoside analogs, small molecules with antiviral activity against infections such as HBV, HIV, and herpes viruses [112, 113].

For nucleosides to become their active metabolites, it is requisite to undergo intracellular phosphorylation [14]. It should be noted that the development of the nucleoside monophosphate is the rate-limiting step for their intracellular activation [114]. Accordingly, nucleosides were modified to phosphoramidate, ester, and monophosphate prodrugs, enhancing both their intracellular delivery and activation [14, 112, 115–117]. With the outbreak of the Ebola virus in West Africa (2013–2016), the library of nucleoside molecules was appraised to find the most potential ones against the virus, resulting in the identification of remdesivir (formerly GS-5734), a monophosphoramidate prodrug of the 1′-cyano-substituted nucleoside analog (GS-441524) [11, 13, 118]. Even though remdesivir was a potential therapeutic agent for the treatment of the Ebola virus and its safety profile in the human population was established, it was outdone by monoclonal antibodies, namely, Zmapp (triple monoclonal antibody cocktail), MAb114 (single monoclonal antibody), and REGN-EB3 (a cocktail of three monoclonal antibodies), in phase 3 clinical trial. Hence, remdesivir is not being utilized in this regard anymore [10, 119, 120]. However, in addition to Ebola virus, remdesivir has demonstrated wide antiviral activities against MERS-CoV, SARS-CoVs, Marburg virus, respiratory syncytial virus, HCV, and several paramyxoviruses [14, 72, 121, 122].

For remdesivir (GS-5734) to be converted into its active metabolite, it undergoes intracellular metabolic conversion [123]. Once remdesivir (GS-5734) enters cells, it is metabolized into an alanine metabolite (GS-704277), processed into the monophosphate derivative, and then it is converted into its active form of nucleoside triphosphate (NTP) [9, 13]. Owing to the fact that the resultant NTP resembles the natural nucleotide, that is, ATP, it could be misleadingly considered by the RdRp as a nucleotide for incorporation into the nascent RNA strand, thereby bringing the replication of RNA to a halt [124–127]. It should be noted that CoVs have a proof-reading ability enabling the virus to remove wrongly incorporated nucleosides [123, 124]. However, remdesivir seems to be capable of suppressing such activities due to the mechanism of its inhibitory effect and delayed RNA chain termination [128].

As it was shown in a series of recent studies on SARS-CoV-2 RdRp and MERS-CoV RdRp, the inhibition of the RNA replication cannot happen immediately after the addition of remdesivir. Rather, it occurs after three nucleotides were added into the nascent RNA [129]. Thus, remdesivir inhibits further growth of the RNA strand by the delayed RNA chain termination phenomenon; meanwhile, the three added nucleotides might account for the protection of inhibitor (remdesivir) from excision by the viral 3'–5' exonuclease activity, which is responsible for the proofreading ability of the CoVs [12, 125].

Many clinical studies are aiming to assess the efficiency of remdesivir for SARS-CoV-2-infected patients. Remdesivir is known to be well tolerated in clinical studies and compassionate use [130–132]. However, its main adverse effects may include multiple organ-dysfunction syndromes, septic shock, acute kidney injury (AKI), and hypotension [133]. In a compassionate use of remdesivir for patients with COVID-19 infection, patients received a 10-day course of treatment with remdesivir (200 mg on day 1 and 100 mg daily for 9 days). The results showed 68% (36 of 53 patients) clinical improvement in patients. However, increased hepatic enzymes, diarrhea, rash, renal impairment, and hypotension were the most common adverse events experienced by patients under treatment, particularly those patients receiving invasive ventilation. Serious adverse events were observed for 12 patients (23%), among which multiple organ-dysfunction syndrome, septic shock, acute kidney injury, and hypotension were most common in patients receiving invasive ventilation at the baseline [132].

In a randomized, double-blind, placebo-controlled, multicenter trial on patients with severe COVID-19 at 10 hospitals in Hubei, China, 237 patients randomly received either placebo or remdesivir (158 remdesivir and 79 placebo). The results showed that patients under treatment with remdesivir had faster clinical improvements compared with those receiving placebo; however, differences were not significant. 66% of patients receiving remdesivir and 64% of those receiving placebo experienced some adverse events [134].

Based on the preliminary results of the first stage of the Adaptive COVID-19 Treatment Trial (ACTT-1), in which patients with COVID-19 were randomly given either remdesivir or placebo as a control group, from among 1059 patients, 538 were assigned to remdesivir and 521 to placebo while the outcome of the study was the time to recovery. The results showed that the patients receiving remdesivir recovered in 11 days; however, recovery time for those receiving placebo was 15 days. Moreover, the Kaplan–Meier estimates of the death rate by 14 days in patients treated with remdesivir and placebo were 7.1% and 11.9%, respectively [131].

In a phase 3 SIMPLE trial, the effect of receiving remdesivir for either 5 or 10 days plus standard of care versus standard of care alone was assessed for patients with moderate COVID-19 pneumonia. The results showed that patients receiving remdesivir for 5 days were 65% more likely to have clinical improvement at day 11 compared with those receiving standard of care alone; however, no significant differences were observed between patients treated with remdesivir for 10 days and those receiving SOC alone.

In the SIMPLE-severe study on patients with SARS-CoV-2 receiving remdesivir for either 5 days or 10 days (200 mg on day 1 and 100 mg daily), from among 397 patients, 200 patients were under treatment for 5 days while 197 patients were under treatment for 10 days. For both groups, the percentage of patients with adverse events was similar (70% in the 5-day group and 74% in the 10-day group). Among all patients, 21% of patients treated for 5 days and 35% of patients treated for 10 days experienced serious adverse events. Also, serious adverse events of Grade 3 or higher for patients receiving remdesivir for 5 days and 10 days were 30% and 43%, respectively. The most common adverse events experienced were nausea (10% in the 5-day group vs. 9% in the 10-day group), increased alanine aminotransferase (6% vs. 8%), acute respiratory failure (6% vs. 11%), and constipation (7% in both groups) while 4% of the 5-day group and 10% of the 10-day group discontinued treatment due to adverse events. The most common serious adverse events in patients receiving remdesivir for 10 days were the acute respiratory failure (9%, vs. 5%) and respiratory failure (5%, vs. 2%). It should be mentioned that for patients with severe COVID-19 independent of mechanical ventilation, no noticeable differences were observed for patients treated with remdesivir for 5 days and 10 days [130]. According to the newest information released from mortality trials, recommended by the WHO expert groups, in hospitalized patients infected with COVID-19, remdesivir had little or no effect on inpatient overall mortality, initiation of ventilation, and duration of hospital stay [135]. Nevertheless, remdesivir, as the first treatment for COVID-19 patients requiring hospitalization, was approved by the FDA after a phase 3 clinical study sponsored by Gilead Sciences (NCT04292899).

2.2.2 Chloroquine and Hydoroxychloroquine

Chloroquine, an amine acidotropic form of quinine, was first synthesized as an anti-malaria drug in 1934 and has been utilized for the treatment and prophylaxis of malaria for many years [15]. In 1946, hydroxychloroquine sulfate, hydroxyl analog of the chloroquine, was synthesized by introducing a hydroxyl group into CQ. Both CQ and HCQ have been utilized for the treatment of malaria, lupus, and rheumatoid arthritis for many years [16, 136]. They share similarities such as pharmacokinetics, mode of action, indications, and type of drug toxicity; however, they slightly differ in the clinical indications and toxic doses [15, 137]. Even though the utilization of both CQ and HCQ for the treatment of malaria is being limited owing to the arisen of chloroquine-resistant P. falciparum strains, they have shown broad-spectrum activities against bacterial, fungal, and viral infections such as autoimmune diseases [15, 138].

Inhibitory effects of CQ and HCQ against CoVs could be fulfilled in various ways [139, 140]. A complete review of their mechanisms of action can be found elsewhere [141]. Despite observed controversy regarding the exact mechanism of action, it was proved that the very mechanism of action, for CoVs entry, is mainly dependent on not only the type of the virus but also the type of the host cells [139, 142, 143]. Since the interaction of COVID-19 S protein with the receptor ACE2 on the host cells is a critical step for initiating the infection process, one of the feasible inhibitory effects of CQ on viral attachment could be through impairing terminal glycosylation of the ACE2 receptor, thereby impeding viral binding and its subsequent entry [18, 144]. On the other hand, another possible way of inhibition could happen through the interaction of CQ and HCQ with viral S proteins, thus preventing the binding of

S proteins on the host cell membrane receptors according to some in silico studies [145]. Moreover, as it is known, for CoVs, the endocytic pathway is one of the chief mechanisms of viral entry into host cells [139]. In this regard, on account of the weak diprotic base nature of CQ and HCQ, their accumulation in acidic organelles such as lysosomes and endosomes increases the pH of their surrounding ambient [5]. Hence, CQ and HCQ are able to prevent the attachment and subsequent entry of the virus mainly dependent on the acidic endo-lysosomal pH, by inhibiting the acidification of the lysosome in that enzymatic protease activities responsible for the cleavage of S protein and subsequent viral entry [146]. In addition, the elevation of pH caused by CQ and HCQ could impair not only the correct maturation and recognition of viral antigens by dendritic cells but also the maturation process of viral proteins completed in the ERGIC and trans-Golgi network (TGN) vesicles both of which require acidic pH [15, 141]. Furthermore, the inhibition of the autophagic process by CQ and HCQ could be involved with the effects of COVID-19 prevention. The viral assembly process occurs in the ERGIC, related directly to autophagosome biogenesis. After the use of CQ/HCQ, the autophagic process could be inhibited by the subsequent pH elevation in lysosomes leading to the SARS-COV-2 halt. Besides, the inhibition of the autophagic process might also associate with the activity suppression of the recycled materials accompanying the autophagic process accounting for the nucleation and replication process of COVID-19 [16].

Moreover, apart from its anti-viral activity, HCQ could act as an anti-inflammatory agent capable of decreasing the production of some cytokines [17, 147]. The secretion of cytokines, such as IL-1β, IL-1RA, IL-6, IL-7, IL-8, IL-2R, TNF-α, known as the cytokine storm, is associated with the disease severity [148, 149]. The possible mechanisms of CQ and HCQ are their involvement in anti-thrombotic activities, and suppressing the release of IL-6, IL-1β, and TNF-α, which are key modulators of inflammation [141, 148].

There are several clinical studies conducted to assess the efficacy of HCQ and CQ on patients infected with COVID-19 [16]. In a study, it was observed that the patients treated with CQ experienced a faster and higher rate of viral suppression compared with those patients in the control group [150]. In another study, the effects of high dosage and low dosage of the CQ on patients infected with severe COVID-19 was assessed, and the results indicated that the higher dosage of CQ should not be used for severe COVID-19 patients, since it might cause a safety hazard, particularly when used with azithromycin and oseltamivir [151]. According to the results of a study, HCQ brought a decreased mortality in critical patients infected with COVID-19 [152]. However, contradictory results were also obtained. For instance, Mahevas et al. found that HCQ could not significantly decrease the admission to ICU, death, or ARDS in COVID-19 patients with hypoxemic pneumonia [153]. According to the findings of another study, it was also demonstrated that CQ cannot prevent the SARS-CoV-2 entry into the lung cells in vitro, in that CQ targets a pathway for viral activation that is not active in the lung cells [154]. Similarly, Mallat et al. indicated that HCQ resulted in a slower viral clearance and mild to moderate disease compared to the control group in patients infected with COVID-19 [155]. On June 15, 2020, the FDA revoked the emergency use authorization for both CQ and HCQ [16] and according to the newest information released from mortality trials, recommended by the WHO expert groups, in hospitalized patients infected with COVID-19, HCQ had little or no effect on inpatient overall mortality, initiation of ventilation, and duration of the hospital stay [135]. Nonetheless, the efficiency of CQ/HCQ as an antiviral treatment for COVID-19 is still assessing in phase 4 clinical studies in the USA (NCT04331600, NCT04382625). HCQ was proved to have 40% less toxicity in animals [156]. However, the most common side effects of both CQ and HCQ at therapeutic doses include myopathy, electrocardiographic changes, bleaching of hair, retinopathy, pruritus, headaches, and gastrointestinal symptoms [5].

2.2.3 Favipiravir (Avigan)

Favipiravir (6-fluoro-3-hydroxy-2-pyrazinecarboxamide, T-705, Avigan), was first discovered by Toyama Chemical Co., Ltd for antiviral activity against the influenza virus and has been approved for the treatment of Influenza in Japan since 2014 because of its proven safety and effectiveness on humans in clinical trials [5, 26, 27]. Concerning COVID-19, favipiravir was approved to be utilized on 15 February 2020, in China against SARS-CoV-2 [157]. For favipiravir to be converted to its active form, that is, favipiravir-RTP (T-705 RTP) undergoes intra-cellularly phosphoribosylation, consequently exerting its antiviral activity as a pro-drug [27]. Also, it was shown that favipiravir-RTP could be efficiently recognized as a guanosine and an adenosine analog by influenza A virus polymerase [158]. Favipiravir triphosphate, a purine nucleoside analog, is believed to directly inhibit the RdRp activity of influenza A virus polymerase [25, 158]. However, the exact mode of action and accurate molecular interaction between the nucleotide and the viral polymerase has yet to be explicated [158]. In a study conducted on the influenza A (H1N1) virus, it was demonstrated that a high rate of mutation is induced with favipiravir generating a non-viable viral phenotype, a lethal mutagenesis which is a key antiviral mechanism of T-705 [159].

Favipiravir has antiviral activity against a great variety of influenza viruses such as A (H1N1) pdm09, A (H5N1), and recently emerged A (H7N9) avian virus. Moreover, favipiravir is capable of inhibiting the influenza strains resisting current antiviral drugs and showing a synergic effect in combination with oseltamivir, thus expanding influenza treatment options [160]. It was shown that its antiviral activity performs in a dose-dependent manner while it has a short half-life of 2–5.5 h [161, 162]. In addition, the metabolism of favipiravir occurs in the liver mainly by aldehyde oxidase (AO), and partially by xanthine oxidase, thereby producing an inactive oxidative metabolite, T-705M1 that is excreted by the kidneys [162]. In a small clinical study conducted on 168 critically ill patients infected with influenza, patients received either a combination of favipiravir and oseltamivir or oseltamivir alone. The results showed that the combination therapy of favipiravir and oseltamivir results in accelerating clinical recovery [163].

Favipiravir, chloroquine, arbidol, and remdesivir are under clinical studies in china to assess their efficacy and safety against SARS-CoV-2 [157]. According to preliminary clinical results obtained from an open-label comparative controlled study of patients infected with COVID-19, patients receiving favipiravir compared with those receiving lopinavir/ritonavir experienced not only faster viral clearance but also better chest computed tomography changes [164]. Furthermore, in an in vitro study conducted on Vero E6 cells, favipiravir inhibited SARS-CoV-2 replication with EC50 values of 61.88 μM (9.4 μg/mL). Nevertheless, another study reported EC50 values > 100 μM (15. 7 μg/mL) for favipiravir [24, 165]. The need for metabolic activation in the host cells for favipiravir could explain the differences between these two studies [24]. In a randomized, controlled, open-label multicenter trial performed on 240 patients infected with COVID-19, patients randomly received arbidol or favipiravir in a 1:1 ratio. According to the results, favipiravir could not considerably improve the clinical recovery rate on day 7 in comparison to arbidol. However, favipiravir appreciably improved the latency to relief for pyrexia and cough and showed mild and manageable adverse effects, including raised serum uric acid, psychiatric symptom reactions, digestive tract reaction, and abnormal LFT [166]. In a double-blinded, placebo-controlled, randomized, phase 3 trial, favipiravir is being administered as a potential therapy for mild to moderate COVID-19 outpatients (NCT04600895).

2.2.4 Lopinavir/Ritonavir (Kaletra)

Lopinavir/ritonavir combination, available under the brand name Kaletra, and developed by Abbott Laboratories, USA, is known as an anti-retroviral drug and was approved by FDA for the treatment of patients infected with HIV in 2000 [5]. Ritonavir, a potent inhibitor of cytochrome P450 3A4, inhibits the metabolism of lopinavir and increases its bioavailability, it was shown that the co-administration of these drugs in healthy volunteers increases the area under the lopinavir plasma concentration–time curve > 100-fold [19, 167]. PLpro, a crucial factor in the protease activity and proper replication of the SARS-CoVs genome has been a target of interest in the treatment of COVID-19 patients [168]. It was demonstrated that lopinavir is a non-covalent, competitive, and potential inhibitor for inhibiting the PLpro of CoVs and subsequently blocking the virus replication [169]. The administration of lopinavir/ritonavir during the early peak viral replication phase (initial 7–10 days) has been reported to be crucial for the efficiency of drugs [170].

In a study, it was demonstrated that after the administration of lopinavir/ritonavir, the viral load and clinical symptoms dramatically decreased [171]. In another study conducted on 36 pediatric patients (aged 0–16 years) infected with COVID-19, all patients received IFN-α by aerosolization twice a day, 14 (39%) patients received lopinavir/ritonavir syrup (twice a day), and 6 patients needed oxygen inhalation. The results indicated that all patients were cured and the hospital stay meantime was 14 days [172]. On the contrary, in a randomized controlled, open-label clinical trial conducted on 199 patients infected with severe COVID-19, no specific difference was observed in patients treated with lopinavir/ritonavir compared to those who received standard care, and gastrointestinal disturbances were more prevalent adverse events between patients treated with lopinavir/ritonavir than patients in the control group [173].

In an open-label, randomized, phase 2 trial in adults infected with COVID-19, patients were assigned to either a 14-day triple combination of IFN-β-1b, lopinavir/ritonavir and ribavirin or a control group (lopinavir/ritonavir); results showed that triple combination therapy was superior to control group regarding decreasing the time of hospital stay and alleviating symptoms in patients with mild to moderate COVID-19 [174]. In another study, four patients infected with COVID-19 underwent treatment with lopinavir/ritonavir (lopinavir 400 mg/ritonavir 100 mg, q12 h through oral route), arbidol (0.2 g, three times in a day through oral route), and Chinese traditional medicine Shufeng Jiedu capsule (SFJDC) (2.08 g, three times in a day through oral route) while the duration of treatment was 6–15 days. According to the obtained results, from among four patients, three patients showed considerable improvement in pneumonia-associated symptoms, and for the other patients suffering from severe pneumonia, signs of improvement were observed [175]. The most common adverse effects of lopinavir/ritonavir have been reported to be diarrhea, nausea, and, vomiting (gastrointestinal adverse effects from mild to moderate). However, less common adverse effects observed in patients treated with lopinavir/ritonavir consist of an allergic reaction, asthenia, malaise, headache, myalgias, arthralgias, myocardial infarction, seizures, and lactic acidosis [20, 167, 176]. Lopinavir/ritonavir is still in phase 4 of a clinical study in China to be evaluated for COVID-19 patients (NCT04252885).

2.2.5 Umifenovir (Arbidol)

Arbidol, or umifenovir, an indole-derivative with broad-spectrum activity against both enveloped and non-enveloped viruses, was initially approved in China and Russia for the treatment of influenza A and B [177, 178]. Arbidol is believed to block the entry of influenza virus (A and B) into the host cells by increasing the stability of the hemagglutinin (HA) and hampering low pH reorganizations necessary for fusion machinery of hemagglutinin with the membrane [5, 21, 23]. Arbidol could interfere with advanced stages of the viral life cycle, in that it is capable of interacting with both viral proteins and lipids [4, 179]. Regarding its structure, the presence of amine in position 4 and the hydroxyl moiety in position 5 is crucial for its antiviral activity [39]. It is reported that 40% of the drug could be excreted unchanged after the administration while its half-life is between 17 and 21 h [22].

In a study conducted on 69 patients infected with SARS-CoV-2 in Wuhan, arbidol therapy led to not only a decrease in the mortality rate but also an improvement in the discharge rate [180]. In another study, the therapeutic efficacy of co-administration of arbidol and lopinavir/ritonavir compared to only lopinavir/ritonavir on COVID-19 patients was evaluated, and the results showed that the combination of arbidol and lopinavir/ritonavir culminates in slowing down the development of lung lesions, decreasing the feasibility of respiratory and gastrointestinal transmission toward decreasing the viral load of COVID-19 [181]. In a clinical trial, 27 COVID-19 patients were recruited, among them, 10 of the patients received chloroquine phosphate, 11 received arbidol, and 6 received lopinavir/ritonavir; the results indicated that both CQ and arbidol decreased the hospitalization time as well as hospitalization expenses and shortened the viral shedding interval [182].

Furthermore, in a study, 200 inpatients infected with common-type COVID-19 received either arbidol hydrochloride capsules (control groups) or a combination of arbidol hydrochloride capsules and Shufeng Jiedu Capsule (SFJDC) (experiment group) for 14 days. The results demonstrated that combining traditional Chinese and western allopathic medicine not only improves recovery time but also has better clinical efficiency and safety [183]. On the contrary, in a clinical trial performed on 141 patients infected with COVID-19, 70 patients received IFN-α-2b, while 71 of them received a combination of arbidol and IFN-α-2b. The outcomes demonstrated that patients receiving co-administration of arbidol and IFN-α-2b experienced neither a decrease in their hospitalization time nor an acceleration in COVID-19 RNA clearance [184]. Likewise, an inefficiency is reported for umifenovir in non-ICU patients [185]. Despite the inconsistent results, arbidol is currently in phase 4 of a clinical trial, which has been conducted on 380 patients with pneumonia caused by SARS-CoV-2 in Ruijin Hospital, Shanghai, China (NCT04260594) [186].

2.2.6 Darunavir

Darunavir, a non-peptidic protease inhibitor (PI) approved by the FDA, is particularly used for the treatment of HIV-1 infection and is majorly utilized in combination with a low boosting dose of ritonavir [38]. Darunavir is more potent compared with other protease inhibitors due to its distinct chemical structure increasing binding affinity and reducing dissociation rate [5]. It has been proved that it is able to prevent viral maturation by inhibiting the cleavage of HIV gag and gag-pol polyproteins alongside inhibiting proteolytic activity and subsequent HIV-1 replication by suppressing dimerization of HIV-1 protease [33]. Therefore, darunavir is recognized as a protease inhibitor while cobicistat could be a supplement for enhancing both pharmacodynamics and pharmacokinetics of darunavir through inhibiting cytochrome P450 (CYP3A) [4, 187]. Darunavir is thoroughly metabolized by hepatic cytochrome P450 (CYP) 3A4 enzymes and is rapidly absorbed after oral intake; moreover, its terminal elimination half-life is 15 h [34]. In a study, it was demonstrated that administration of darunavir is accompanied by an increase in the risk of myocardial infarction in patients infected with HIV. Hence, employing darunavir as a potential therapeutic may be associated with enhancing the risk of cardiovascular diseases [188].

In an open-label trial conducted on 30 patients infected with COVID-19, patients randomly received either darunavir/cobicistat for 5 days on top of IFN-α-2b inhaling or IFN-α-2b inhaling alone. The results showed that darunavir/cobicistat therapy did not change the viral clearance rate at day 7 in comparison to the control group; furthermore, for patients receiving darunavir/cobicistat the median duration of viral shedding from randomization was 8 days, while 7 days in the control group. However, no statistical significance was observed, and the recurrence of adverse events in both groups was similar. On the other hand, one of the patients receiving darunavir/cobicistat developed anemia (a decrease in the level of hemoglobin from 11.3 to 9.9 g/dL). Other observed adverse events were elevated transaminase levels and renal dysfunction. It should be noted that all of the adverse events were mild [189]. In a phase 3 clinical study, the efficacy and safety of darunavir and cobicistat are evaluating on COVID-19 patients in China (NCT04252274).

2.2.7 Ribavirin

Ribavirin, a guanosine analog, is an antiviral drug used for the treatment of patients infected with HCV and respiratory syncytial virus [5]. Its mechanisms of action could be divided into indirect and direct mechanisms. The direct mechanisms consist of interfering with RNA capping, polymerase inhibition, as well as lethal mutagenesis, and indirect mechanisms are comprised of inosine monophosphate dehydrogenase inhibition and immunomodulatory effects [190]. Ribavirin has established a good reputation for being utilized in emergency clinical plans against CoVs infection due to its availability and low cost. The most convincing outcomes generally have been obtained with early administration upon presentation with pneumonia and before sepsis or organ system failure [30]. Its half-life time is estimated to be 3.7 h, with an oral bioavailability of 52%, which could be because of the first-pass metabolism in the liver [31, 191]. Even though ribavirin is known as a potential therapeutic for the treatment of HCV, it is highly toxic. Hence, it is recommended to be used in combination therapy with IFNs or lopinavir/ritonavir in the Diagnosis and Treatment Guidelines of COVID-19 in China [39].

A combination of ribavirin and IFN-α-2b was utilized for the treatment of MERS-CoV infected rhesus macaques and demonstrated a decrease in viral replication, moderating the host response, and improving clinical results [192]. In addition, according to an open-label, randomized, phase 2 trial conducted on patients infected with COVID-19, triple combination therapy of patients with interferon-β-1b, lopinavir/ritonavir, as well as ribavirin was much safer and superior to the administration of only lopinavir/ritonavir regarding the decreasing symptoms, reducing the time of hospital stay, and viral shedding in patients infected with mild to moderate COVID-19 [174]. In order to compare the efficacy and safety of three antivirals, namely, ribavirin, lopinavir/ritonavir, and IFN-α-1b for the treatment of patients infected with COVID-19, three different therapeutic regimes were applied in a clinical trial, that is, ribavirin plus IFN-α1b or lopinavir/ritonavir plus IFN-α1b and or ribavirin plus lopinavir/ritonavir plus IFN-α1b. According to the obtained results, the combination of ribavirin plus lopinavir/ritonavir caused a considerable increase in gastrointestinal adverse effects [193]. A combination of ribavirin, nitazoxanide, and ivermectin for a duration of 7 days is assessed for COVID-19 treatment at Mansoura University in Egypt (NCT04392427).

2.2.8   Oseltamivir (Tamiflu)

Oseltamivir (Tamiflu), a neuraminidase inhibitor (NAIs) licensed for the treatment of both influenzas A and B, was synthesized through employing two natural products from plants, namely, quinic acid, and shikimic acid [29, 194]. Oseltamivir prodrug is known as oseltamivir phosphate [28]. In the liver, oseltamivir is metabolized and converted to its active metabolite, that is, oseltamivir carboxylate [28]. It is able to prevent the release of viral particles from the host cells by binding to influenza viral neuraminidase, thereby decreasing the spread of the virus in the respiratory tract [4, 28]. Nevertheless, according to the result of a study performed on patients infected with COVID-19 in china, no positive results were obtained for patients receiving tamiflu [195]. However, the administration of oseltamivir and its combination with other drugs such as CQ, arbidol, lopinavir/ritonavir, and favipiravir are under clinical studies to evaluate their potential in the treatment of SARS-CoV-2 infection [85, 196]. In an open, prospective/retrospective, randomized controlled cohort study, the efficiency of three antiviral drugs including oseltamivir, arbidol hydrochloride, and lopinavir/ritonavir is compared for COVID-19 treatment in China (NCT04255017).

2.2.9   Ivermectin

Ivermectin, approved as both an anti-parasitic and anthelmintic agent, is a macrolide endectocide macrocyclic lactone that was originally derived from an actinomycete (*streptomyces avermitilis*) [5, 38]. Its antiviral activity was initially found by its capability in inhibiting the interaction between the nuclear transport receptor importin α/β (IMP) and integrase molecule of HIV [4, 48]. In fact, its antiviral mechanism of action involves the dissociation of the preformed IMPα/β1 heterodimer, responsible for the transport of viral proteins to the nuclear [77, 197].

   According to the result of a study conducted in Australia, ivermectin demonstrated antiviral activity against SARS-CoV-2 in clinical isolate in vitro Vero-hSLAM cells with the addition of a single dose 2 h post-infection, and it was able to reduce viral RNA around 5,000 times. Moreover, the hypothesized mechanism of action for ivermectin was observed to be likely through inhibiting IMPα/β1-mediated nuclear import of viral proteins as anticipated [198]. This drug is currently under a phase 3 clinical trial against COVID-19 in the USA, Pennsylvania, Temple University (NCT04530474).

2.2.10   Tenofovir

Tenofovir, a nucleotide analog (NA) of adenosine 5'-monophosphate, is a reverse transcriptase inhibitor with two different formulations, namely, tenofovir disoproxil fumarate (TDF) and tenofovir alafenamide (TAF) [35]. They are commercially available prodrugs of tenofovir capable of improving their oral bioavailability and membrane permeability [35, 36]. Tenofovir alafenamide is able to selectively activate presenting preferential distribution in lymphatic tissues, and it is formulated to reduce adverse events associated with the administration of tenofovir disoproxil fumarate [199]. Both of them are vital components for the treatment of HIV and HBV [35]. Tenofovir is one of the potential nucleotide analogs under investigation for the treatment of SARS-CoV-2 [200]. A combinational administration of tenofovir/emtricitabine in addition to the use of personal protective equipment (PPE) is currently under phase 2/3 of a clinical trial for COVID-19 patients by Hospital Universitario San Ignacio, Colombia (NCT04519125).

2.2.11 Camostat Mesylate

Camostat mesylate, a serine protease inhibitor first used for the treatment of dystrophic epidermolysis, chronic pancreatitis, and oral squamous cell carcinoma, was initially manufactured by the Nichi-Iko Pharmaceutical Co., Ltd. in contribution with Ono Pharmaceutical, Japan [38, 201]. It should be noted that the S protein of human CoVs is primed by TMPRSS2, which is a serine protease [37]. In this regard, camostat mesylate may be able to inhibit the SARS-CoV-2 entry into the host cell, since it is a serine protease inhibitor blocking TMPRSS2 activity [39, 202]. In a study conducted on a pathogenic animal model of SARS-CoV-1 infection, it was observed that camostat has the potential to prevent viral spread and pathogenesis of SARS-CoV-1 [203]. Camostat mesylate is currently under phase 3 of a clinical trial for the treatment of COVID-19 patients in French (NCT04608266).

2.2.12 Nafamostat Mesylate

Nafamostat, a synthetic serine protease inhibitor that is known as an anticoagulant in nature, was first brought to the Japanese market in 1986 for the treatment of acute symptoms of pancreatitis and for applying to certain bleeding complications. This drug is capable of inhibiting different enzymatic systems such as complement, kallikrein-kinin, fibrinolytic systems, and coagulation. It has been also utilized for the prevention of liver transplantation and post-transplant syndrome [5, 38, 204]. Nafamostat is able to prevent viral entry through the host cell surface membrane. Hence, it is considered as one of the potential repurposing drugs against COVID-19 [205]. Its mechanism of action is anticipated to be through inhibiting the human protein TMPRSS2, the S protein-dependent enzyme that cleaves and thereby activates the S protein for binding to ACE2 [206, 207]. In an in vitro study, nafamostat prevented the entry of MERS-CoV, and it was demonstrated as the most potential protease inhibitor among all 1000 drugs screened [40]. This drug is currently under phase 3 of the clinical trial on patients infected with COVID-19 by the University of Edinburgh, the UK (NCT04473053).

2.2.13 Molnupiravir

Molnupiravir, or MK-4482/EIDD-2801, pro-drug of the nucleoside analog N4-hydroxycytidine (NHC), is an RNA polymerase inhibitor that is orally available and was originally developed for the treatment of influenza [43–45]. It has shown appreciable anti-influenza activity in ferrets, mice, guinea pigs, and human airway epithelium organoids [44, 208, 209]. As a result of the collaboration between Ridgeback Biotherapeutics and Merck, molnupiravir is developing for the treatment of COVID-19 patients [43]. In a study, the effect of molnupiravir in a Syrian hamster SARS-CoV-2 infection model was investigated, and the results showed that molnupiravir considerably decreased not only infectious virus titers but also viral RNA loads in the lungs, thereby improving lung histopathology [210]. Moreover, in another in vivo study conducted on animals infected with SARS-CoV-2, molnupiravir was proved to be a potential oral drug capable of considerably decreasing the viral load in the upper respiratory tract and preventing the spread to untreated contact animals [45]. Molnupiravir administration is currently in phase 2/3 of a multicenter clinical trial by Merck Sharp & Dohme Corp (NCT04575584).

2.2.14 Sofosbuvir

Sofosbuvir, a direct-acting antiviral drug that was initially approved as an anti-HCV, could be utilized as a repurposed antiviral drug for the treatment of COVID-19 [46, 47]. Among the studies, it was predicted that sofosbuvir might be capable of binding to the SARS-CoV-2 RdRp enzyme, thereby inhibiting its activity [64, 211, 212]. In a single-center, randomized controlled trial in patients infected with moderate COVID-19, patients received either a combination therapy of sofosbuvir/daclatasvir/ribavirin or standard care. The results demonstrated that the combinational administration of these three drugs engendered recovery and lower mortality rates for patients. Nevertheless, an imbalance was observed in the baseline characteristics between the arms. Thus, larger randomized trials are needed to prove these results [213]. Additionally, according to a molecular docking study, ribavirin, remdesivir, sofosbuvir, galidesivir, and tenofovir are potent drugs against COVID-19 that tightly bind to the RdRp of the SARS-CoV-2 strain, thereby preventing its function [214]. Sofosbuvir is currently under phase 2/3 of a clinical trial in Egypt by Tanta University (NCT04497649).

2.2.15 Famotidine

Famotidine, a histamine-2 receptor antagonist (H2RA), reduces the production of gastric acid [50, 51]. An in vitro study of this drug demonstrated that H2RA has anti-viral activity against HIV replication [215]. Regarding the treatment of COVID-19, according to the results obtained from in silico molecular docking, Famotidine could inhibit PLpro enzyme activity in the viral replication cycle [216]. Hence, this drug is capable of inhibiting vital enzymes in the life cycle of SARS-CoV-2 and consequently mediating the maturation of non-structural proteins [51]. In a multi-site, randomized, double-blind phase 3 clinical study, the efficiency of famotidine is evaluating for COVID-19 patients in the USA (NCT04370262).

2.2.16 Nitazoxanide

Nitazoxanide, a synthetic nitrothiazolyl-salicylamide derivative, is a broad-spectrum antiviral agent used for the treatment of a wide range of viruses, including influenza A, B, and Ebola viruses [3, 6, 54]. Nitazoxanide demonstrated in vitro antiviral activity against MERS-CoV and other CoVs; also, this drug suppresses the production of pro-inflammatory cytokines in peripheral blood mononuclear cells and IL-6 in mice [217]. Moreover, the antiviral activity of this drug could be attributed not to the virus-specific pathways, but rather to its interference with host-regulated pathways involved in viral replication [5, 217, 218]. Nitazoxanide is currently in phase 4 of clinical trials for COVID-19 treatment in Mexico by Laboratorios Liomont (NCT04406246).

2.2.17 Nelfinavir

Nelfinavir, a non-peptidic, competitive HIV protease inhibitor, is considered as one of the potential drugs against COVID-19 [55]. This drug was approved by the FDA for the treatment of HIV infection in 1997 [219]. According to the results of a study in which HIV protease inhibitors were screened to find potential drugs, CoVs, it was indicated that nelfinavir is capable of inhibiting the replication cycle of SARS-CoV-1 [56]. Similarly, in another in vitro study, it was shown that nelfinavir could inhibit 3CLpro of the virus and consequently suppressing the replication cycle of SARS-CoV-2 with EC50 and EC90 of 1.13 µM and 1.76 µM, respectively [57, 220]. According to an in silico study, based on the combinational administration of nelfinavir and cepharanthine, nelfinavir could bind the SARS-COV-2 main protease, thereby inhibiting the viral replication cycle, while cepharanthine is able to prevent viral attachment and entry into cells. Hence, their combination could be a potent multidrug for the treatment of COVID-19 [221].

2.2.18 Auranofin

Auranofin is a gold-containing triethyl phosphine that has been explored for therapeutic applications against a wide range of diseases, including cancer, neurodegenerative disorders, HIV, as well as parasitic and bacterial infections [58, 59]. According to the results of one in vitro study, auranofin inhibits the replication of SARS-COV-2 in human cells at a low micromolar concentration by reducing the viral RNA up to 95% at 48 h after infection [58]. Its mechanism of action consists of inhibiting the redox enzymes as well as induction of ER stress, thereby interfering with the protein synthesis of SARS-CoV-2 [58–61, 222].

2.2.19 Carmofur

Carmofur, an approved antineoplastic drug that was derived from 5-fluorouracil (5-FU) and was explored for the treatment of breast, gastric, bladder, and colorectal cancers, is shown to have inhibitory effects against SARS-CoV-2 [222–224]. It was demonstrated that carmofur inhibits the main protease (3CLpro) activity of the SARS-CoV-2 with an IC50 value of 1.82 μM and prevents viral replication in cells with an EC50 value of 24.30 μM [63, 64].

2.2.20 Galidesivir

Galidesivir (BCX4430, Immucillin-A), an adenosine analog as well as RdRp inhibitor that was first developed for the treatment of HCV, has shown antiviral activity against a wide variety of viruses, including togaviruses, filoviruses, arenaviruses, paramyxoviruses, orthomyxovirus bunyaviruses, CoVs, picornavirus, and flaviviruses [38, 66, 67]. This drug is currently under phase 1 of clinical trials for COVID-19 treatment in Brazil (NCT03891420).

2.2.21 Azvudine

Azvudine or 2′-deoxy-2′-β-fluoro-4′-azidocytidine (FNC) was first developed for the treatment of HIV and has antiviral activity against HBV and HCV [68, 225]. Azvudine might be able to inhibit the reverse enzyme transcriptase vital in viral transcription, thereby interfering with the replication of the CoVs [38]. This drug is currently under phase 3 of a randomized clinical trial for patients infected with COVID-19 in Brazil (NCT04668235).

2.3   Immunomodulators and Anti-inflammatory Drugs

The immune (innate and adaptive) system includes cells, molecules, and processes working together to provide the body protection against aggressive viruses, bacteria, toxins, parasites, fungi, and cancer cells [7]. However, the immune system may be weakened in individuals owing to high age or immunodeficiency disorders [1]. On the other hand, the inflammatory phase, the third phase after the mild infection and pulmonary phases, is initiated and accompanied by cytokine storm due to the excessive immune response of the host upon infection and may create complications like ARDS leading to death in many cases [148]. Considering the biology of SARS-CoV-2 and by exploring the molecular mechanisms employed by the virus regarding its interactions with host cells, the development of host immune response could be illuminated, which may lead to proposing efficient drugs for inhibiting COVID-19 [7]. The medications that have the potential to interact with the host immune system generally can fall into two main categories, the remedies with the aim of boosting the immune system and the therapeutics that intervene in the host immune response and play their role in immunomodulation or alleviating damages caused by the dysregulated inflammatory responses [1]. In this respect, many immunomodulatory therapies and anti-inflammatory drugs such as NKs, IFNs, MSCs, convalescent plasma, interleukin inhibitors, anticitokines, anticoagulants, corticosteroids, and monoclonal antibodies have been administered to control the symptoms, modulate the immune system leading to COVID-19 treatment (Table 2.1) [1, 8, 85].

2.3.1   Natural Killer Cells

The higher mortality rate of elderly patients compared to other generation individuals infected with COVID-19 could be explained by the weakening of the immune system with age and considered somehow as aging-associated diseases in chronic disease states [77, 101, 226]. NK cells as practical components of the innate immune system are against viral infections. NK cells are able to rapidly release granzymes and perforins inducing cell lysis. In addition, they are crucial sources of interferon-gamma (IFN-γ) capable of mobilizing antigen-presenting cells (APCs) and activating antiviral immunity [7]. Hence, the injection of NK cells into elder and fragile patients' bodies could be efficacious in SARS-CoV-2 clearance with no severe side effects [1, 77]. Chen et al. indicated that macrophages and NK cells have a crucial function in the clearance of SARS-CoV-1 [227]. The safety and efficacy of CYNK-001, an immunotherapy containing NK cells derived from human placental CD34 + cells on moderate COVID-19 patients, are currently assessing in a phase 2/3 clinical study at multicenters in the USA (NCT04365101).

2.3.2   Mesenchymal Stem Cells

In severe patients, SARS-CoV-2 infection may generate a threatening cytokine storm in the lungs. MSCs, as a safe and well-tolerated therapeutic option, can be used to benefit through their immunomodulatory, antimicrobial, antiapoptotic, and regenerative effects [70]. The anti-inflammatory activity of MSCs has been associated with decreasing pro-inflammatory cytokines and producing paracrine factors leading to regenerative medicine in pulmonary epithelial cells [1]. MSCs are able to regulate the function of both innate and adaptive immune systems through either passive or active cell–cell interaction, secretion of trophic factors, or activation of regulatory T cells [2, 70]. Clinically, the employing of intravenous transplantation of MSCs for severe SARS-CoV-2 patients has been reported to be safe and effective [71, 228, 229]. It has been reported that MSC therapy for COVID-19 patients exerts no adverse effects on the patients [230, 231]. The efficacy and safety of the MSC remestemcel-L administration are evaluating on COVID-19 patients with ARDS in phase 3 multicenter clinical study in the USA (NCT04371393).

2.3.3   Interferons

IFNs are soluble endogenous signaling proteins with high antiviral activity secreted by cells including cells with hematopoietic origin upon viral or bacterial infection. The interferon-stimulating genes (ISG) generally associated with immunomodulation, signaling, and inflammation are activated by the INF fixation on interferon α/β receptor (IFNAR), the receptors found at most cells plasma membrane [4, 232]. IFNs have been utilized as a therapeutic option against autoimmune disorders, different cancers, and viral infections including hepatitis B and C [1, 232]. IFNs play significant roles in enhancing the immune system by restricting the spread of infectious viral, adjusting innate immunity responses, and activation of adaptive immune responses [7, 233, 234]. Hence, employing recombinant human interferons (rhIFNs) has been considered as a potential treatment method against COVID-19 while SARS-CoV-2 has shown sensitivity to some human type I IFNs like IFN-α and IFN-β [94, 235, 236]. However, the adverse reactions of fever, myalgias, headaches, leukopenia, lymphopenia, and autoimmune hepatitis may be associated with the administration of IFNs [72].

2.3.3.1   Interferon-α

IFN-α is a cytokine secreted by the immune cells in the body capable of eliciting a practical host-mediated immune cell response for various cancer treatments and inhibition of replication of viruses like SARS-CoVs, HIV, and HCV while SARS-CoV-2 has shown significantly high sensitivity to IFN-α [8, 72, 237]. IFN-α has been exploited as a favorable antiviral activity due to low toxicity and its crucial roles in the inhibition of the virus replication at the early stage of infection, moderating the symptoms of the acute phase of the disease, shortening the disease duration leading to the survival of severe patient [8]. To enhance the stability of IFN-α and prolong its half-life from 4.6 to 22–60 h, Pegylated IFN-α-2b has been utilized providing a situation in which the lower dosing could be injected frequently [72]. In a retrospective multicenter cohort clinical trial, 242 of 446 COVID-19 patients received IFN-α-2b as a treatment in Hubei, China. It was observed that early administration (≤5 days after admission) of IFN-α-2b was responsible for the reduced in-hospital mortality while peculiarly, late administration of IFN-α-2b was involved with increased mortality [238]. In a phase 3 clinical study, the efficiency of rhIFN-α-1b in preventing COVID-19 is currently assessing on a large number of patients at Taihe Hospital Shiyan, Hubei, China (NCT04320238).

2.3.3.2 Interferon-β

IFN-β is a signaling cytokine that has a broad range of applications against viral infections like HCV and HBV. It is able to activate cytoplasmic enzymes stimulating them to prevent viral replication [38, 74]. IFN-β with the ability of maintaining endothelial barrier activity, pro-inflammatory response, and defensive function in the lungs has demonstrated the highest potency among the IFNs in prophylactic protection and antiviral potential post-infection effects [74, 234, 239]. IFN-β-1 has demonstrated efficacy against SARS-CoV-1 [240]. Consequently, on account of the high similarity of SARS-CoV-1 with SARS-CoV-2, IFN-β-1 has been introduced as a potential therapeutic in COVID-19 treatment [241]. Despite inhibiting the production of IFN-β and obstructing the innate immune system response by SARS-CoV-2, the virus has shown sensitivity to the antiviral activity of externally administered type I IFNs [236].

In a phase 2 clinical study (NCT04385095), safety and efficacy of inhaled nebulized IFN-β-1-a (SNG001) were assessed for the treatment of patients admitted to hospital with COVID-19, and it was demonstrated that in comparison with patients who received placebo, treated patients with SNG001 had a greater chance of treatment with more rapid recovery [73]. Likewise, IFN-β-1-b has indicated potency in inhibiting SARS-CoV-1, MERS-CoV, and SARS-CoV-2 infections [241]. Also, a combination therapy of IFN-β-1-b, lopinavir/ritonavir, and ribavirin was evaluated on COVID-19 patients admitted to hospital in a phase 2 clinical study (NCT04276688) and demonstrated a high potential in alleviating symptoms and reducing the disease duration and hospital stay in patients with mild to moderate SARS-CoV-2 infection [174].

2.3.4   Convalescent Plasma or Intravenous Immunoglobulin

The pooled plasma or hyperimmune immunoglobulins derived from recovered patients of a disease, termed convalescent plasma (CP) has been widely employed to treat many infectious diseases such as MERS-CoV, SARS-CoV-1, Ebola, and SARS-CoV-2 with favorable outcomes through passive immunity delivery and replacement therapy for antibody deficiencies [38, 242, 243]. On account of the fact that CP or intravenous immunoglobulin (IVIG) possesses neutralizing antibodies to SARS-CoV-2, which could be exploited as a potential therapy to directly neutralize the virus, modulate the inflammatory response, and control the overactive immune system (i.e., cytokine storm) [75, 76, 243]. In the hope of minimizing morbidity and mortality, all these benefits of CP are expected to be attained if used in early administration and non-critically hospitalized patients [8, 76].

Improvement in the clinical status of small sample sizes of critically ill patients was reported in some studies by the administration of CP with low serious adverse reactions [243–247]. Despite the risks involved with CP and IVIG administration for COVID-19 patients, including transfusion-associated lung injury and circulatory overload, allergic/anaphylactic reactions and less common risks like transmission of infections and red blood cell alloimmunization [41], very low adverse events have been reported in a safety study of CP for 20,000 hospitalized patients implying that CP is safe in hospitalized patients with COVID-19 [246]. In a phase 3 clinical trial, CP has been employed for treating hospitalized COVID-19 patients in New York, the USA (NCT04418518).

2.3.5   Anticitokines, Immunosuppressants, and JAK Inhibitors

Since COVID-19 is associated with a significant increase in the level of serum cytokines, that is, the cytokine storm, repurposing of available anticytokines with proven safety has come to the fore [248, 249]. In this regard, considering the role of the Janus kinase signal transducer and activator of transcription (JAK-STAT) pathway in Angiotensin II type 1 receptor, the receptor that is expressed on peripheral tissues and immune cells as cytokine receptors with the role of the renin-angiotensin system (RAS) signal transduction, targeting of this pathway in hospitalized patients by employing JAK and Adaptor-associated kinase (AAK) inhibitors like sirolimus, baricitinib, ruxolitinib, imatinib, cyclosporine, and tofacitinib capable of inhibiting type I/II cytokine receptors could not only reduce the clinical symptoms in organs like lung, kidney, and heart but also inhibit the cytokine storm in ARDS condition associated with severe SARS-CoV-2 infection [3, 250, 251].

2.3.5.1 Sirolimus

Sirolimus, known as rapamycin (trademark name: Rapamune), is a natural product isolated from the bacterium *Streptomyces hygroscopicus in* Easter Island. Sirolimus was initially isolated as an antifungal agent with potential anticandida activity [92, 252]. Nevertheless, its antitumor/antiproliferative and immunosuppressive properties were proved by further studies [252]. Also, sirolimus is capable of weakening the immune system, surprisingly strengthening T cells activity in the course of pathogenic invasions, delaying age-related illnesses in humans, and having an inhibitory effect on the mammalian target of rapamycin complex 1 (mTORC1) receptor [91].

On account of the fact that mTORC1 has a pivotal role in the viral replication of different viruses, including orthohantavirus and CoVs, sirolimus could be a potential therapeutic agent for repurposing against COVID-19 [85, 91, 253]. According to the results obtained from an in vitro study, sirolimus affected PI3K/AKT/mTOR pathway, thereby inhibiting MERS-CoV activity [254]. Moreover, an in silico study showed that sirolimus could be a potential drug for the treatment of patients infected with COVID-19 using a network-based drug repurposing model [253]. This drug is currently in phase 2 of a clinical trial for patients infected with COVID-19 performing in the USA by the University of Cincinnati (NCT04341675).

2.3.5.2 Baricitinib

Baricitinib under the brand name olumiant, an inhibitor of cytokine-release approved in 2018 for the treatment of rheumatoid arthritis, is a potential JAK inhibitor that selectively inhibits the JAK1 and 2, consequently reducing inflammation in patients infected with COVID-19 [3, 255–257]. Moreover, baricitinib is capable of inhibiting AAK1 and Cyclin G-associated kinase (GAK) [3]. Both AAK1 and GAK are important regulators of endocytosis. Hence, targeting AAK1 and GAK makes baricitinib also a potential candidate for not only inhibiting the viral entry but also interfering with the virus assembly [258]. The effectiveness and safety of baricitinib have been shown through conclusive results such as lower fatality rate and higher discharge rate obtained from clinical studies conducted on COVID-19 patients receiving baricitinib [256, 259]. This drug is undergoing phase 3 of a clinical study performing on 1400 COVID-19 patients in the USA (NCT04421027).

2.3.5.3 Ruxolitinib

Ruxolitinib, also known as INC424 or INCB18424, is a potent inhibitor of JAK1 and 2 that was approved by the FDA against myelofibrosis, polycythemia vera, and acute graft-versus-host disease [38, 93, 260, 261]. Its main mechanism of action includes interfering with the JAK-STAT, one of the chief regulator cell signaling pathways, through interacting with JAK and preventing the activation of STAT, thereby reducing the elevated levels of cytokines [93, 257]. Given these points, ruxolitinib could be considered as one of the potent drugs against a wide range of diseases, including COVID-19. According to studies conducted on COVID-19 patients receiving ruxolitinib, obtained results demonstrated that this drug successfully reduced both the inflammatory blood cytokine levels such as IL-6 and the acute phase protein ferritin; moreover, the administration of ruxolitinib brought about rapid respiratory and cardiac improvement, significant chest computed tomography (CT) improvement, faster recovery from lymphopenia, clinical stabilization, as well as favorable side-effect profile [262, 263]. Ruxolitinib is under phase 3, randomized, double-blind, placebo-controlled, multicenter clinical study on patients infected with COVID-19 in the USA (NCT04377620).

2.3.5.4 Fingolimod

Fingolimod, known as FTY720, is an orally administered compound acting as the modulator of sphingosine-1-phosphate (S1P) receptors and was chemically derived from an immunosuppressive metabolite (myriocin) isolated from a fungus (*Isaria sinclairii*) [38, 264, 265]. In fact, fingolimod could play a role as a potent functional antagonist of S1P1 receptors on T cells subsequently sequestering lymphocytes in lymph nodes [1]. This drug has shown conclusive results in the treatment of multiple sclerosis (MS) on account of its capability to reduce the inflammatory damages and effect on the central nervous system (CNS) [95, 266]. Based on pathological findings, besides ventilator support, immune modulators such as fingolimod should be taken into consideration, in that, their combination might prevent the progression of ARDS [196].

2.3.5.5 Thalidomide

Thalidomide, which is an antiangiogenic, anti-inflammatory, as well as anti-fibrotic agent, was initially synthesized by the CIBA pharmaceutical company in 1954 and used as a sedative, antiemetic, and tranquilizer for morning sickness [267]. Thalidomide has inhibitory effects on TNF-α synthesis and is used for the treatment of multiple inflammatory diseases, including Behçets' disease and Crohn's disease [77, 96]. Even though the exact mechanism of action for the anti-inflammatory effects of thalidomide has yet to be found, researchers have attributed its anti-inflammatory effects to its ability for accelerating the degradation of messenger RNA in blood cells, thereby reducing the blood serum level of TNF-α, which is a cytokine involved in systemic inflammation and cytokine storm [268, 269]. Due to its properties, clinical studies are conducting to assess the immunomodulatory effects of thalidomide on reducing lung damage caused by SARS-CoV-2 [77]. Thalidomide is under phase 2 of a randomized, multicenter, placebo-controlled, double-blind clinical trial on 100 patients infected with COVID-19 performing by Wenzhou Medical University in China (NCT04273529).

2.3.6   Non-steroidal Anti-inflammatory Drugs (NSAIDs)

NSAIDs have been widely utilized for controlling acute and chronic inflammatory circumstances [270]. Generally, NSAIDs act by curbing prostaglandin synthases 1 and 2, known as cyclooxygenase enzymes (COX-1 and COX-2) that are responsible for producing prostaglandins (PGs) and provoking pain and fever [271–273]. NSAIDs have been employed to reduce fever and muscle pain caused by COVID-19. However, there is a heated controversy as to whether these drugs are safe [270]. On account of the fact that most studies postulating not protective effects of NSAIDs have been majorly in vitro or on animals rather than on humans [272], further studies are needed to determine the role of NSAIDs in the context of SARS-CoV-2 infection.

2.3.6.1   Naproxen

Naproxen a member of NSAIDs is a propionic acid derivative, which is administered orally and rectally, has been widely used against rheumatic diseases and non-rheumatic circumstances [97, 98]. It was demonstrated by Zheng et al. that naproxen is capable of exerting antiviral activity against influenza A and B viruses, in that this drug obstructs the nuclear export of the viral nucleoproteins, hampering influenza replication [274]. Additionally, in another study conducted on COVID-19 patients receiving ibuprofen and naproxen, conclusive results such as diminishing the probability of hospitalization and requiring mechanical ventilation were obtained [275]. Currently, the efficacy of naproxen on hospitalized COVID-19 patients is assessing through phase 3 of a randomized clinical study performing by Hôpitaux de Paris (NCT04325633).

2.3.6.2   Ibuprofen

Ibuprofen is NSAID with analgesic, antipyretic, and anti-inflammatory properties, which was first introduced in the UK in 1969 and has been used against symptoms of acute pain, inflammation, fever, osteoarthritis, rheumatoid arthritis, ankylosing spondylitis, gout, and Bartter's syndromes [38, 276, 277]. Its mechanisms of action consist of reducing the activity of COX enzyme, thereby inhibiting the production of prostaglandins [278]. Regarding the effectiveness of ibuprofen against COVID-19 infection, there are quite a few contradictory suggestions made by studies. For example, whether ibuprofen could facilitate the cleavage of the ACE2 receptor on host cells, thereby interfering with the viral entry, or whether ibuprofen may decrease the excess inflammation or cytokine release in COVID-19 infection has been discussing. However, due to a lack of substantiated evidence, these claims are just possible protective effects of ibuprofen against CoV infection [279, 280]. Moreover, even though it is hypothesized that ibuprofen might increase the severity of COVID-19 infection, according to one study performed on 430 patients infected with COVID-19,

ibuprofen was not associated with worse clinical outcomes, compared with paracetamol or no antipyretic. Nevertheless, further clinical studies are required to confirm
their results [281]. This drug is currently under phase 4 of a clinical study conducting
on 230 participants infected with COVID-19 by King's College London in the UK
(NCT04334629).

2.3.6.3   Paracetamol

Paracetamol (acetaminophen), which was initially synthesized through its precursor,
that is, phenacetin in 1878, has been used to relieve acute and chronic pain worldwide [282, 283]. This drug is currently the most prevalent analgesic in the world and
possesses weak inhibitory effects on the synthesis of prostaglandins [283, 284].
According to suggested warnings regarding the administration of ibuprofen for
COVID-19 patients, paracetamol was recommended as a safer option. Although
paracetamol has been reported to have no or insignificant anti-inflammatory and
antiplatelet activity, this drug has been constantly used for the control of COVID-
19 [285, 286]. Nonetheless, paracetamol might cause glutathione (GSH) depletion,
which might result in developing severe COVID-19, particularly in more vulnerable
groups. Therefore, clinical studies are needed to investigate the efficacy and adverse
effects of this drug on patients infected with COVID-19 [286].

2.3.7   Corticosteroids

Corticosteroids have been proposed to be utilized for the suppression of lung inflammation in SARS-CoV-1 and MERS-CoV owing to their anti-inflammatory effects
and the potential in reducing mortality [72, 287]. Prognosis improvement and clinical recovery promotion have been reported in a systematic review of corticosteroid
therapy for severe COVID-19 patients [288]. Nevertheless, the efficacy and safety
of the administration of corticosteroids like methylprednisolone, dexamethasone,
and budesonide for the management of SARS-CoV-2 infection are still controver-
sial [200, 287, 289]. The administration of corticosteroids in the management of
COVID-19 may bring the risk of damages like prolonged mechanical ventilation,
delayed viral clearance, and avascular necrosis. Thus, it demonstrates the need for
a high consideration in corticosteroid administration for COVID-19 patients while
also requiring more clinical data [289].

2.3.7.1   Methylprednisolone

Whether methylprednisolone could be a potential drug for the suppression of
unwanted immune reactions is questionable [1]. However, it is believed by many
medical researchers that methylprednisolone could improve the deregulation of the

host immune response and increase the blood pressure where is low due to the cytokine storm [85]. Wu et al. reported that the risk of mortality was decreased by the administration of methylprednisolone for severe patients with ARDS. In fact, 23 of 50 (46%) patients who received methylprednisolone died while the mortality rate in patients with no methylprednisolone treatment was higher (61.8% (21 of 34 patients)) [101]. The routine use of corticosteroids including methylprednisolone is opposed by the Infectious Diseases Society of American. On the other hand, they do recommend the administration of corticosteroids for the patients with developed ARDS in order to set the cytokine storm in the context of a clinical trial [290]. The efficacy of different hormone doses of methylprednisolone is evaluating on severe COVID-19 patients in a phase 4 clinical study in Hubei, China (NCT04263402).

2.3.7.2 Dexamethasone

Dexamethasone is on the list of essential medicine of the WHO, which is available worldwide at low cost [291]. It is published that early treatment of ARDS with dexamethasone could reduce the ventilator days and mortality in patients generally with established moderate-to-severe ARDS [102]. A lower 28-day mortality was reported by employing dexamethasone treatment at a dose of 6 mg once daily for up to 10 days in hospitalized patients with COVID-19 who were receiving respiratory support (NCT04381936) [291]. In a phase 3 clinical study (NCT04327401), administration of intravenous dexamethasone plus standard care, compared with standard care alone for the COVID-19 patients with moderate or severe ARDS, showed promising results and caused a significant increase in the number of days alive and free of mechanical ventilation over 28 days [103]. In a phase 4 clinical trial in Brno, Czechia, the effect of two different doses of dexamethasone is assessing on COVID-19 patients with ARDS (NCT04663555).

2.3.8 Antibiotics

2.3.8.1 Azithromycin

Azithromycin, a broad-spectrum antibiotic, is an orally administered acid-stable antibacterial drug that is structurally related to erythromycin with analogous antimicrobial activity [38, 292, 293]. This drug is known for its antimicrobial activity against some gram-negative organisms, especially Haemophilus influenza that is associated with respiratory tract infections [293]. Besides its antibacterial activity, azithromycin has shown a wide variety of antiviral and immunomodulatory activities. Hence, this drug could be a potent candidate in suppressing viral infections, particularly COVID-19 [294]. Further, regarding SARS-CoV-2, it was shown that azithromycin inhibits the viral entry into the host cells by interacting with SARS-CoV-2 S protein and ACE2 [4, 100]. Also, it should be noted that, in the context of COVID-19, the combination of HCQ and azithromycin has been associated with serious adverse events, including a higher risk of cardiac toxicity and arrhythmias [294]. However, in one randomized-controlled clinical trial, results showed that treatment with a combination of azithromycin and HCQ was associated with a reduction in mortality of COVID-19 patients, and while administered alone, azithromycin did not show a higher risk of adverse events compared with the administration of the combination of HCQ and azithromycin or HCQ alone [295]. Azithromycin versus usual care is under phase 3 of a multicenter open-label clinical trial in ambulatory care of COVID19 by the collaboration of Pfizer in the UK (NCT04381962).

2.3.8.2 Teicoplanin

Teicoplanin is a glycopeptide antibiotic that besides its antibacterial activities against gram-positive bacteria, including staphylococci, streptococci, and enterococci, has shown conclusive results against the Ebola virus, influenza virus, flavivirus, HCV, HIV, and CoVs such as MERS-CoV and SARS-CoV-1 [62, 296]. As for CoVs, including SARS CoV-2, teicoplanin is capable of preventing the replication of the virus-cell cycle by inhibiting the viral RNA release. This could happen simply because teicoplanin inhibits the low pH cleavage of viral S protein by cathepsin L in the endosome [4]. Moreover, the concentration of teicoplanin needed in vitro for inhibiting 50% of SARS-CoV-2 viruses (IC50) was 1.66 μM, which was far lower than the IC50 reached in human blood (8.78 μM for a daily dose of 400 mg) [297, 298]. Nonetheless, these results are required to be confirmed by further clinical studies.

2.3.8.3 Tetracyclines

Tetracyclines (doxycycline, tetracyclines, minocycline) are polyketide antibiotics that have a broadspectrum antimicrobial activity [4, 299]. Tetracycline's mechanism of action, that is, blocking protein synthesis in *staphylococcus aureus* cells and inhibiting cell growth in a bacteriostatic manner was first delineated in 1953; further studies showed that these drugs act through binding to bacterial ribosomes [300]. Studies on the skin showed that tetracyclines are also capable of reducing the levels of inflammatory cytokines [301]. Thus, due to their anti-inflammatory effects, tetracyclines could be considered for the treatment of COVID-19 patients. Another possible mechanism of action of tetracyclines against COVID-19 is related to its chelating activity. In better words, tetracyclines are capable of chelating zinc from host matrix metalloproteinases (MMPs), thereby limiting viral replication, and this happens because CoVs bind to the host MMPs for viral survival [302–304]. Doxycycline is under phase 3 of a multicenter, randomized, clinical study performed on 330 COVID-19 patients receiving either doxycycline or placebo conducting by Nantes University Hospital in France (NCT04371952).

2.3.9 Low Molecular Weight Heparins as Anticoagulants

Owing to the fact that coagulopathy has been one of the major causes of morbidity and mortality in COVID-19 patients, anticoagulation therapy has been considered as one of the potential ways of combating COVID-19 [305]. In this regard, heparins are clinically approved anti-coagulants, and low molecular weight heparins (LMWHs), derived from unfractionated heparins either by chemical or by enzymatic depolymerization, are glycosaminoglycans that have been utilized in the prophylaxis of post-surgical venous thromboembolism as well as non-surgical patients with acute pathology and reduced mobility [77, 99, 306]. Due to the fact that unfractionated heparins and LMWHs have inhibitory effects on proteases, including factor Xa, thrombin, furin, and cathepsin-L, it has been hypothesized that LMWH and unfractionated heparin could be considered as potential drugs for not only targeting protease cleavage but also the viral entry of SARS-CoV-2 [307]. Nevertheless, these suggestions need to be confirmed by in vitro and clinical studies. According to a retrospective clinical study, from among 42 patients with COVID-19, 21 underwent LMWH treatment, and 21 were assigned to the control group during hospitalization. Results showed that LMWH treatment not only caused improvement in the coagulation dysfunction of patients but also exerts anti-inflammatory effects through reducing IL-6 and increasing lymphocyte percent [308]. Another retrospective study was performed to assess the safety of intermediate-dose regimens of one LWMH, that is, enoxaparin in COVID-19 patients with pneumonia, especially in older patients. Their results proposed that the use of an intermediate dose of LWMH seems to be safe and possible for COVID-19 patients, but further clinical studies are needed to substantiate these suggestions [305]. LMWH is currently under phase 4 of a randomized clinical trial performing by Ain Shams University of Egypt (NCT04584580). Moreover, another phase 3 ongoing clinical trial, in the USA, is performing to compare the effects of full dose administration of enoxaparin vs. prophylactic or intermediate dose of enoxaparin in high-risk COVID-19 patients (NCT04401293).

2.3.10 Adjunctive Supplements and Vitamins

2.3.10.1 Vitamin D

Vitamin D, a crucial group of fat-soluble secosteroids, is generally known for its functions in the maintenance of bone health and calcium-phosphorus metabolism [309]. A wide range of antioxidant, immunomodulatory, anti-inflammatory, and antifibrotic functions have been recently attributed to vitamin D. It has been considered to be able to inhibit cytokine storm in SARS-CoV-2 infection and decrease the expression levels of pro-inflammatory type 1 cytokines like IL-12, IL-16, IL-8, TNF-α, IFN-γ while increasing regulatory T cells and type 2 cytokines including IL-4, IL-5, IL-10

[69, 104, 309]. The elderly patients and individuals with common variable immunodeficiency and bronchiectasis who are recognized with vitamin D deficiency are reported to be at a high risk of viral respiratory tract infections, acute lung injury, and particularly COVID-19 infection [104, 310]. On the contrary, no relationship between vitamin D concentration and the severity of COVID-19 in hospitalized patients was reported by Hernández et al. [311]. However, considering the several beneficial functions of vitamin D and its effects on immune cell proliferation and activity, pulmonary ACE2 expression, and priming effects against the viral replication, it has been proposed to employ high-dose vitamin D as a safe adjuvant therapeutic intervention to reduce the risk of COVID-19 severity and mortality [38, 104, 309]. Nevertheless, further studies are still needed to validate this association between vitamin D and COVID-19. Two clinical trials are currently in phase 4 to evaluate the efficiency of high doses of cholecalciferol (vitamin D3) on morbidity and mortality of COVID-19 patients in Spain and Argentina (NCT04552951, NCT04411446).

2.3.10.2 Vitamin C

Vitamin C (ascorbic acid) as an essential micronutrient and a potent antioxidant agent could play significant roles in neutralizing free radicals, preventing cellular damage, and associating with immune health [105]. Vitamin C has been reported to be effective against viruses like influenza viruses and reducing the duration and severity of upper respiratory infections [69, 85]. Many studies have demonstrated that vitamin C could involve with the development and maturation of T lymphocytes and NKs in the immune defense. Generally, it accumulates in phagocytic cells like neutrophils and can contribute to enhancing chemotaxis, phagocytosis, inhibition of reactive oxygen species (ROS) generation, preventing the neutrophils accumulation in the lung, and modulation of cytokine production network in the host inflammation response. Hence, employing vitamin C appears to be effective in preventing and treating respiratory and systemic infections [85, 105, 312]. The high dosage administration of vitamin C, a safe supportive treatment with no major side effects, has been considered as a potential treatment for reducing the cytokine storm and recovering COVID-19 patients [38, 85, 313, 314]. Currently, in a phase 3 clinical trial, the effect of high-dose intravenous vitamin C is evaluated on the mortality or persistent organ dysfunction of COVID-19 patients in Canada (NCT04401150).

2.3.10.3 Zinc

Zinc, as a micronutrient food supplement, has anti-inflammatory and antioxidant activities with an evident function in immunity on account of its roles as a cofactor, signaling molecule, and a structural element [69, 315]. Zinc deficiency has been reported to be responsible for the up-regulation of TNF-α, IFN-γ, JAK signaling in the lungs, cytokine production, and induction of apoptosis in lung epithelial cells [316]. Zinc could play its role in preventing viral pathogenesis through inhibiting viral entry, fusion, and replication alongside attenuating the risk of hyper-inflammation, preserving natural tissue barriers while protecting cells and tissues from oxidative damage and dysfunction [69, 106, 315]. The administration of zinc as a potential well-tolerated supplementary therapeutic against COVID-19 has been considered in several studies owing to its possible anti-inflammatory, antioxidant, immunomodulatory, and direct antiviral effects [69]. The efficacy of zinc in higher risk COVID-19 outpatients is currently assessing in a phase 4 clinical placebo-controlled trial in the USA (NCT04621461).

2.3.11 Miscellaneous Therapies

2.3.11.1 Nitric Oxide

Nitric oxide (NO) is an important cellular signaling molecule that is produced by nitric oxide synthase (NOS) by converting arginine and oxygen into citrulline and nitric oxide [317–319]. NOS exists in a wide range of cells such as neurons, macrophages, airway epithelial cells, and vascular endothelial cells, and mediate neurotransmission, smooth muscle contraction, and mucin secretions [2]. It was demonstrated that NO possesses broadspectrum antiviral activity against ectromelia virus, vaccinia virus, herpes simplex type 1 viruses, CoVs, and influenza A and B viruses [319–321]. Besides, different inflammatory stimuli like cytokines can bring about high and sustained production of NO by inducible nitric oxide synthase (iNOS). Furthermore, iNOS causes anti-inflammatory or pro-inflammatory responses, cytotoxicity, or cytoprotection [319]. Concerning SARS-CoV-2, inhalation of nitric oxide is being evaluated for use against COVID-19, because inhaled NO has a chief role in pulmonary and cardiovascular physiology [322]. Inhaled nitric oxide gas is under phase 2 of a randomized clinical trial performing on 470 COVID-19 patients by Massachusetts General Hospital (NCT04312243).

2.3.11.2 Statins

Statins are inhibitors of 3-hydroxy-3-methylglutaryl coenzyme A (HMG-CoA) with anti-inflammatory and immunomodulatory, anti-thrombotic, and antioxidant activities. Therefore, they could be considered as repurposing drugs for restoring viral infection-induced endothelial dysfunction, decreasing the severity of the lung injury and mortality rate caused by SARS-CoV-2, and maintaining the homeostasis of the patients [5, 323]. Moreover, it was postulated that stains might decrease the fatality rate caused by MERS-CoV [324]. Nevertheless, studies on animal models infected with SARS and MERS infections proposed that the administration of statins that suppress the myeloid differentiation primary response protein (MYD88) signaling might worsen the disease condition [5, 325]. It should be noted that since statins have a potency for drug–drug interaction with some protease inhibitor drugs, their co-administration is contraindicated [326]. Additionally, myopathy and severe rhabdomyolysis are two major side effects of statins, and less prevalent adverse effects include peripheral neuropathy, hepatotoxicity, impaired myocardial contractility, and autoimmune diseases [327]. Phase 2 of a randomized clinical study is currently conducting on COVID-19 patients to assess the efficacy of atorvastatin as an adjunctive treatment by Mount Auburn hospital in the USA (NCT04380402).

2.3.11.3 Losartan

Losartan, an angiotensin-receptor antagonist without agonist properties, is a selectively and orally available ACE2 inhibitor that plays a role by blocking the vasoconstrictor and aldosterone-secreting effects of angiotensin 2 through inhibiting the binding of it to the angiotensin II type 1 receptor [38, 328]. Thus, it is considered as a repurposing drug against COVID-19 infection. While orally administered, roughly 14% of the losartan is converted to its metabolite, E 3174, which is 10-to-40-fold more active compared with its original compound, with an estimated terminal half-life from 6 to 9 h [328]. Losartan is currently under phase 3 of a multicenter clinical trial for assessing its protective effect against COVID-19 (NCT04606563).

2.4   Recombinant Proteins and Monoclonal Antibodies

Three phases of infection have been proposed for COVID-19, the first one is mild infection requiring only symptomatic treatment. The pulmonary phase is the second phase necessitating mostly antiviral treatment where recombinant proteins like APN01, meplazumab and novaferon could play a vital role by inhibiting the viral entry and replication. The third phase of the SARS-CoV-2 infection is the inflammatory response phase mainly experienced by severe COVID-19 patients. The third phase is generally associated with complications and immune-inflammatory response accompanied by abundant macrophages, neutrophils, lymphocytes, immune mediators, and pro-inflammatory cytokines. IL-1, IL-6, and TNF-α are the most prominent pro-inflammatory cytokines in the body [62, 148].

IL-1 binds to the IL-1 receptor to modulate its function leading to the production of other pro-inflammatory cytokines. IL-6, another key pro-inflammatory cytokine, binds to the IL-6 receptor expressed on monocytes, neutrophils, macrophages, and other leukocytes and interact with membrane-bound gp130 to activate its downstream JAK signal. The excessive IL-6 signaling may cause a faster decline of lung elasticity, more severe bronchoalveolar inflammation, and organ damages. The need for mechanical ventilation has been recently reported to be strongly connected to the elevated IL-6 [1, 148, 329]. In this regard, the administration of monoclonal antibodies (mAbs) and recombinant proteins could be highly effective. Several mAbs and proteins such as tocilizumab, sarilumab, bevacizumab, and anakinra have been utilized to reduce hyper inflammation and the risk of ARDS and organ dysfunction (Table 2.1) [38, 330].

2.4.1 APN01

APN01, a recombinant human ACE2 (rhACE2) that was originally developed by Apeiron Biologics, is currently utilized for the treatment of patients infected with COVID-19 [38, 41]. APN01, which is a soluble recombinant ACE2, may prevent SARS-CoV-2 entry, as it imitates the innate human enzyme ACE2 on the surface of the host cells employed by the virus for entering the cells [3, 42]. By so doing, the S protein of the virus binds not to the ACE2 on the host cell, but rather to the soluble ACE2, (APN01), thereby preventing viral infection and increasing the viral load [77]. In the meantime, APN01 decreases damaging inflammatory reactions in the lungs and reduces lung injuries [3, 331]. It was demonstrated that the administration of APN01 as a therapeutic could decrease the level of Angiotensin II, thus preventing the ACE enzyme from accessing its substrate. This mechanism has the potential to inhibit further activations of the ACE2 angiotensin receptor [77, 332]. In a study, it was demonstrated that APN01 can reduce SARS-CoV-2 recovery from Vero cells by a factor of 1,000–5,000 in a dose-dependent manner [333]. APN01 is currently under phase 2 in a randomized, double-blind clinical trial for COVID-19 therapy, which is performing in a multicenter in Austria, Denmark, Germany, Russian Federation, and the UK by Apeiron Biologics (NCT04335136).

2.4.2 Novaferon

Novaferon (Nova) is a novel recombinant IFN-α like protein with 176 amino acids. Novaferon has exhibited anticancer and antiviral activities. In China, this drug has been approved for the treatment of chronic HBV. Li et al. in 2014, by studying the antitumor effects of novaferon and comparing it with recombinant humanized IFN-α-2b (rhIFN-α-2b), demonstrated that novaferon has stronger antitumor effects than rhIFN-α-2b [334, 335]. Novafen is able to block the virus replication in COVID-19-infected cells and also prevent the virus from entering healthy cells.

Zhang et al. reported a significantly higher clearance rate of SARS-CoV-2 employing novafron alone or by its combination with lopinavir/ritonavir compared with lopinavir/ritonavir alone [52, 53, 336]. Novaferon is currently under Phase 3 of randomized, double-blind clinical trials for hospitalized COVID-19 patients' treatment (NCT04669015).

2.4.3   Tocilizumab

Tocilizumab or atlizumab is a recombinant humanized monoclonal antibody (rhmAb) against IL-6 under the trade name Actemra [337]. It has FDA approval for the treatment of rheumatoid and polyarticular juvenile idiopathic arthritis and systematic juvenile idiopathic. Tocilizumab plays its role through the membrane and soluble IL-6 receptor blockade and inhibiting the cytokine release syndrome process (CRS) [38]. Actemra is able to restrain the cytokine storm induced by SARS-CoV-2 [330]. Several studies have shown that a single dose of 400 ml administration of tocilizumab could render benefits through better breathing, faster fever reduction while it is also capable of reducing the inflammation response, vasopressor support, and mortality rate in COVID-19 [8, 77, 338]. Nevertheless, some side effects regarding the administration of tocilizumab have been reported including increased upper respiratory infections, hypertension, hematological effects, gastrointestinal perforation and hepatotoxicity [77]. For the management of the severe SARS-CoV-2 infection, tocilizumab has been employed in phase 4 and 3 clinical trials (NCT04377750, NCT04320615).

2.4.4   Sarilumab

Sarilumab, developed by Regeneron Pharmaceuticals and Sanofi, is another immunomodulatory drug. In May 2017, the FDA approved Sarilumab for the treatment of rheumatoid arthritis under the Kevzara brand name. Sarilumab has the potential to suppress the growth of some xenograft prostate and lung tumors either as a single drug or in combination with other therapeutics [339, 340]. Sarilumab, like tocilizumab, is an IL-6 rhmAb that could suppress COVID-19-associated overactive inflammatory immune responses and cytokine storm. The most common side effects of this drug include cough or sore throat, thrombocytopenia, blocked or runny nose, urinary and respiratory tract infections, neutropenia, hypercholesterolemia, mild hepatotoxicity, and cold sores [78, 341]. In a randomized, embedded, multifactorial, adaptive platform trial for community-acquired pneumonia (REMAP-CAP), a clinical phase 4 study, the efficiency of a range of interventions including sarilumab has been evaluating on ICU admitted COVID-19 patients (NCT02735707).

2.4.5   Eculizumab

Eculizumab has the FDA approval for the treatment of atypical hemolytic uremic syndrome (aHUS), paroxysmal nocturnal hemoglobinuria (PNH) diseases [342]. One of the most effective methods for preventing tissue damage is to suppress the production of excessive inflammation responses and cytokines caused by SARS CoV-2. Eculizumab with the brand name of soliris as a rhmAb against the complement protein C5 could inhibit the cleavage of C5 into C5a and C5b [343]. Consequently, it is able to prevent the terminal complement complex C5b-9 and subsequently the membrane attack complex formation, production of reactive oxygen species, and initiation of releasing cytokine storm. In addition, it could inhibit the formation of the C5a, responsible for the development of acute lung injury [80]. Despite the benefits of eculizumab, it may exhibit some side effects such as bradycardia, atrioventricular block, and hypertension in some patients [79]. Eculizumab is currently under phase 2 clinical trial against COVID-19 in France (NCT04346797).

2.4.6 Bevacizumab

Bevacizumab (under brand name avastin) is a humanized monoclonal antivascular endothelial growth factor (anti-VEGF) antibody that has been approved for the treatment of several cancers such as breast, brain, and renal cancers by the FDA. Studies have shown that upon ARDS the amount of VEGF produced by epithelial and inflammatory cells increases in patients. The increase in VEGF causes vascular permeability and pulmonary edema [1, 344, 345]. Bevacizumab with a specific ability in binding to VEGF and subsequent inhibition of its linking to VEGF receptor on the surface of endothelial cells could be a potential therapeutic for the treatment of ARDS and ALI caused by COVID-19 [81, 346]. However, the administration of bevacizumab is associated with an increased risk of cardiovascular events in some cases [347]. In multiple cohort randomized controlled trials, bevacizumab is under phase 2 for COVID-19 treatment at Hôpitaux de Paris in France (NCT04344782).

2.4.7 Infliximab

Infliximab is a chimeric monoclonal anti-tumor necrosis factor (anti-TNF) alpha antibody (IgG1), which could inhibit TNF from binding to its receptors by targeting it [82, 84]. TNF is a cytokine that releases in the acute phase of inflammation and binds to its receptors (TNFRI and TNFRII) in all cells except erythrocytes. Several signaling pathways including transcription factor activation (nuclear factor-κB), proteases (caspases), and protein kinases (MAP kinase, c-Jun N-terminal kinase) are activated by connecting TNF to its receptor leading to the activation of the target cells for immune and inflammatory responses by the release of apoptotic pathway initiation and several cytokines. TNF also plays role in the activation of lymphocytes (B, T) and macrophages, production of proinflammatory cytokines such as IL-1, IL-6, and expression of adhesion molecules (ICAM-1, E-selectin) [83]. Studies have shown that TNF could induce cytokine cascade in rheumatoid arthritis and promote pathogenesis in SARS-CoV-2 [348]. Infliximab has been approved in the USA since

1988 as a drug for the treatment of some autoimmune inflammatory diseases. Infliximab could be a proper option to reduce inflammation, cytokine storm, and subsequent organ failure due to COVID-19 [82, 84, 349]. Some reactions may occur upon infliximab Infusion, which could be prevented by employing antihistamines, acetaminophen, and corticosteroids as pre-medications [83]. Infliximab is currently under phase 3 of clinical trials for severe COVID-19 patients in multi centers of the USA (NCT04593940).

2.4.8 Anakinra

Anakinra under the brand name of kinert is a modified recombinant human IL-1 receptor antagonist with a short half-life of about 3–4 h and a proper safety profile that is approved for use in rheumatoid arthritis and neonatal-onset multisystem inflammatory treatment [85, 86]. IL-1 receptors induce the innate immune response and are associated with excessive inflammation response of the host [350]. It has been hypothesized that anakinra could assist in neutralizing the SARS-CoV-2-related hyperinflammatory state, one of the main causes of ARDS in COVID-19 patients [86, 351]. Anakinra was administered for severe COVID-19 patients in a cohort study and it was observed that it can reduce both the need for invasive mechanical ventilation and mortality without serious side effects [86]. As an anti-proinflammatory cytokine drug, it has been employed in several clinical studies against COVID-19 with encouraging results and it is currently under phase 3 in multi centers of Greece (NCT04680949).

2.4.9 Emapalumab

Emapalumab is a humanized monoclonal anti-IFN-γ antibody, which is known as gamifant brand name. FDA approved Emapalumab for the treatment of hemophagocytic lymphohistiocytosis (HLH) (an illness caused by an overactive immune system) [87, 148, 352, 353]. Emapalumab could prevent the binding of IFN-γ to its cell surface receptors subsequently inhibiting the activation of inflammatory signals and cytokine release syndrome caused by SARS-COV-2 [88, 148]. In order to minimize the rate of inflammation, and the needing mechanical ventilation emapalumab was utilized in combination with anakinra in a phase 2/3 multicenter randomized clinical trial against COVID-19 (NCT04324021). Immunosuppression is reported as one of the side effects of utilizing emapalumab in some patients. Thus, patients with weak immune systems should take this drug with caution [87].

2.4.10 Meplazumab

The main entry route of the virus is to bind host cells through ACE2 receptors at the surface of the host cells. SARS-CoV-2 can also enter the host cells through the cluster of differentiation 147 (CD147). In fact, CD147 can act as a receptor for the SARS-CoV-2 spike protein [89, 354]. Moreover, CD147 could function as a mediator in the inflammatory response as a receptor for cyclophilin A (CyPA), the activator of the intracellular antiviral response, and a potent chemotactic factor for inflammatory leukocytes [90, 355]. Meplazumab is a humanized monoclonal anti-CD147 antibody (IgG2) that inhibits SARS-CoV-2 from entering the cell by blocking the expression of CD147 and reducing the infection caused by the virus. Meplazumab also plays a critical role in reducing the cytokine storm caused by COVID-19 by suppressing cyclophilin A from linking to CD147 [356, 357]. In a phase 2/3 multicenter clinical study, the safety and efficacy of meplazumab are assessing for hospitalized COVID-19 patients (NCT04586153).

2.5 Bioactive Natural Compounds and Herbal Medicines

Natural compounds as highly safe and available products have exhibited promising biological and pharmacological activities including anticancer, antiviral, antimicrobial, anti-inflammatory, and antioxidant properties. Medicinal plant-based natural compounds and traditional herbal medicines have demonstrated antiviral properties against several viruses like the influenza virus, HBV, HCV, SARS-CoV-1, and MERS-CoV. The intervention in both the viral life cycle and host response is attributed to the antiviral functions of natural compounds [358, 359]. Due to the high similarity between SARS-CoV-2 and SARS-CoV-1 in respect to genomics, epidemiologic, and pathogenesis, some herbal and natural medicines are used for the treatment of SARS-CoV-1 could be employed for inhibiting SARS-CoV-2 as well [85]. In this regard, natural compounds and herbal medicines such as theaflavin, cepharanthine, lectins, silvestrol, tryptanthrin, hirsutenoone, tanshinones I-VII, celastrol, pristimererin, iguesterin, tingenone have indicated the potential to prevent SARS-CoV-2 infection through inhibiting RdRp, ACE2, PLpro, or 3CLpro [94, 358, 359].

On the other hand, curcumin and piperine, quercetin, emodin, and scutellarein have been reported to be able to associate with the inhibition of COVID-19 while rendering their anti-inflammatory activities [358]. Particularly, extracted food supplements from plants like curcumin, piperine, and quercetin have the potential to interfere in cellular entry and replication of SARS-CoV-2 and play their roles by immune-boosting, antioxidant and anti-inflammatory functions and by repairing the tissue damages induced by COVID-19. The immunomodulatory and anticytokine effects are also proposed for these agents. Furthermore, these drugs are highly potent to be employed as adjuvants to enhance the bioavailability of other drugs by rendering their multidrug resistance (MDR) effect [69, 360–362]. The nanoencapsulation of quercetin, curcumin/piperine has been practiced and developed due to their hydrophobicity and utilized for the treatment of cancer cells with the capability of suppressing MDR [363, 364]. Likewise, despite the uncertainty of the precise mechanism of many natural compounds, Traditional Chinese Medicines (TCMs) like *Glycyrrhiza uralensis, Saposhnikoviae divaricata, Astragalus membranaceus, Rhizoma Atractylodis Macrocephalae* have been reported to be effective in inhibiting COVID-19 and its subsequent lung inflammation or acute lung injury [85, 94].

2.6   Combination Therapy Approach for COVID-19

Numerous drugs have been reported to be effective against SARS-CoV-2 infection employing the drug repurposing approach that targets viral entry, fusion, replication, and translation alongside regulating immunity and inflammatory response attenuating [234, 365]. In this respect, antivirals, immunomodulators, and anti-inflammatory drugs possessing different mechanisms of action could be exploited in combination to simultaneously inhibit viral functions while providing support and symptomatic treatment for SARS-CoV-2 patients. Besides, the treatment of the severe COVID-19 patients, most at the risk of dying due to the cytokine storm should be practiced with utmost importance by the combination therapy while considering drug–drug interactions and side effects [234, 366]. Some promising combinational administrations of drugs for COVID-19 therapy are presented in Table 2.2.

2.7   Perspectives and Conclusion

SARS-CoV-2 infection is associated with both direct damages induced by the virus and host inflammatory and immune response. In this regard, many antivirals have been administered to inhibit the virus while immunomodulators and anti-inflammatory drugs, as well as biological and natural compounds, have been utilized to either enhance the innate immune system or manage the deregulated inflammatory responses and control the symptoms leading to quick recovery of patients and reducing mortality. Accordingly, combination therapy could be more effective against SARS-CoV-2 infection in the case being utilized timely by taking drug-drug interactions into account. In many clinical trials, combinational administration of antivirals, immunomodulators, and anti-inflammatory drugs has been proposed considering different targets to inhibit the infection.

Alternatively, nanotechnology as a promising strategy could be applied to the COVID-19 treatment principle. Highly biocompatible natural-based vehicles such as proteins and polysaccharides are highly potent to be employed to encapsulate the potential COVID-19 therapeutics and deliver them in an efficient way by enhancing the stability and bioavailability of drugs like favipiravir alongside reducing their

Table 2.2 Potential combination therapy undergoing clinical trials for the treatment of COVID-19

Therapeutics	Targets	Proposed effects	Clinical trial phase (NCT number) (https://clinicaltrials.gov)	Refs.
Remdesivir (antiviral) + baricitinib (immunomodulator)	RdRP + JAK	Inhibiting viral replication + reducing cytokine storm	Phase 3 (NCT04401579)	[1, 7, 367]
Remdesivir (antiviral) + tocilizumab (immunomodulator)	RdRP + IL-6	Inhibiting viral replication + Reducing cytokine storm	Phase 3 (NCT04409262), phase 3 (NCT04678739)	[1, 65, 365]
Remdesivir (antiviral) + IFN-β (immunomodulator)	RdRP + IFNAR signaling	Inhibiting viral replication + boosting immunity against viral infection	Phase 3 (NCT04492475), phase 2 (NCT04647695)	[1, 7]
Lopinavir/ritonavir (antiviral) + ribavirin (antiviral) + IFN-β-1b (immunomodulator)	3CLpro + RdRP + IFNAR signaling	Inhibiting viral replication + boosting immunity against viral infection	Phase 2 (NCT04276688)	[174, 366]
Favipiravir (antiviral) + tocilizumab (immunomodulator)	RdRp + IL-6	Inhibiting viral replication + Reducing cytokine storm	N/A phase (NCT04310228)	[365, 366, 368]
HCQ (antiviral) + azithromycin (immunomodulator)	Endosoamal pH + S protein/ACE2 interaction (feasible)	Inhibiting viral entry and post entry	Phase 3 (NCT04321278), Phase 3 (NCT04347512/withdrawn)	[365, 369]
HCQ (antiviral) + nitazoxanide (antiviral/immunomodulator)	Endosoamal pH, ACE2 + Immune interferon response (feasible)	Inhibiting viral entry + reducing cytokine storm (feasible)	Phase 2/3 (NCT04361318)	[94, 365]
HCQ (antiviral) + ribavirin (antiviral) + nitazoxanide (antiviral/immunomodulator)	Endosoamal pH, ACE2 + RdRP + Immune interferon response (feasible)	Inhibiting viral entry + inhibiting viral replication + reducing cytokine storm (feasible)	Phase 2 (NCT04605588)	[6, 62]

(continued)

Table 2.2 (continued)

Therapeutics	Targets	Proposed effects	Clinical trial phase (NCT number) (https://clinicaltrials.gov)	Refs.
HCQ (antiviral) + azithromycin (immunomodulator) + tocilizumab (immunomodulator)	Endosoamal pH + S protein/ACE2 interaction (feasible) + IL-6	Inhibiting viral entry and post entry + reducing cytokine storm	Phase 2 (NCT04332094)	[4, 7, 100]
HCQ (antiviral) + azithromycin (immunomodulator) + lopinavir/ritonavir (antiviral)	Endosoamal pH + S protein/ACE2 interaction (feasible) + 3CLpro	Inhibiting viral entry and post entry + inhibiting viral replication	Phase 2 (NCT04459702)	[4, 65, 100]
Ivermectin (antiviral) + nitazoxanide (antiviral/immunomodulator)	Nuclear transport process + immune interferon response	Inhibiting viral replication + reducing cytokine storm (feasible)	Phase 2/3 (NCT04360356)	[5, 365]
HCQ (antiviral) + favipiravir (antiviral)	Endosoamal pH, ACE2 + RdRP	Inhibiting viral entry + inhibiting viral replication	Phase 3 (NCT04411433), N/A phase (NCT04392973)	[7, 170, 365]
Ribavirin (antiviral) + ivermectin (antiviral) + nitazoxanide (antiviral/immunomodulator)	RdRP + nuclear transport process + immune interferon response	Inhibiting viral replication + reducing cytokine storm (feasible)	Phase 3 (NCT04392427)	[5, 7, 94]
Danoprevir (antiviral) + ritonavir (antiviral)	Protease	Inhibiting protease activity in the replication cycle	Phase 4 (NCT04345276)	[38, 370]
CP (immunomodulator) + MSC (immunomodulator)	S protein + immune system	Boosting immunity against viral entry and pathogenesis + regenerating tissues and reducing cytokine storm	N/A phase (NCT04492501)	[1, 371]
Tocilizumab (immunomodulator) + anakinra (immunomodulator)	IL-6 + IL-1	Reducing cytokine storm	Phase 3 (NCT04330638)	[7, 65]

(continued)

Table 2.2 (continued)

Therapeutics	Targets	Proposed effects	Clinical trial phase (NCT number) (https://clinicaltrials.gov)	Refs.
Quercetin (antiviral) + bromelain (immunomodulator) + zinc (immunomodulator) + vitamin C (immunomodulator)	S protein/ACE2 interaction + immune system	Inhibiting viral entry and pathogenesis + boosting immunity and reducing cytokine storm	Phase 4 (NCT04468139)	[69, 372]

side effects. Besides, by utilizing the targeted delivery, the nanoparticle-based therapeutics could be triggered toward susceptible alveolar cells prone to be infected by SARS-CoV-2 to either protect them against COVID-19 or provide them with inhibitory drugs. It is worth noting that the nanoparticle-based vaccines have already been taken into consideration for the control of the COVID-19 pandemic and prevent its higher outbreak.

As we obtain more information about the potency of the drug formulations against COVID-19 with respect to their mechanism of action particularly in severe patients, we will be better equipped to optimize therapeutic strategies.

Acknowledgements The authors would like to thank Maria Tajbakhsh Rigi (MD) for sharing her experience in the treatment of COVID-19 patients.

Financial support

This research did not receive any specific fundings.

Conflict of interest

The authors declare no conflict of interests.

References

1. Tu Y-F, Chien C-S, Yarmishyn AA, Lin Y-Y, Luo Y-H, Lin Y-T, Lai W-Y, Yang D-M, Chou S-J, Yang Y-P (2020) A review of SARS-CoV-2 and the ongoing clinical trials. Int J Mol Sci 21(7):2657
2. Oroojalian F, Haghbin A, Baradaran B, Hemat N, Shahbazi M-A, Baghi HB, Mokhtarzadeh A, Hamblin MR (2020) Novel insights into the treatment of SARS-CoV-2 infection: an overview of current clinical trials. Int J Biol Macromol
3. Gil C, Ginex T, Maestro I, Nozal V, Barrado-Gil L, Cuesta-Geijo MA, Urquiza J, Ramírez D, Alonso C, Campillo NE (2020) COVID-19: drug targets and potential treatments. J Med Chem
4. Pandey A, Nikam AN, Shreya AB, Mutalik SP, Gopalan D, Kulkarni S, Padya BS, Fernandes G, Mutalik S, Prassl R (2020) Potential therapeutic targets for combating SARS-CoV-2: drug repurposing, clinical trials and recent advancements. Life Sci 117883
5. Singh TU, Parida S, Lingaraju MC, Kesavan M, Kumar D, Singh RK (2020) Drug repurposing approach to fight COVID-19. Pharmacol Rep 1–30
6. Santos IdA, Grosche VR, Bergamini FRG, Sabino-Silva R, Jardim ACG (2020) Antivirals against coronaviruses: candidate drugs for SARS-coV-2 treatment? Front Microbiol 11:1818
7. Alnefaie A, Albogami S (2020) Current approaches used in treating COVID-19 from a molecular mechanisms and immune response perspective. Saudi Pharmaceut J

8. Yang X, Liu Y, Liu Y, Yang Q, Wu X, Huang X, Liu H, Cai W, Ma G (2020) Medication therapy strategies for the coronavirus disease 2019 (COVID-19): recent progress and challenges. Exp Rev Clin Pharmacol 13(9):957–975
9. Eastman RT, Roth JS, Brimacombe KR, Simeonov A, Shen M, Patnaik S, Hall MD (2020) Remdesivir: a review of its discovery and development leading to emergency use authorization for treatment of COVID-19. ACS Central Sci
10. Nili A, Farbod A, Neishabouri A, Mozafarihashjin M, Tavakolpour S, Mahmoudi H (2020) Remdesivir: a beacon of hope from Ebola virus disease to COVID-19. Rev Med Virol e2133
11. Santoro MG, Carafoli E (2020) Remdesivir: from ebola to COVID-19. Biochem Biophys Res Commun
12. Gordon CJ, Tchesnokov EP, Woolner E, Perry JK, Feng JY, Porter DP, Götte M (2020) Remdesivir is a direct-acting antiviral that inhibits RNA-dependent RNA polymerase from severe acute respiratory syndrome coronavirus 2 with high potency. J Biol Chem 295(20):6785–6797
13. Jorgensen SC, Kebriaei R, Dresser LD (2020) Remdesivir: review of pharmacology, preclinical data and emerging clinical experience for COVID-19. Pharmacother J Human Pharmacol Drug Ther
14. Malin JJ, Suárez I, Priesner V, Fätkenheuer G, Rybniker J (2020) Remdesivir against COVID-19 and other viral diseases. Clin Microbiol Rev 34(1)
15. Devaux CA, Rolain J-M, Colson P, Raoult D (2020) New insights on the antiviral effects of chloroquine against coronavirus: what to expect for COVID-19? Int J Antimicrob Agents 105938
16. Hoffmann M, Schroeder S, Kleine-Weber H, Müller MA, Drosten C, Pöhlmann S (2020) Nafamostat mesylate blocks activation of SARS-CoV-2: new treatment option for COVID-19. Antimicrob Agents Chemother
17. Naghipour S, Ghodousi M, Rahsepar S, Elyasi S (2020) Repurposing of well-known medications as antivirals: hydroxychloroquine and chloroquine–from HIV-1 infection to COVID-19. Exp Rev Anti-Infect Ther 18(11):1119–1133
18. Vincent MJ, Bergeron E, Benjannet S, Erickson BR, Rollin PE, Ksiazek TG, Seidah NG, Nichol ST (2005) Chloroquine is a potent inhibitor of SARS coronavirus infection and spread. Virology J 2(1):1–10
19. Hurst M, Faulds D (2000) Lopinavir. Drugs 60(6):1371–1379
20. Cvetkovic RS, Goa KL (2003) Lopinavir/ritonavir. Drugs 63(8):769–802
21. Blaising J, Polyak SJ, Pécheur E-I (2014) Arbidol as a broad-spectrum antiviral: an update. Antiviral Res 107:84–94
22. Deng P, Zhong D, Yu K, Zhang Y, Wang T, Chen X (2013) Pharmacokinetics, metabolism, and excretion of the antiviral drug arbidol in humans. Antimicrob Agents Chemother 57(4):1743–1755
23. Villalaín J (2010) Membranotropic effects of arbidol, a broad anti-viral molecule, on phospholipid model membranes. J Phys Chem B 114(25):8544–8554
24. Boretti A (2020) Favipiravir use for SARS CoV-2 infection. Pharmacol Rep 72(6):1542–1552
25. Coomes EA, Haghbayan H (2020) Favipiravir, an antiviral for COVID-19? J Antimicrob Chemother
26. Shiraki K, Daikoku T (2020) Favipiravir, an anti-influenza drug against life-threatening RNA virus infections. Pharmacol Therapeut 107512
27. Furuta Y, Komeno T, Nakamura T (2017) Favipiravir (T-705), a broad spectrum inhibitor of viral RNA polymerase. Proc Jpn Acad Series B 93(7):449–463
28. Uyeki TM (2018) Oseltamivir treatment of influenza in children. Oxford University Press, US
29. Yousefi B, Valizadeh S, Ghaffari H, Vahedi A, Karbalaei M, Eslami M (2020) A global treatments for coronaviruses including COVID-19. J Cell Physiol
30. Khalili JS, Zhu H, Mak NSA, Yan Y, Zhu Y (2020) Novel coronavirus treatment with ribavirin: groundwork for an evaluation concerning COVID-19. J Med Virol
31. Preston SL, Drusano GL, Glue P, Nash J, Gupta S, McNamara P (1999) Pharmacokinetics and absolute bioavailability of ribavirin in healthy volunteers as determined by stable-isotope methodology. Antimicrob Agents Chemother 43(10):2451–2456

32. Back D, Sekar V, Hoetelmans R (2008) Darunavir: pharmacokinetics and drug interactions. Antiviral Ther 13(1):1
33. Deeks ED (2014) Darunavir: a review of its use in the management of HIV-1 infection. Drugs 74(1):99–125
34. Rittweger M, Arasteh K (2007) Clinical pharmacokinetics of darunavir. Clin Pharmacokinet 46(9):739–756
35. Wassner C, Bradley N, Lee Y (2020) A review and clinical understanding of tenofovir: tenofovir disoproxil fumarate versus tenofovir alafenamide. J Int Associat Provid AIDS Care (JIAPAC) 19:2325958220919231
36. Clososki GC, Soldi RA, Silva RMd, Guaratini T, Lopes JN, Pereira PR, Lopes JL, Santos Td, Martins RB, Costa CS (2020) Tenofovir disoproxil fumarate: new chemical developments and encouraging in vitro biological results for SARS-CoV-2. J Brazil Chem Soc 31(8):1552–1556
37. Uno Y (2020) Camostat mesilate therapy for COVID-19. Int Emerg Med 1–2
38. Sarkar C, Mondal M, Torequl Islam M, Martorell M, Docea AO, Maroyi A, Sharifi-Rad J, Calina D (2020) Potential therapeutic options for COVID-19: current status, challenges, and future perspectives. Front Pharmacol 11:1428
39. Zheng L, Zhang L, Huang J, Nandakumar KS, Liu S, Cheng K (2020) Potential treatment methods targeting 2019-nCoV infection. Eur J Med Chem 112687
40. Yamamoto M, Matsuyama S, Li X, Takeda M, Kawaguchi Y, Inoue J-I, Matsuda Z (2016) Identification of nafamostat as a potent inhibitor of Middle East respiratory syndrome coronavirus S protein-mediated membrane fusion using the split-protein-based cell-cell fusion assay. Antimicrob Agents Chemother 60(11):6532–6539
41. Chary MA, Barbuto AF, Izadmehr S, Hayes BD, Burns MM (2020) COVID-19: therapeutics and their toxicities. J Med Toxicol 16(3):10.1007
42. Yang P, Gu H, Zhao Z, Wang W, Cao B, Lai C, Yang X, Zhang L, Duan Y, Zhang S (2014) Angiotensin-converting enzyme 2 (ACE2) mediates influenza H7N9 virus-induced acute lung injury. Sci Rep 4:7027
43. Corum J, Wu KJ, Zimmer C (2020) Coronavirus drug and treatment tracker. The New York Times
44. Shang Z, Chan SY, Liu WJ, Li P, Huang W (2020) Recent Insights into emerging coronavirus: SARS-CoV-2. ACS Infect Dis
45. Cox RM, Wolf JD, Plemper RK (2020) Therapeutically administered ribonucleoside analogue MK-4482/EIDD-2801 blocks SARS-CoV-2 transmission in ferrets. Nat Microbiol 1–8
46. Nourian A, Khalili H (2020) Sofosbuvir as a potential option for the treatment of COVID-19. Acta Bio Medica: Atenei Parmensis 91(2):239
47. Sayad B, Sobhani M, Khodarahmi R (2020) Sofosbuvir as repurposed antiviral drug against COVID-19: why were we convinced to evaluate the drug in a registered/approved clinical trial? Arch Med Res 51(6):577–581
48. Wagstaff KM, Sivakumaran H, Heaton SM, Harrich D, Jans DA (2012) Ivermectin is a specific inhibitor of importin α/β-mediated nuclear import able to inhibit replication of HIV-1 and dengue virus. Biochem J 443(3):851–856
49. Altay O, Mohammadi E, Lam S, Turkez H, Boren J, Nielsen J, Mardinoglu A (2020) Current status of COVID-19 therapies and drug repositioning applications. iScience 23(7):101303
50. Sethia R, Prasad M, Jagannath S, Nischal N, Soneja M, Garg P (2020) Efficacy of famotidine for COVID-19: a systematic review and meta-analysis, medRxiv
51. Aguila EJT, Cua IHY (2020) Repurposed GI drugs in the treatment of COVID-19. Dig Dis Sci 65(8):2452–2453
52. Kumari P, Singh A, Ngasainao MR, Shakeel I, Kumar S, Lal S, Singhal A, Sohal SS, Singh IK, Hassan MI (2020) Potential diagnostics and therapeutic approaches in COVID-19. Clinica Chimica Acta; Int J clinical Chem 510:488–497
53. Zheng F, Zhou Y, Zhou Z, Ye F, Huang B, Huang Y, Ma J, Zuo Q, Tan X, Xie J (2020) Novel protein drug, novaferon, as the potential antiviral drug for COVID-19, medRxiv
54. Rossignol J-F (2014) Nitazoxanide: a first-in-class broad-spectrum antiviral agent. Antiviral Res 110:94–103

55. Bardsley-Elliot A, Plosker GL (2000) Nelfinavir. Drugs 59(3):581–620
56. Yamamoto N, Yang R, Yoshinaka Y, Amari S, Nakano T, Cinatl J, Rabenau H, Doerr HW, Hunsmann G, Otaka A (2004) HIV protease inhibitor nelfinavir inhibits replication of SARS-associated coronavirus. Biochem Biophys Res Commun 318(3):719–725
57. Bolcato G, Bissaro M, Pavan M, Sturlese M, Moro S (2020) Targeting the coronavirus SARS-CoV-2: computational insights into the mechanism of action of the protease inhibitors lopinavir, ritonavir and nelfinavir. Sci Rep 10(1):20927
58. Rothan HA, Stone S, Natekar J, Kumari P, Arora K, Kumar M (2020) The FDA-approved gold drug Auranofin inhibits novel coronavirus (SARS-COV-2) replication and attenuates inflammation in human cells. Virology
59. Harbut MB, Vilchèze C, Luo X, Hensler ME, Guo H, Yang B, Chatterjee AK, Nizet V, Jacobs WR, Schultz PG (2015) Auranofin exerts broad-spectrum bactericidal activities by targeting thiol-redox homeostasis. Proc Natl Acad Sci 112(14):4453–4458
60. Thangamani S, Mohammad H, Abushahba MF, Sobreira TJ, Seleem MN (2016) Repurposing auranofin for the treatment of cutaneous staphylococcal infections. Int J Antimicrob Agents 47(3):195–201
61. May HC, Yu J-J, Guentzel MN, Chambers JP, Cap AP, Arulanandam BP (2018) Repurposing auranofin, ebselen, and PX-12 as antimicrobial agents targeting the thioredoxin system. Front Microbiol 9:336
62. Ahamad S, Branch S, Harrelson S, Hussain MK, Saquib M, Khan S (2020) Primed for global coronavirus pandemic: emerging research and clinical outcome. Eur J Med Chem 112862
63. Jin Z, Zhao Y, Sun Y, Zhang B, Wang H, Wu Y, Zhu Y, Zhu C, Hu T, Du X (2020) Structural basis for the inhibition of SARS-CoV-2 main protease by antineoplastic drug carmofur. Nat Struct Mol Biol 27(6):529–532
64. Chien M, Anderson TK, Jockusch S, Tao C, Li X, Kumar S, Russo JJ, Kirchdoerfer RN, Ju J (2020) Nucleotide analogues as inhibitors of SARS-CoV-2 polymerase, a key drug target for COVID-19. J Proteome Res 19(11):4690–4697
65. Asselah T, Durantel D, Pasmant E, Lau G, Schinazi RF (2020) COVID-19: discovery, diagnostics and drug development. J Hepatol
66. Ataei M, Hosseinjani H (2020) Molecular mechanisms of galidesivir as a potential antiviral treatment for COVID-19. J Pharmaceut Care 8(3):150–151
67. Westover JB, Mathis A, Taylor R, Wandersee L, Bailey KW, Sefing EJ, Hickerson BT, Jung K-H, Sheridan WP, Gowen BB (2018) Galidesivir limits Rift Valley fever virus infection and disease in Syrian golden hamsters. Antiviral Res 156:38–45
68. Yu B, Chang J (2020) Azvudine (FNC): a promising clinical candidate for COVID-19 treatment. Signal Transduct Target Ther 5(1):1–2
69. Mrityunjaya M, Pavithra V, Neelam R, Janhavi P, Halami P, Ravindra P (2020) Immune-boosting, antioxidant and anti-inflammatory food supplements targeting pathogenesis of COVID-19. Front Immunol 11
70. Durand N, Mallea J, Zubair AC (2020) Insights into the use of mesenchymal stem cells in COVID-19 mediated acute respiratory failure. npj Regenerat Med 5(1):1–9
71. Sadeghi S, Soudi S, Shafiee A, Hashemi SM (2020) Mesenchymal stem cell therapies for COVID-19: current status and mechanism of action. Life Sci 262:118493
72. Barlow A, Landolf KM, Barlow B, Yeung SYA, Heavner JJ, Claassen CW, Heavner MS (2020) Review of emerging pharmacotherapy for the treatment of coronavirus disease 2019. Pharmacother J Human Pharmacol Drug Ther 40(5):416–437
73. Monk PD, Marsden RJ, Tear VJ, Brookes J, Batten TN, Mankowski M, Gabbay FJ, Davies DE, Holgate ST, Ho L-P (2020) Safety and efficacy of inhaled nebulised interferon beta-1a (SNG001) for treatment of SARS-CoV-2 infection: a randomised, double-blind, placebo-controlled, phase 2 trial. Lancet Respirat Med
74. Hensley LE, Fritz EA, Jahrling PB, Karp C, Huggins JW, Geisbert TW (2004) Interferon-β 1a and SARS coronavirus replication. Emerg Infect Dis 10(2):317
75. Chen L, Xiong J, Bao L, Shi Y (2020) Convalescent plasma as a potential therapy for COVID-19. Lancet Infect Dis 20(4):398–400

76. Rojas M, Rodríguez Y, Monsalve DM, Acosta-Ampudia Y, Camacho B, Gallo JE, Rojas-Villarraga A, Ramírez-Santana C, Díaz-Coronado JC, Manrique R (2020) Convalescent plasma in Covid-19: possible mechanisms of action. Autoimmun Rev 102554
77. Nittari G, Pallotta G, Amenta F, Tayebati SK (2020) Current pharmacological treatments for SARS-COV-2: a narrative review. Eur J Pharmacol 173328
78. Montesarchio V, Parella R, Iommelli C, Bianco A, Manzillo E, Fraganza F, Palumbo C, Rea G, Murino P, De Rosa R (2020) Outcomes and biomarker analyses among patients with COVID-19 treated with interleukin 6 (IL-6) receptor antagonist sarilumab at a single institution in Italy. J Immunother Cancer 8(2)
79. Liu J, Virani SS, Alam M, Denktas AE, Hamzeh I, Khalid U (2020) Coronavirus disease-19 and cardiovascular disease: a risk factor or a risk marker? Rev Med Virol e2172
80. Annane D, Heming N, Grimaldi-Bensouda L, Frémeaux-Bacchi V, Vigan M, Roux A-L, Marchal A, Michelon H, Rottman M, Moine P (2020) Eculizumab as an emergency treatment for adult patients with severe COVID-19 in the intensive care unit: a proof-of-concept study. EClinicalMedicine 28:100590
81. Zhang J, Xie B, Hashimoto K (2020) Current status of potential therapeutic candidates for the COVID-19 crisis. Brain Behav Immun
82. Stallmach A, Kortgen A, Gonnert F, Coldewey SM, Reuken P, Bauer M (2020) Infliximab against severe COVID-19-induced cytokine storm syndrome with organ failure—a cautionary case series. Crit Care 24(1):1–3
83. Gerriets V, Bansal P, Khaddour K (2019) Tumor necrosis factor (TNF) inhibitors. StatPearls [Internet], StatPearls Publishing
84. Feldmann M, Maini RN, Woody JN, Holgate ST, Winter G, Rowland M, Richards D, Hussell T (2020) Trials of anti-tumour necrosis factor therapy for COVID-19 are urgently needed. Lancet 395(10234):1407–1409
85. Wu R, Wang L, Kuo H-CD, Shannar A, Peter R, Chou PJ, Li S, Hudlikar R, Liu X, Liu Z (2020) An update on current therapeutic drugs treating COVID-19. Curr Pharmacol Rep 1
86. Huet T, Beaussier H, Voisin O, Jouveshomme S, Dauriat G, Lazareth I, Sacco E, Naccache J-M, Bézie Y, Laplanche S (2020) Anakinra for severe forms of COVID-19: a cohort study. Lancet Rheumatol
87. Henderson LA, Canna SW, Schulert GS, Volpi S, Lee PY, Kernan KF, Caricchio R, Mahmud S, Hazen MM, Halyabar O (2020) On the alert for cytokine storm: immunopathology in COVID-19. Arthrit Rheumatol
88. Boettler T, Newsome PN, Mondelli MU, Maticic M, Cordero E, Cornberg M, Berg T (2020) Care of patients with liver disease during the COVID-19 pandemic: EASL-ESCMID position paper. JHEP Rep 100113
89. Ulrich H, Pillat MM (2020) CD147 as a target for COVID-19 treatment: suggested effects of azithromycin and stem cell engagement. Stem Cell Rev Rep 1–7
90. Xia P, Dubrovska A (2020) Tumor markers as an entry for SARS-CoV-2 infection? FEBS J 287(17):3677–3680
91. Naserifar M, Hosseinjani H (2020) Novel immunological aspects of sirolimus as a new targeted therapy for COVID-19. J Pharmaceut Care 8(3):152–153
92. Seto B (2012) Rapamycin and mTOR: a serendipitous discovery and implications for breast cancer. Clin Translat Med 1(1):1–7
93. Bagca BG, Avci CB (2020) Overview of the COVID-19 and JAK/STAT pathway inhibition: ruxolitinib perspective. Cytokine Growth Factor Rev (2020)
94. Maurya VK, Kumar S, Bhatt ML, Saxena SK (2020) Therapeutic development and drugs for the treatment of COVID-19. In: Coronavirus disease 2019 (COVID-19). Springer, pp 109–126
95. Chun J, Hartung H-P (2010) Mechanism of action of oral fingolimod (FTY720) in multiple sclerosis. Clin Neuropharmacol 33(2):91
96. Vargesson N (2015) Thalidomide-induced teratogenesis: history and mechanisms. Birth Def Res Part C Embryo Today Rev 105(2):140–156
97. Brogden R, Heel R, Speight T, Avery G (1979) Naproxen up to date: a review of its pharmacological properties and therapeutic efficacy and use in rheumatic diseases and pain states. Drugs 18(4):241–277

98. Todd PA, Clissold SP (1990) Naproxen. Drugs 40(1):91–137
99. Jamwal S, Gautam A, Elsworth J, Kumar M, Chawla R, Kumar P (2020) An updated insight into the molecular pathogenesis, secondary complications and potential therapeutics of COVID-19 pandemic. Life Sci 118105
100. Damle B, Vourvahis M, Wang E, Leaney J, Corrigan B (2020) Clinical pharmacology perspectives on the antiviral activity of azithromycin and use in COVID-19. Clin Pharmacol Therapeut
101. Wu C, Chen X, Cai Y, Zhou X, Xu S, Huang H, Zhang L, Zhou X, Du C, Zhang Y (2020) Risk factors associated with acute respiratory distress syndrome and death in patients with coronavirus disease 2019 pneumonia in Wuhan, China. JAMA Int Med
102. Villar J, Ferrando C, Martínez D, Ambrós A, Muñoz T, Soler JA, Aguilar G, Alba F, González-Higueras E, Conesa LA (2020) Dexamethasone treatment for the acute respiratory distress syndrome: a multicentre, randomised controlled trial. Lancet Respirat Med 8(3):267–276
103. Tomazini BM, Maia IS, Cavalcanti AB, Berwanger O, Rosa RG, Veiga VC, Avezum A, Lopes RD, Bueno FR, Silva MVA (2020) Effect of dexamethasone on days alive and ventilator-free in patients with moderate or severe acute respiratory distress syndrome and COVID-19: the CoDEX randomized clinical trial. JAMA 324(13):1307–1316
104. Ebadi M, Montano-Loza AJ (2020) Perspective: improving vitamin D status in the management of COVID-19. Eur J Clin Nutrit 1–4
105. Carr AC, Maggini S (2017) Vitamin C and immune function. Nutrients 9(11):1211
106. Wessels I, Rolles B, Rink L (2020) The potential impact of zinc supplementation on COVID-19 pathogenesis. Front Immunol 11:1712
107. Jassim SAA, Naji MA (2003) Novel antiviral agents: a medicinal plant perspective. J Appl Microbiol 95(3):412–427
108. Antonelli G, Turriziani O (2012) Antiviral therapy: old and current issues. Int J Antimicrob Agents 40(2):95–102
109. Safrin S (2001) Antiviral agents. Basic Clin Pharmacol 11:845–875
110. Chan S-W (2020) Current and future direct-acting antivirals against COVID-19. Front Microbiol 11:2880
111. Ita K (2020) Coronavirus disease (COVID-19): current status and prospects for drug and vaccine development. Arch Med Res
112. Jordan PC, Stevens SK, Deval J (2018) Nucleosides for the treatment of respiratory RNA virus infections. Antiviral Chem Chemother 26:2040206618764483
113. De Clercq E (2011) A 40-year journey in search of selective antiviral chemotherapy. Annu Rev Pharmacol Toxicol 51:1–24
114. Siegel D, Hui HC, Doerffler E, Clarke MO, Chun K, Zhang L, Neville S, Carra E, Lew W, Ross B (2017) Discovery and synthesis of a phosphoramidate prodrug of a pyrrolo [2, 1-f][triazin-4-amino] adenine C-nucleoside (GS-5734) for the treatment of ebola and emerging viruses. ACS Publications
115. De Clercq E, Herdewijn P (2010) Strategies in the design of antiviral drugs. Pharmaceut Sci Encycl Drug Discov Develop Manufact 1–56
116. Seley-Radtke KL, Yates MK (2018) The evolution of nucleoside analogue antivirals: a review for chemists and non-chemists. Part 1: Early structural modifications to the nucleoside scaffold. Antiviral Res 154:66–86
117. Lo MK, Jordan R, Arvey A, Sudhamsu J, Shrivastava-Ranjan P, Hotard AL, Flint M, McMullan LK, Siegel D, Clarke MO (2017) GS-5734 and its parent nucleoside analog inhibit Filo-, Pneumo-, and Paramyxoviruses. Sci Rep 7:43395
118. Warren TK, Jordan R, Lo MK, Ray AS, Mackman RL, Soloveva V, Siegel D, Perron M, Bannister R, Hui HC (2016) Therapeutic efficacy of the small molecule GS-5734 against Ebola virus in rhesus monkeys. Nature 531(7594):381–385
119. Mulangu S, Dodd LE, Davey Jr RT, Tshiani Mbaya O, Proschan M, Mukadi D, Lusakibanza Manzo M, Nzolo D, Tshomba Oloma A, Ibanda A (2019) A randomized, controlled trial of Ebola virus disease therapeutics. N Engl J Med 381(24):2293–2303
120. Lamb YN (2020) Remdesivir: first approval. Drugs 1–9

121. Sheahan TP, Sims AC, Graham RL, Menachery VD, Gralinski LE, Case JB, Leist SR, Pyrc K, Feng JY, Trantcheva I (2017) Broad-spectrum antiviral GS-5734 inhibits both epidemic and zoonotic coronaviruses. Sci Translat Med 9(396)
122. Kaddoura M, AlIbrahim M, Hijazi G, Soudani N, Audi A, Alkalamouni H, Haddad S, Eid A, Zaraket H (2020) COVID-19 therapeutic options under investigation. Front Pharmacol 11
123. Zhu W, Chen CZ, Gorshkov K, Xu M, Lo DC, Zheng W (2020) RNA-dependent RNA polymerase as a target for COVID-19 drug discovery. SLAS DISCOVERY Adv Sci Drug Discov 2472555220942123
124. Amirian ES, Levy JK (2020) Current knowledge about the antivirals remdesivir (GS-5734) and GS-441524 as therapeutic options for coronaviruses. One Health 100128
125. Gordon CJ, Tchesnokov EP, Feng JY, Porter DP, Götte M (2020) The antiviral compound remdesivir potently inhibits RNA-dependent RNA polymerase from Middle East respiratory syndrome coronavirus. J Biol Chem 295(15):4773–4779
126. Cao Y-c, Deng Q-x, Dai S-x (2020) Remdesivir for severe acute respiratory syndrome coronavirus 2 causing COVID-19: an evaluation of the evidence. Travel Med Infect Dis 101647
127. Ramezankhani R, Solhi R, Memarnejadian A, Nami F, Hashemian SM, Tricot T, Vosough M, Verfaillie C (2020) Therapeutic modalities and novel approaches in regenerative medicine for COVID-19. Int J Antimicrob Agents 106208
128. Liu W, Morse JS, Lalonde T, Xu S (2020) Learning from the past: possible urgent prevention and treatment options for severe acute respiratory infections caused by 2019-nCoV. Chembiochem
129. Kirchdoerfer RN (2020) Halting coronavirus polymerase. J Biol Chem 295(15):4780–4781
130. Goldman JD, Lye DC, Hui DS, Marks KM, Bruno R, Montejano R, Spinner CD, Galli M, Ahn M-Y, Nahass RG (2020) Remdesivir for 5 or 10 days in patients with severe Covid-19. N Engl J Med
131. Beigel JH, Tomashek KM, Dodd LE, Mehta AK, Zingman BS, Kalil AC, Hohmann E, Chu HY, Luetkemeyer A, Kline S (2020) Remdesivir for the treatment of Covid-19—preliminary report. N Engl J Med
132. Grein J, Ohmagari N, Shin D, Diaz G, Asperges E, Castagna A, Feldt T, Green G, Green ML, Lescure F-X (2020) Compassionate use of remdesivir for patients with severe Covid-19. N Engl J Med 382(24):2327–2336
133. Khan Z, Karataş Y, Ceylan AF, Rahman H (2020) COVID-19 and therapeutic drugs repurposing in hand: the need for collaborative efforts. Le Pharmacien Hospitalier et Clinicien
134. Wang Y, Zhang D, Du G, Du R, Zhao J, Jin Y, Fu S, Gao L, Cheng Z, Lu Q (2020) Remdesivir in adults with severe COVID-19: a randomised, double-blind, placebo-controlled, multicentre trial. Lancet
135. WST Consortium (2020) Repurposed antiviral drugs for COVID-19—interim WHO SOLIDARITY trial results. N Engl J Med
136. Schrezenmeier E, Dörner T (2020) Mechanisms of action of hydroxychloroquine and chloroquine: implications for rheumatology. Nat Rev Rheumatol 1–12
137. Khuroo MS, Sofi AA, Khuroo M (2020) Chloroquine and Hydroxychloroquine in Coronavirus Disease 2019 (COVID-19). Facts, fiction & the hype. A critical appraisal. Int J Antimicrob Agents 106101
138. Rolain JM, Colson P, Raoult D (2007) Recycling of chloroquine and its hydroxyl analogue to face bacterial, fungal and viral infections in the 21st century. Int J Antimicrob Agents 30(4):297–308
139. Yang N, Shen H-M (2020) Targeting the endocytic pathway and autophagy process as a novel therapeutic strategy in COVID-19. Int J Biolog Sci 16(10):1724
140. Iyer M, Jayaramayya K, Subramaniam MD, Lee SB, Dayem AA, Cho S-G, Vellingiri B (2020) COVID-19: an update on diagnostic and therapeutic approaches. BMB Rep 53(4):191
141. Roldan EQ, Biasiotto G, Magro P, Zanella I (2020) The possible mechanisms of action of 4-aminoquinolines (chloroquine/hydroxychloroquine) against Sars-Cov-2 infection (COVID-19): a role for iron homeostasis? Pharmacol Res 104904

142. Wang H, Yang P, Liu K, Guo F, Zhang Y, Zhang G, Jiang C (2008) SARS coronavirus entry into host cells through a novel clathrin-and caveolae-independent endocytic pathway. Cell Res 18(2):290–301
143. Inoue Y, Tanaka N, Tanaka Y, Inoue S, Morita K, Zhuang M, Hattori T, Sugamura K (2007) Clathrin-dependent entry of severe acute respiratory syndrome coronavirus into target cells expressing ACE2 with the cytoplasmic tail deleted. J Virol 81(16):8722–8729
144. Yan R, Zhang Y, Li Y, Xia L, Guo Y, Zhou Q (2020) Structural basis for the recognition of SARS-CoV-2 by full-length human ACE2. Science 367(6485):1444–1448
145. Fantini J, Di Scala C, Chahinian H, Yahi N (2020) Structural and molecular modeling studies reveal a new mechanism of action of chloroquine and hydroxychloroquine against SARS-CoV-2 infection. Int J Antimicrob Agents 105960
146. Simmons G, Reeves JD, Rennekamp AJ, Amberg SM, Piefer AJ, Bates P (2004) Characterization of severe acute respiratory syndrome-associated coronavirus (SARS-CoV) spike glycoprotein-mediated viral entry. Proc Natl Acad Sci 101(12):4240–4245
147. Liu J, Cao R, Xu M, Wang X, Zhang H, Hu H, Li Y, Hu Z, Zhong W, Wang M (2020) Hydroxychloroquine, a less toxic derivative of chloroquine, is effective in inhibiting SARS-CoV-2 infection in vitro. Cell Discov 6(1):1–4
148. Magro G (2020) COVID-19: Review on latest available drugs and therapies against SARS-CoV-2. Coagulation and inflammation cross-talking. Vir Res 198070
149. Qin C, Zhou L, Hu Z, Zhang S, Yang S, Tao Y, Xie C, Ma K, Shang K, Wang W (2020) Dysregulation of immune response in patients with COVID-19 in Wuhan, China. Clin Infect Dis
150. Huang M, Li M, Xiao F, Pang P, Liang J, Tang T, Liu S, Chen B, Shu J, You Y (2020) Preliminary evidence from a multicenter prospective observational study of the safety and efficacy of chloroquine for the treatment of COVID-19. Natl Sci Rev 7(9):1428–1436
151. Borba MGS, Val FFA, Sampaio VS, Alexandre MAA, Melo GC, Brito M, Mourão MPG, Brito-Sousa JD, Baía-da-Silva D, Guerra MVF (2020) Effect of high vs low doses of chloroquine diphosphate as adjunctive therapy for patients hospitalized with severe acute respiratory syndrome coronavirus 2 (SARS-CoV-2) infection: a randomized clinical trial. JAMA Netw Open 3(4):e208857–e208857
152. Yu B, Wang DW, Li C (2020) Hydroxychloroquine application is associated with a decreased mortality in critically ill patients with COVID-19, medRxiv
153. Mahevas M, Tran V-T, Roumier M, Chabrol A, Paule R, Guillaud C, Gallien S, Lepeule R, Szwebel T-A, Lescure X (2020) No evidence of clinical efficacy of hydroxychloroquine in patients hospitalized for COVID-19 infection with oxygen requirement: results of a study using routinely collected data to emulate a target trial. MedRxiv
154. Hoffmann M, Mösbauer K, Hofmann-Winkler H, Kaul A, Kleine-Weber H, Krüger N, Gassen NC, Müller MA, Drosten C, Pöhlmann S (2020) Chloroquine does not inhibit infection of human lung cells with SARS-CoV-2. Nature 585(7826):588–590
155. Mallat J, Hamed F, Balkis M, Mohamed MA, Mooty M, MalikA, Nusair A, Bonilla F (2020) Hydroxychloroquine is associated with slower viral clearance in clinical COVID-19 patients with mild to moderate disease: a retrospective study, medRxiv
156. McChesney EW (1983) Animal toxicity and pharmacokinetics of hydroxychloroquine sulfate. Am J Med 75(1):11–18
157. Dong L, Hu S, Gao J (2020) Discovering drugs to treat coronavirus disease 2019 (COVID-19). Drug Discov Therapeut 14(1):58–60
158. Jin Z, Smith LK, Rajwanshi VK, Kim B, Deval J (2013) The ambiguous base-pairing and high substrate efficiency of T-705 (favipiravir) ribofuranosyl 5′-triphosphate towards influenza A virus polymerase. PLoS ONE 8(7):e68347
159. Baranovich T, Wong S-S, Armstrong J, Marjuki H, Webby RJ, Webster RG, Govorkova EA (2013) T-705 (favipiravir) induces lethal mutagenesis in influenza A H1N1 viruses in vitro. J Virol 87(7):3741–3751
160. Furuta Y, Gowen BB, Takahashi K, Shiraki K, Smee DF, Barnard DL (2013) Favipiravir (T-705), a novel viral RNA polymerase inhibitor. Antiviral Res 100(2):446–454

161. de Mello CPP, Tao X, Kim TH, Vicchiarelli M, Bulitta JB, Kaushik A, Brown AN (2018) Clinical regimens of favipiravir inhibit Zika virus replication in the hollow-fiber infection model. Antimicrob Agents Chemother 62(9)

162. Du YX, Chen XP (2020) Favipiravir: pharmacokinetics and concerns about clinical trials for 2019-nCoV infection. Clin Pharmacol Therapeut

163. Wang Y, Fan G, Salam A, Horby P, Hayden FG, Chen C, Pan J, Zheng J, Lu B, Guo L (2020) Comparative effectiveness of combined favipiravir and oseltamivir therapy versus oseltamivir monotherapy in critically ill patients with influenza virus infection. J Infect Dis 221(10):1688–1698

164. Cai Q, Yang M, Liu D, Chen J, Shu D, Xia J, Liao X, Gu Y, Cai Q, Yang Y (2020) Experimental treatment with favipiravir for COVID-19: an open-label control study. Engineering

165. Choy K-T, Wong AY-L, Kaewpreedee P, Sia S-F, Chen D, Hui KPY, Chu DKW, Chan MCW, Cheung PP-H, Huang X (2020) Remdesivir, lopinavir, emetine, and homoharringtonine inhibit SARS-CoV-2 replication in vitro. Antiviral Res 104786

166. Chen C, Huang J, Cheng Z, Wu J, Chen S, Zhang Y, Chen B, Lu M, Luo Y, Zhang J (2020) Favipiravir versus arbidol for COVID-19: a randomized clinical trial. MedRxiv

167. Chandwani A, Shuter J (2008) Lopinavir/ritonavir in the treatment of HIV-1 infection: a review. Ther Clin Risk Manag 4(5):1023

168. Maciorowski D, Idrissi SZE, Gupta Y, Medernach BJ, Burns MB, Becker DP, Durvasula R, Kempaiah P (2020) A review of the preclinical and clinical efficacy of remdesivir, hydroxychloroquine, and lopinavir-ritonavir treatments against COVID-19. SLAS DISCOVERY Advan Sci Drug Discov 2472555220958385

169. Ratia K, Pegan S, Takayama J, Sleeman K, Coughlin M, Baliji S, Chaudhuri R, Fu W, Prabhakar BS, Johnson ME (2008) A noncovalent class of papain-like protease/deubiquitinase inhibitors blocks SARS virus replication. Proc Natl Acad Sci 105(42):16119–16124

170. Sanders JM, Monogue ML, Jodlowski TZ, Cutrell JB (2020) Pharmacologic treatments for coronavirus disease 2019 (COVID-19): a review. JAMA 323(18):1824–1836

171. Lim J, Jeon S, Shin H-Y, Kim MJ, Seong YM, Lee WJ, Choe K-W, Kang YM, Lee B, Park S-J (2020) Case of the index patient who caused tertiary transmission of COVID-19 infection in Korea: the application of lopinavir/ritonavir for the treatment of COVID-19 infected pneumonia monitored by quantitative RT-PCR. J Kor Med Sci 35(6)

172. Qiu H, Wu J, Hong L, Luo Y, Song Q, Chen D (2020) Clinical and epidemiological features of 36 children with coronavirus disease 2019 (COVID-19) in Zhejiang, China: an observational cohort study. Lancet Infect Dis

173. Cao B, Wang Y, Wen D, Liu W, Wang J, Fan G, Ruan L, Song B, Cai Y, Wei M (2020) A trial of lopinavir–ritonavir in adults hospitalized with severe Covid-19. N Engl J Med

174. Hung IF-N, Lung K-C, Tso EY-K, Liu R, Chung TW-H, Chu M-Y, Ng Y-Y, Lo J, Chan J, Tam AR (2020) Triple combination of interferon beta-1b, lopinavir–ritonavir, and ribavirin in the treatment of patients admitted to hospital with COVID-19: an open-label, randomised, phase 2 trial. Lancet 395(10238):1695–1704

175. Wang Z, Chen X, Lu Y, Chen F, Zhang W (2020) Clinical characteristics and therapeutic procedure for four cases with 2019 novel coronavirus pneumonia receiving combined Chinese and Western medicine treatment. Biosci Trends

176. Bongiovanni M, Cicconi P, Landonio S, Meraviglia P, Testa L, Di Biagio A, Chiesa E, Tordato F, Bini T, Monforte AdA (2005) Predictive factors of lopinavir/ritonavir discontinuation for drug-related toxicity: results from a cohort of 416 multi-experienced HIV-infected individuals. Int J Antimicrob Agents 26(1):88–91

177. Wang X, Cao R, Zhang H, Liu J, Xu M, Hu H, Li Y, Zhao L, Li W, Sun X (2020) The anti-influenza virus drug, arbidol is an efficient inhibitor of SARS-CoV-2 in vitro. Cell Discov 6(1):1–5

178. Jomah S, Asdaq SMB, Al-Yamani MJ (2020) Clinical efficacy of antivirals against novel coronavirus (COVID-19): a review. J Infect Public Health

179. Zeng L-Y, Yang J, Liu S (2017) Investigational hemagglutinin-targeted influenza virus inhibitors. Expert Opin Investig Drugs 26(1):63–73

180. Wang Z, Yang B, Li Q, Wen L, Zhang R (2020) Clinical features of 69 cases with coronavirus disease 2019 in Wuhan, China. Clin Infect Dis
181. Deng L, Li C, Zeng Q, Liu X, Li X, Zhang H, Hong Z, Xia J (2020) Arbidol combined with LPV/r versus LPV/r alone against corona virus disease 2019: a retrospective cohort study. J Infect
182. Huang H, Guan L, Yang Y, Le Grange JM, Tang G, Xu Y, Yuan J, Lin C, Xue M, Zhang X (2020) Chloroquine, arbidol (umifenovir) or lopinavir/ritonavir as the antiviral monotherapy for COVID-19 patients: a retrospective cohort study
183. Chen J, Lin S, Niu C, Xiao Q (2020) Clinical evaluation of Shufeng Jiedu Capsules combined with umifenovir (Arbidol) in the treatment of common-type COVID-19: a retrospective study. Exp Rev Respirat Med 1–9
184. Xu P, Huang J, Fan Z, Huang W, Qi M, Lin X, Song W, Yi L (2020) Arbidol/IFN-α2b therapy for patients with corona virus disease 2019: a retrospective multicenter cohort study. Microbes Infect 22(4–5):200–205
185. Lian N, Xie H, Lin S, Huang J, Zhao J, Lin Q (2020) Umifenovir treatment is not associated with improved outcomes in patients with coronavirus disease 2019: a retrospective study. Clin Microbiol Infect
186. Jieming Q (2020) Clinical study of arbidol hydrochloride tablets in the treatment of pneumonia caused by novel coronavirus. NCT04260594
187. Santos JR, Curran A, Navarro-Mercade J, Ampuero MF, Pelaez P, Perez-Alvarez N, Clotet B, Paredes R, Molto J (2019) Simplification of antiretroviral treatment from darunavir/ritonavir monotherapy to darunavir/cobicistat monotherapy: effectiveness and safety in routine clinical practice. AIDS Res Hum Retroviruses 35(6):513–518
188. Triant VA, Siedner MJ (2020) Darunavir and cardiovascular risk: evaluating the data to inform clinical care. J Infect Dis 221(4):498–500
189. Chen J, Xia L, Liu L, Xu Q, Ling Y, Huang D, Huang W, Song S, Xu S, Shen Y (2020) Antiviral activity and safety of darunavir/cobicistat for the treatment of COVID-19, Open forum infectious diseases. Oxford University Press, US, p ofaa241
190. Graci JD, Cameron CE (2006) Mechanisms of action of ribavirin against distinct viruses. Rev Med Virol 16(1):37–48
191. Loustaud-Ratti V, Stanke-Labesque F, Marquet P, Gagnieu M-C, Maynard M, Babany G, Trépo C (2009) Optimizing ribavirin dosage: a new challenge to improve treatment efficacy in genotype 1 hepatitis C patients. Gastroentérologie clinique et biologique 33(6–7):580–583
192. Falzarano D, De Wit E, Rasmussen AL, Feldmann F, Okumura A, Scott DP, Brining D, Bushmaker T, Martellaro C, Baseler L (2013) Treatment with interferon-α2b and ribavirin improves outcome in MERS-CoV–infected rhesus macaques. Nat Med 19(10):1313–1317
193. Zeng Y-M, Xu X-L, He X-Q, Tang S-Q, Li Y, Huang Y-Q, Harypursat V, Chen Y-K (2020) Comparative effectiveness and safety of ribavirin plus interferon-alpha, lopinavir/ritonavir plus interferon-alpha, and ribavirin plus lopinavir/ritonavir plus interferon-alpha in patients with mild to moderate novel coronavirus disease 2019: study protocol. Chin Med J 133(9):1132–1134
194. Zhang ZJ, Morris-Natschke SL, Cheng YY, Lee KH, Li RT (2020) Development of anti-influenza agents from natural products. Med Res Rev 40(6):2290–2338
195. Wang D, Hu B, Hu C, Zhu F, Liu X, Zhang J, Wang B, Xiang H, Cheng Z, Xiong Y (2020) Clinical characteristics of 138 hospitalized patients with 2019 novel coronavirus–infected pneumonia in Wuhan, China. Jama 323(11):1061–1069
196. Rosa SGV, Santos WC (2020) Clinical trials on drug repositioning for COVID-19 treatment. Revista Panamericana de Salud Pública 44:e40
197. Caly L, Wagstaff KM, Jans DA (2012) Nuclear trafficking of proteins from RNA viruses: potential target for antivirals? Antiviral Res 95(3):202–206
198. Caly L, Druce JD, Catton MG, Jans DA, Wagstaff KM (2020) The FDA-approved drug ivermectin inhibits the replication of SARS-CoV-2 in vitro. Antiviral Res 104787
199. De Salazar PM, Ramos J, Cruz VL, Polo R, Del Amo J, Martínez-Salazar J (2020) Tenofovir and remdesivir ensemble docking with the SARS-CoV-2 polymerase and template-nascent RNA. Authorea Preprints

200. Drożdżal S, Rosik J, Lechowicz K, Machaj F, Kotfis K, Ghavami S, Łos MJ (2020) FDA approved drugs with pharmacotherapeutic potential for SARS-CoV-2 (COVID-19) therapy. Drug Resist Updates 100719

201. Ohkoshi M, Oka T (1984) Clinical experience with a protease inhibitor [N, N-dimethylcarbamoylmethyl 4-(4-guanidinobenzoyloxy)-phenylacetate] methanesulfate for prevention of recurrence of carcinoma of the mouth and in treatment of terminal carcinoma. J Maxillofacial Surg 12:148–152

202. Hoffmann M, Kleine-Weber H, Schroeder S, Krüger N, Herrler T, Erichsen S, Schiergens TS, Herrler G, Wu N-H, Nitsche A (2020) SARS-CoV-2 cell entry depends on ACE2 and TMPRSS2 and is blocked by a clinically proven protease inhibitor. Cell

203. Zhou Y, Vedantham P, Lu K, Agudelo J, Carrion R Jr, Nunneley JW, Barnard D, Pöhlmann S, McKerrow JH, Renslo AR (2015) Protease inhibitors targeting coronavirus and filovirus entry. Antiviral Res 116:76–84

204. Bittmann S, Luchter E, Weissenstein A, Villalon G, Moschuring-Alieva E (2020) TMPRSS2-inhibitors play a role in cell entry mechanism of COVID-19: an insight into camostat and nefamostat. J Regen Biol Med 2(2):1–3

205. Jang S, Rhee J-Y (2020) Three cases of treatment with Nafamostat in elderly patients with COVID-19 pneumonia who need oxygen therapy. Int J Inf Dis

206. Ragia G, Manolopoulos VG (2020) Inhibition of SARS-CoV-2 entry through the ACE2/TMPRSS2 pathway: a promising approach for uncovering early COVID-19 drug therapies. Eur J Clin Pharmacol 1–8

207. Hempel T, Raich L, Olsson S, Azouz NP, Klingler AM, Rothenberg ME, Noé F (2020) Molecular mechanism of SARS-CoV-2 cell entry inhibition via TMPRSS2 by Camostat and Nafamostat mesylate. BioRxiv

208. Toots M, Yoon J-J, Cox RM, Hart M, Sticher ZM, Makhsous N, Plesker R, Barrena AH, Reddy PG, Mitchell DG (2019) Characterization of orally efficacious influenza drug with high resistance barrier in ferrets and human airway epithelia. Sci Translat Med 11(515)

209. Toots M, Yoon J-J, Hart M, Natchus MG, Painter GR, Plemper RK (2020) Quantitative efficacy paradigms of the influenza clinical drug candidate EIDD-2801 in the ferret model. Translat Res 218:16–28

210. Abdelnabi R, Foo CS, Kaptein SJ, Zhang X, Langendries L, Vangeel L, Vergote V, Heylen E, Dallmeier K, Chatterjee A (2020) Molnupiravir (EIDD-2801) inhibits SARS-CoV2 replication in Syrian hamsters model

211. Eslami G, Mousaviasl S, Radmanesh E, Jelvay S, Bitaraf S, Simmons B, Wentzel H, Hill A, Sadeghi A, Freeman J (2020) The impact of sofosbuvir/daclatasvir or ribavirin in patients with severe COVID-19. J Antimicrob Chemother 75(11):3366–3372

212. Ju J, Li X, Kumar S, Jockusch S, Chien M, Tao C, Morozova I, Kalachikov S, Kirchdoerfer R, Russo JJ (2020) Nucleotide analogues as inhibitors of SARS-CoV polymerase. BioRxiv

213. Abbaspour Kasgari H, Moradi S, Shabani AM, Babamahmoodi F, Davoudi Badabi AR, Davoudi L, Alikhani A, Hedayatizadeh Omran A, Saeedi M, Merat S (2020) Evaluation of the efficacy of sofosbuvir plus daclatasvir in combination with ribavirin for hospitalized COVID-19 patients with moderate disease compared with standard care: a single-centre, randomized controlled trial. J Antimicrob Chemother 75(11):3373–3378

214. Elfiky AA (2020) Ribavirin, remdesivir, sofosbuvir, galidesivir, and tenofovir against SARS-CoV-2 RNA dependent RNA polymerase (RdRp): a molecular docking study, Life Sci 117592

215. Bourinbaiar AS, Fruhstorfer EC (1996) The effect of histamine type 2 receptor antagonists on human immunodeficiency virus (HIV) replication: identification of a new class of antiviral agents. Life Sci 59(23):PL365–PL370

216. Wu C, Liu Y, Yang Y, Zhang P, Zhong W, Wang Y, Wang Q, Xu Y, Li M, Li X (2020) Analysis of therapeutic targets for SARS-CoV-2 and discovery of potential drugs by computational methods. Acta Pharmaceutica Sinica B

217. Rossignol J-F (2016) Nitazoxanide, a new drug candidate for the treatment of Middle East respiratory syndrome coronavirus. J Infect Public Health 9(3):227–230

218. Yavuz S, Ünal S (2020) Antiviral treatment of COVID-19. Turk J Med Sci 50(SI-1):611–619

219. Xu Z, Yao H, Shen J, Wu N, Xu Y, Lu X, Li L-J (2020) Nelfinavir is active against SARS-CoV-2 in Vero E6 cells
220. Yamamoto N, Matsuyama S, Hoshino T, Yamamoto N (2020) Nelfinavir inhibits replication of severe acute respiratory syndrome coronavirus 2 in vitro. BioRxiv
221. Ohashi H, Watashi K, Saso W, Shionoya K, Iwanami S, Hirokawa T, Shirai T, Kanaya S, Ito Y, Kim KS (2020) Multidrug treatment with nelfinavir and cepharanthine against COVID-19, bioRxiv
222. Cui W, Yang K, Yang H (2020) Recent progress in the drug development targeting SARS-CoV-2 main protease as treatment for COVID-19. Front Mole Biosci 7
223. Sakamoto J, Hamada C, Rahman M, Kodaira S, Ito K, Nakazato H, Ohashi Y, Yasutomi M (2005) An individual patient data meta-analysis of adjuvant therapy with carmofur in patients with curatively resected colon cancer. Jpn J Clin Oncol 35(9):536–544
224. Morimoto K, Koh M (2003) Postoperative adjuvant use of carmofur for early breast cancer. Osaka City Med J 49(2):77–84
225. Wang R-R, Yang Q-H, Luo R-H, Peng Y-M, Dai S-X, Zhang X-J, Chen H, Cui X-Q, Liu Y-J, Huang J-F (2014) Azvudine, a novel nucleoside reverse transcriptase inhibitor showed good drug combination features and better inhibition on drug-resistant strains than lamivudine in vitro. PLoS ONE 9(8):e105617
226. Phillip JM, Wu P-H, Gilkes DM, Williams W, McGovern S, Daya J, Chen J, Aifuwa I, Lee JS, Fan R (2017) Biophysical and biomolecular determination of cellular age in humans. Nat Biomed Eng 1(7):0093
227. Chen J, Lau YF, Lamirande EW, Paddock CD, Bartlett JH, Zaki SR, Subbarao K (2010) Cellular immune responses to severe acute respiratory syndrome coronavirus (SARS-CoV) infection in senescent BALB/c mice: CD4 + T cells are important in control of SARS-CoV infection. J Virol 84(3):1289–1301
228. Leng Z, Zhu R, Hou W, Feng Y, Yang Y, Han Q, Shan G, Meng F, Du D, Wang S (2020) Transplantation of ACE2-mesenchymal stem cells improves the outcome of patients with COVID-19 pneumonia. Aging Dis 11(2):216
229. Canham MA, Campbell JD, Mountford JC (2020) The use of mesenchymal stromal cells in the treatment of coronavirus disease 2019. J Translat Med 18(1):1–15
230. Cao Y, Wu H, Zhai W, Wang Y, Li M, Li M, Yang L, Tian Y, Song Y, Li J (2020) A safety consideration of mesenchymal stem cell therapy on COVID-19. Stem Cell Res 49:102066
231. Rajarshi K, Chatterjee A, Ray S (2020) Combating COVID-19 with Mesenchymal stem cell therapy. Biotechnol Rep e00467
232. Lin F-C, Young HA (2014) Interferons: success in anti-viral immunotherapy. Cytokine Growth Factor Rev 25(4):369–376
233. Ivashkiv LB, Donlin LT (2014) Regulation of type I interferon responses. Nat Rev Immunol 14(1):36–49
234. Nile SH, Nile A, Qiu J, Li L, Jia X, Kai G (2020) COVID-19: pathogenesis, cytokine storm and therapeutic potential of interferons. Cytokine Growth Factor Rev
235. Sallard E, Lescure F-X, Yazdanpanah Y, Mentre F, Peiffer-Smadja N, Florence A, Yazdanpanah Y, Mentre F, Lescure F-X, Peiffer-Smadja N (2020) Type 1 interferons as a potential treatment against COVID-19. Antiviral Res 104791
236. Schreiber G (2020) The role of type I interferons in the pathogenesis and treatment of COVID-19. Front Immunol 11
237. Lokugamage KG, Hage A, Schindewolf C, Rajsbaum R, Menachery VD (2020) SARS-CoV-2 is sensitive to type I interferon pretreatment. BioRxiv
238. Wang N, Zhan Y, Zhu L, Hou Z, Liu F, Song P, Qiu F, Wang X, Zou X, Wan D (2020) Retrospective multicenter cohort study shows early interferon therapy is associated with favorable clinical responses in COVID-19 patients. Cell Host Microbe 28(3):455–464, e2
239. Cinatl J, Morgenstern B, Bauer G, Chandra P, Rabenau H, Doerr H (2003) Treatment of SARS with human interferons. Lancet 362(9380):293–294
240. Kindler E, Thiel V, Weber F (2016) Interaction of SARS and MERS coronaviruses with the antiviral interferon response. Adv Virus Res 96:219–243

241. Abdolvahab MH, Moradi-Kalbolandi S, Zarei M, Bose D, Majidzadeh-A K, Farahmand L (2020) Potential role of interferons in treating COVID-19 patients. Int Immunopharmacol 107171
242. Bloch EM, Shoham S, Casadevall A, Sachais BS, Shaz B, Winters JL, van Buskirk C, Grossman BJ, Joyner M, Henderson JP (2020) Deployment of convalescent plasma for the prevention and treatment of COVID-19. J Clin Investig 130(6):2757–2765
243. Shen C, Wang Z, Zhao F, Yang Y, Li J, Yuan J, Wang F, Li D, Yang M, Xing L (2020) Treatment of 5 critically ill patients with COVID-19 with convalescent plasma. JAMA 323(16):1582–1589
244. Duan K, Liu B, Li C, Zhang H, Yu T, Qu J, Zhou M, Chen L, Meng S, Hu Y (2020) Effectiveness of convalescent plasma therapy in severe COVID-19 patients. Proc Natl Acad Sci 117(17):9490–9496
245. Zhang B, Liu S, Tan T, Huang W, Dong Y, Chen L, Chen Q, Zhang L, Zhong Q, Zhang X (2020) Treatment with convalescent plasma for critically ill patients with SARS-CoV-2 infection. Chest
246. Joyner MJ, Bruno KA, Klassen SA, Kunze KL, Johnson PW, Lesser ER, Wiggins CC, Senefeld JW, Klompas AM, Hodge DO (2020) Safety update: COVID-19 convalescent plasma in 20,000 hospitalized patients. In: Mayo Clinic Proceedings, Elsevier, pp 1888–1897
247. Ye M, Fu D, Ren Y, Wang F, Wang D, Zhang F, Xia X, Lv T (2020) Treatment with convalescent plasma for COVID-19 patients in Wuhan, China. J Med Virol
248. Mangalmurti N, Hunter CA (2020) Cytokine storms: understanding COVID-19. Immunity
249. Channappanavar R, Perlman S (2017) Pathogenic human coronavirus infections: causes and consequences of cytokine storm and immunopathology, Seminars in immunopathology, Springer, pp 529–539
250. Mehta P, McAuley DF, Brown M, Sanchez E, Tattersall RS, Manson JJ, HAS Collaboration (2020) COVID-19: consider cytokine storm syndromes and immunosuppression. Lancet (London, England) 395(10229):1033
251. Seif F, Aazami H, Khoshmirsafa M, Kamali M, Mohsenzadegan M, Pornour M, Mansouri D (2020) JAK inhibition as a new treatment strategy for patients with COVID-19. Int Arch Allergy Immunol 181(6):467–475
252. Sehgal S (2003) Sirolimus: its discovery, biological properties, and mechanism of action. In: Transplantation proceedings. Elsevier, pp S7–S14
253. Zhou Y, Hou Y, Shen J, Huang Y, Martin W, Cheng F (2020) Network-based drug repurposing for novel coronavirus 2019-nCoV/SARS-CoV-2. Cell Discov 6(1):1–18
254. Kindrachuk J, Ork B, Hart BJ, Mazur S, Holbrook MR, Frieman MB, Traynor D, Johnson RF, Dyall J, Kuhn JH (2015) Antiviral potential of ERK/MAPK and PI3K/AKT/mTOR signaling modulation for Middle East respiratory syndrome coronavirus infection as identified by temporal kinome analysis. Antimicrob Agents Chemother 59(2):1088–1099
255. Mogul A, Corsi K, McAuliffe L (2019) Baricitinib: the second FDA-approved JAK inhibitor for the treatment of rheumatoid arthritis. Ann Pharmacother 53(9):947–953
256. Cantini F, Niccoli L, Matarrese D, Nicastri E, Stobbione P, Goletti D (2020) Baricitinib therapy in COVID-19: a pilot study on safety and clinical impact. J Infect
257. Stebbing J, Phelan A, Griffin I, Tucker C, Oechsle O, Smith D, Richardson P (2020) COVID-19: combining antiviral and anti-inflammatory treatments. Lancet Infect Dis 20(4):400–402
258. Richardson P, Griffin I, Tucker C, Smith D, Oechsle O, Phelan A, Stebbing J (2020) Baricitinib as potential treatment for 2019-nCoV acute respiratory disease. Lancet, Lancet Publishing Group
259. Tang JW, Young S, May S, Bird P, Bron J, Mohamedanif T, Bradley C, Patel D, Holmes CW, Kwok KO (2020) Comparing hospitalised, community and staff COVID-19 infection rates during the early phase of the evolving COVID-19 epidemic. J Infect
260. Mesa RA, Yasothan U, Kirkpatrick P, Ruxolitinib, Nature Publishing Group (2012)
261. Harrison C, Kiladjian J-J, Al-Ali HK, Gisslinger H, Waltzman R, Stalbovskaya V, McQuitty M, Hunter DS, Levy R, Knoops L (2012) JAK inhibition with ruxolitinib versus best available therapy for myelofibrosis. N Engl J Med 366(9):787–798

262. Neubauer A, Wiesmann T, Vogelmeier CF, Mack E, Skevaki C, Gaik C, Keller C, Figiel J, Sohlbach K, Rolfes C (2020) Ruxolitinib for the treatment of SARS-CoV-2 induced acute respiratory distress syndrome (ARDS). Leukemia 34(8):2276–2278

263. Li H, Liu H (2020) Whether the timing of patient randomization interferes with the assessment of the efficacy of Ruxolitinib for severe COVID-19. J Allergy Clin Immunol 146(6):1453

264. Fujita T, Inoue K, Yamamoto S, Ikumoto T, Sasaki S, Toyama R, Chiba K, Hoshino Y, Okumoto T (1994) Fungal metabolites. Part 11. A potent immunosuppressive activity found in Isaria sinclairii metabolite. J Antibiot 47(2):208–215

265. Brinkmann V (2009) FTY720 (fingolimod) in multiple sclerosis: therapeutic effects in the immune and the central nervous system. Br J Pharmacol 158(5):1173–1182

266. Ingwersen J, Aktas O, Kuery P, Kieseier B, Boyko A, Hartung H-P (2012) Fingolimod in multiple sclerosis: mechanisms of action and clinical efficacy. Clin Immunol 142(1):15–24

267. Franks ME, Macpherson GR, Figg WD (2004) Thalidomide. The Lancet 363(9423):1802–1811

268. Newfield C (2018) New medical indications for thalidomide and its derivatives. Sci J Lander Coll Arts Sci 12(1):3

269. Paravar T, Lee DJ (2008) Thalidomide: mechanisms of action. Int Rev Immunol 27(3):111–135

270. Robb CT, Goepp M, Rossi AG, Yao C (2020) Non-steroidal anti-inflammatory drugs, prostaglandins, and COVID-19. Br J Pharmacol 177(21):4899–4920

271. FitzGerald GA (2020) Misguided drug advice for COVID-19. Science 367(6485):1434

272. Capuano A, Scavone C, Racagni G, Scaglione F (2020) NSAIDs in patients with viral infections, including Covid-19: victims or perpetrators? Pharmacol Res 104849

273. Crosby JC, Heimann MA, Burleson SL, Anzalone BC, Swanson JF, Wallace DW, Greene CJ (2020) COVID-19: a review of therapeutics under investigation. J Am Coll Emer Phys Open

274. Zheng W, Fan W, Zhang S, Jiao P, Shang Y, Cui L, Mahesutihan M, Li J, Wang D, Gao GF (2019) Naproxen exhibits broad anti-influenza virus activity in mice by impeding viral nucleoprotein nuclear export. Cell Rep 27(6):1875–1885, e5

275. Castro VM, Ross RA, McBride SM, Perlis RH (2020) Identifying common pharmacotherapies associated with reduced COVID-19 morbidity using electronic health records, medRxiv

276. Rainsford K (2009) Ibuprofen: pharmacology, efficacy and safety. Inflammopharmacology 17(6):275–342

277. Mititelu RR, Pădureanu R, Băcănoiu M, Pădureanu V, Docea AO, Calina D, Barbulescu AL, Buga AM (2020) Inflammatory and oxidative stress markers—mirror tools in rheumatoid arthritis. Biomedicines 8(5):125

278. Cole GM, Frautschy SA (2010) Mechanisms of action of non-steroidal anti-inflammatory drugs for the prevention of Alzheimer's disease. CNS Neurol Disorders-Drug Targets (Formerly Current Drug Targets-CNS & Neurological Disorders) 9(2):140–148

279. Smart L, Fawkes N, Goggin P, Pennick G, Rainsford K, Charlesworth B, Shah N A narrative review of the potential pharmacological influence and safety of ibuprofen on coronavirus disease 19 (COVID-19), ACE2, and the immune system: a dichotomy of expectation and reality. Inflammopharmacology 1–12

280. Batlle D, Wysocki J, Satchell K (2020) Soluble angiotensin-converting enzyme 2: a potential approach for coronavirus infection therapy? Clin Sci 134(5):543–545

281. Rinott E, Kozer E, Shapira Y, Bar-Haim A, Youngster I (2020) Ibuprofen use and clinical outcomes in COVID-19 patients. Clin Microbiol Inf

282. McCrae J, Morrison E, MacIntyre I, Dear J, Webb D (2018) Long-term adverse effects of paracetamol–a review. Br J Clin Pharmacol 84(10):2218–2230

283. Graham GG, Scott KF (2005) Mechanism of action of paracetamol. Am J Ther 12(1):46–55

284. Roberts E, Nunes VD, Buckner S, Latchem S, Constanti M, Miller P, Doherty M, Zhang W, Birrell F, Porcheret M (2016) Paracetamol: not as safe as we thought? A systematic literature review of observational studies. Ann Rheumat Dis 75(3):552–559

285. Leont'ev D, Babaev B, Shishkov M, Ostreĭkov I (2005) Effect of nonsteroidal anti-inflammatory drugs and paracetamol on hemodynamic changes during postoperative analgesia in children. Anesteziologiia i reanimatologiia (1):22

286. Sestili P, Fimognari C (2020) Paracetamol use in COVID-19: friend or enemy?
287. Ye Z, Wang Y, Colunga-Lozano LE, Prasad M, Tangamornsuksan W, Rochwerg B, Yao L, Motaghi S, Couban RJ, Ghadimi M (2020) Efficacy and safety of corticosteroids in COVID-19 based on evidence for COVID-19, other coronavirus infections, influenza, community-acquired pneumonia and acute respiratory distress syndrome: a systematic review and meta-analysis. CMAJ
288. Cheng W, Li Y, Cui L, Chen Y, Shan S, Xiao D, Chen X, Chen Z, Xu A (2020) Efficacy and safety of corticosteroid treatment in patients with COVID-19: a systematic review and meta-analysis. Front Pharmacol 11:1378
289. Russell CD, Millar JE, Baillie JK (2020) Clinical evidence does not support corticosteroid treatment for 2019-nCoV lung injury. Lancet 395(10223):473–475
290. Bhimraj A, Morgan RL, Shumaker AH, Lavergne V, Baden L, Cheng VC-C, Edwards KM, Gandhi R, Muller WJ, O'Horo JC (2020) Infectious diseases Society of America guidelines on the treatment and management of patients with COVID-19. Clin Inf Dis
291. Dexamethasone in hospitalized patients with Covid-19—preliminary report. N Engl J Med (2020)
292. Peters DH, Friedel HA, McTavish D (1992) Azithromycin. Drugs 44(5):750–799
293. Dunn CJ, Barradell LB (1996) Azithromycin. Drugs 51(3):483–505
294. Echeverría-Esnal D, Martin-Ontiyuelo C, Navarrete-Rouco ME, De-Antonio Cuscó M, Ferrández O, Horcajada JP, Grau S (2020) Azithromycin in the treatment of COVID-19: a review. Exp Rev Anti-inf Ther 1–17
295. Arshad S, Kilgore P, Chaudhry ZS, Jacobsen G, Wang DD, Huitsing K, Brar I, Alangaden GJ, Ramesh MS, McKinnon JE (2020) Treatment with hydroxychloroquine, azithromycin, and combination in patients hospitalized with COVID-19. Int J Infect Dis 97:396–403
296. Campoli-Richards DM, Brogden RN, Faulds D (1990) Teicoplanin. Drugs 40(3):449–486
297. Baron SA, Devaux C, Colson P, Raoult D, Rolain J-M (2020) Teicoplanin: an alternative drug for the treatment of coronavirus COVID-19. Int J Antimicrob Agents 105944(10.1016)
298. Zhang J, Ma X, Yu F, Liu J, Zou F, Pan T, Zhang H (2020) Teicoplanin potently blocks the cell entry of 2019-nCoV. BioRxiv
299. Klein NC, Cunha BA (1995) Tetracyclines. Med Clin North Am 79(4):789–801
300. Nelson ML, Levy SB (2011) The history of the tetracyclines. Ann N Y Acad Sci 1241(1):17–32
301. Henehan M, Montuno M, De Benedetto A (2017) Doxycycline as an anti-inflammatory agent: updates in dermatology. J Eur Acad Dermatol Venereol 31(11):1800–1808
302. Aggarwal HK, Jain D, Talapatra P, Yadav RK, Gupta T, Kathuria KL (2010) Evaluation of role of doxycycline (a matrix metalloproteinase inhibitor) on renal functions in patients of diabetic nephropathy. Ren Fail 32(8):941–946
303. Phillips JM, Gallagher T, Weiss SR (2017) Neurovirulent murine coronavirus JHM. SD uses cellular zinc metalloproteases for virus entry and cell-cell fusion. J Virol 91(8)
304. Poinas A, Boutoille D, Vrignaud F, Nguyen J-M, Bonnet F, Rat C, Garcia G, Dompmartin A, Leccia M-T, Piroth L (2020) Impact of doxycycline on Covid-19 patients with risk factors of disease degradation: dynamic, a randomised controlled double-blind trial
305. Mattioli M, Benfaremo D, Mancini M, Mucci L, Mainquà P, Polenta A, Baldini PM, Fulgenzi F, Dennetta D, Bedetta S (2020) Safety of intermediate dose of low molecular weight heparin in COVID-19 patients. J Thromb Thrombol 1–7
306. Cosmi B, Hirsh J (1994) Low molecular weight heparins. Curr Opin Cardiol 9(5):612–618
307. Belen-Apak FB, Sarialioglu F (2020) The old but new: Can unfractioned heparin and low molecular weight heparins inhibit proteolytic activation and cellular internalization of SARS-CoV2 by inhibition of host cell proteases? Med Hypotheses 109743
308. Shi C, Wang C, Wang H, Yang C, Cai F, Zeng F, Cheng F, Liu Y, Zhou T, Deng B (2020) The potential of low molecular weight heparin to mitigate cytokine storm in severe COVID-19 patients: a retrospective clinical study. Medrxiv
309. Panfili FM, Roversi M, D'Argenio P, Rossi P, Cappa M, Fintini D (2020) Possible role of vitamin D in Covid-19 infection in pediatric population. J Endocrinol Invest

310. Hansdottir S, Monick MM (2011) Vitamin D effects on lung immunity and respiratory diseases. Vit Hormon, Elsevier 217–237
311. Hernández JL, Nan D, Fernandez-Ayala M, García-Unzueta M, Hernández-Hernández MA, López-Hoyos M, Muñoz-Cacho P, Olmos JM, Gutiérrez-Cuadra M, Ruiz-Cubillán JJ (2020) Vitamin D status in hospitalized patients with SARS-CoV-2 infection. J Clin Endocrinol Metabol
312. Van Gorkom GN, Klein Wolterink RG, Van Elssen CH, Wieten L, Germeraad WT, Bos GM (2018) Influence of vitamin C on lymphocytes: an overview. Antioxidants 7(3):41
313. Feyaerts AF, Luyten W (2020) Vitamin C as prophylaxis and adjunctive medical treatment for COVID-19? Nutrition 79:110948
314. Cheng RZ (2020) Can early and high intravenous dose of vitamin C prevent and treat coronavirus disease 2019 (COVID-19)? Med Drug Discov 5:100028
315. Pal A, Squitti R, Picozza M, Pawar A, Rongioletti M, Dutta AK, Sahoo S, Goswami K, Sharma P, Prasad R (2020) Zinc and COVID-19: basis of current clinical trials. Biol Trace Element Res 1–11
316. Bao S, Knoell DL (2006) Zinc modulates cytokine-induced lung epithelial cell barrier permeability. Am J Physiol-Lung Cell Molecul Physiol 291(6):L1132–L1141
317. Lowenstein CJ, Dinerman JL, Snyder SH (1994) Nitric oxide: a physiologic messenger. Ann Intern Med 120(3):227–237
318. Bruckdorfer R (2005) The basics about nitric oxide. Mol Aspects Med 26(1–2):3–31
319. Darwish I, Miller C, Kain KC, Liles WC (2012) Inhaled nitric oxide therapy fails to improve outcome in experimental severe influenza. Int J Med Sci 9(2):157
320. De Groote MA, Fang FC (1995) NO inhibitions: antimicrobial properties of nitric oxide. Clin Inf Dis 21(Supplement_2):S162–S165
321. Croen KD (1993) Evidence for antiviral effect of nitric oxide. Inhibition of herpes simplex virus type 1 replication. J Clin Investigat 91(6):2446–2452
322. Martel J, Ko Y-F, Young JD, Ojcius DM (2020) Could nasal nitric oxide help to mitigate the severity of COVID-19?
323. Subir R (2020) Pros and cons for use of statins in people with coronavirus disease-19 (COVID-19). Diab Metabol Syn Clin Res Rev 14(5):1225–1229
324. Yuan S (2015) Statins may decrease the fatality rate of Middle East respiratory syndrome infection. MBio 6(4)
325. Totura AL, Baric RS (2015) Reply to "statins may decrease the fatality rate of MERS infection. Mbio 6(5)
326. Chauvin B, Drouot S, Barrail-Tran A, Taburet A-M (2013) Drug–drug interactions between HMG-CoA reductase inhibitors (statins) and antiviral protease inhibitors. Clin Pharmacokinet 52(10):815–831
327. Bełtowski J, Wójcicka G, Jamroz-Wiśniewska A (2009) Adverse effects of statins—mechanisms and consequences. Curr Drug Saf 4(3):209–228
328. Sica DA, Gehr TW, Ghosh S (2005) Clinical pharmacokinetics of losartan. Clin Pharmacokinet 44(8):797–814
329. Salvi R, Patankar P (2020) Emerging pharmacotherapies for COVID-19. Biomed Pharmacother 110267
330. Alzghari SK, Acuña VS (2020) Supportive treatment with tocilizumab for COVID-19: a systematic review. J Clin Virol 127:104380
331. Wösten-van Asperen RM, Lutter R, Specht PA, Moll GN, van Woensel JB, van der Loos CM, van Goor H, Kamilic J, Florquin S, Bos AP (2011) Acute respiratory distress syndrome leads to reduced ratio of ACE/ACE2 activities and is prevented by angiotensin-(1–7) or an angiotensin II receptor antagonist. J Pathol 225(4):618–627
332. Khan A, Benthin C, Zeno B, Albertson TE, Boyd J, Christie JD, Hall R, Poirier G, Ronco JJ, Tidswell M (2017) A pilot clinical trial of recombinant human angiotensin-converting enzyme 2 in acute respiratory distress syndrome. Crit Care 21(1):1–9
333. Monteil V, Kwon H, Prado P, Hagelkrüys A, Wimmer RA, Stahl M, Leopoldi A, Garreta E, Del Pozo CH, Prosper F (2020) Inhibition of SARS-CoV-2 infections in engineered human tissues using clinical-grade soluble human ACE2. Cell

334. Li M, Rao C, Pei D, Wang L, Li Y, Gao K, Wang M, Wang J (2014) Novaferon, a novel recombinant protein produced by DNA-shuffling of IFN-α, shows antitumor effect in vitro and in vivo. Cancer Cell Int 14(1):8

335. Mousavi SM, Hashemi SA, Parvin N, Gholami A, Ramakrishna S, Omidifar N, Moghadami M, Chiang W-H, Mazraedoost S (2020) Recent biotechnological approaches for treatment of novel COVID-19: from bench to clinical trial. Drug Metabol Rev 1–30

336. Zheng F, Zhou Y, Zhou Z, Ye F, Huang B, Huang Y, Ma J, Zuo Q, Tan X, Xie J (2020) SARS-CoV-2 clearance in COVID-19 patients with Novaferon treatment: a randomized, open-label, parallel-group trial. Int J Infect Dis 99:84–91

337. Luo P, Liu Y, Qiu L, Liu X, Liu D, Li J (2020) Tocilizumab treatment in COVID-19: a single center experience. J Med Virol 92(7):814–818

338. Fu B, Xu X, Wei H (2020) Why tocilizumab could be an effective treatment for severe COVID-19? J Translat Med 18(1):1–5

339. Yao X, Huang J, Zhong H, Shen N, Faggioni R, Fung M, Yao Y (2014) Targeting interleukin-6 in inflammatory autoimmune diseases and cancers. Pharmacol Ther 141(2):125–139

340. Yousefi H, Mashouri L, Okpechi SC, Alahari N, Alahari SK (2020) Repurposing existing drugs for the treatment of COVID-19/SARS-CoV-2 infection: a review describing drug mechanisms of action. Biochem Pharmacol 114296

341. López RL, Fernández SC, Pérez LL, Palacios AR, Fernández-Roldán MC, Alonso EA, Camacho IP, Rodriguez-Baño J, Merchante N, Olalla J (2020) Efficacy and safety of early treatment with sarilumab in hospitalised adults with COVID-19 presenting cytokine release syndrome (SARICOR STUDY): protocol of a phase II, open-label, randomised, multicentre, controlled clinical trial. BMJ Open 10(11):e039951

342. Mahajan R, Lipton M, Broglie L, Jain NG, Uy NS (2020) Eculizumab treatment for renal failure in a pediatric patient with COVID-19. J Nephrol 33(6):1373–1376

343. Dixit SB, Zirpe KG, Kulkarni AP, Chaudhry D, Govil D, Mehta Y, Jog SA, Khatib KI, Pandit RA, Samavedam S (2020) Current approaches to COVID-19: therapy and prevention. Ind J Crit Care Med: Peer-Reviewed, Official Publication of Indian Society of Critical Care Medicine 24(9):838

344. Scavone C, Brusco S, Bertini M, Sportiello L, Rafaniello C, Zoccoli A, Berrino L, Racagni G, Rossi F, Capuano A (2020) Current pharmacological treatments for COVID-19: what's next? Br J Pharmacol

345. Rudrapal M, Khairnar SJ, Borse LB, Jadhav AG (2020) Coronavirus disease-2019 (COVID-19): an updated review. Drug Res 70(9):389

346. Pang J, Xu F, Aondio G, Li Y, Fumagalli A, Lu M, Valmadre G, Wei J, Bian Y, Canesi M (2020) Efficacy and tolerability of bevacizumab in patients with severe Covid-19, medRxiv

347. Kumar A, Dey AD, Behl T, Chadha S, Aggarwal V (2020) Exploring the multifocal therapeutic approaches in COVID-19: a ray of hope. Int Immunopharmacol

348. Abdin SM, Elgendy SM, Alyammahi SK, Alhamad DW, Omar HA (2020) Tackling the cytokine storm in COVID-19, challenges, and hopes. Life Sci 118054

349. Nicolela Susanna F, Pavesio C (2020) A review of ocular adverse events of biological anti-TNF drugs. J Ophthal Inflamm Infect 10:1–9

350. Dinarello CA (2018) Overview of the IL-1 family in innate inflammation and acquired immunity. Immunol Rev 281(1):8–27

351. Monteagudo LA, Boothby A, Gertner E (2020) Continuous intravenous anakinra infusion to calm the cytokine storm in macrophage activation syndrome. ACR Open Rheumatol 2(5):276–282

352. Al-Salama ZT (2019) Emapalumab: first global approval. Drugs 79(1):99–103

353. Liu B, Li M, Zhou Z, Guan X, Xiang Y (2020) Can we use interleukin-6 (IL-6) blockade for coronavirus disease 2019 (COVID-19)-induced cytokine release syndrome (CRS)? J Autoimmun 102452

354. Jin Y, Yang H, Ji W, Wu W, Chen S, Zhang W, Duan G (2020) Virology, epidemiology, pathogenesis, and control of COVID-19. Viruses 12(4):372

355. Landras A, Reger de Moura C, Jouenne F, Lebbe C, Menashi S, Mourah S (2019) CD147 is a promising target of tumor progression and a prognostic biomarker. Cancers 11(11):1803

356. Trivedi N, Verma A, Kumar D (2020) Possible treatment and strategies for COVID-19: review and assessment. Eur Rev Med Pharmacol Sci 24:12593–12608

357. Bian H, Zheng Z-H, Wei D, Zhang Z, Kang W-Z, Hao C-Q, Dong K, Kang W, Xia J-L, Miao J-L (2020) Meplazumab treats COVID-19 pneumonia: an open-labelled, concurrent controlled add-on clinical trial. MedRxiv

358. Xian Y, Zhang J, Bian Z, Zhou H, Zhang Z, Lin Z, Xu H (2020) Bioactive natural compounds against human coronaviruses: a review and perspective. Acta Pharmaceutica Sinica B

359. Mani JS, Johnson JB, Steel JC, Broszczak DA, Neilsen PM, Walsh KB, Naiker M (2020) Natural product-derived phytochemicals as potential agents against coronaviruses: a review. Virus Res 197989

360. Soni VK, Mehta A, Ratre YK, Tiwari AK, Amit A, Singh RP, Sonkar SC, Chaturvedi N, Shukla D, Vishvakarma NK (2020) Curcumin, a traditional spice component, can hold the promise against COVID-19? Eur J Pharmacol 173551

361. Abolhassani H, Shojaosadati SA (2019) A comparative and systematic approach to desolvation and self-assembly methods for synthesis of piperine-loaded human serum albumin nanoparticles. Colloids Surf, B 184:110534

362. Srinivasan K (2007) Black pepper and its pungent principle-piperine: a review of diverse physiological effects. Crit Rev Food Sci Nutr 47(8):735–748

363. Abolhassani H, Safavi MS, Handali S, Nosrati M, Shojaosadati SA (2020) Synergistic effect of self-assembled curcumin and piperine co-loaded human serum albumin nanoparticles on suppressing cancer cells. Drug Dev Ind Pharm 46(10):1647–1655

364. Kumari A, Yadav SK, Pakade YB, Singh B, Yadav SC (2010) Development of biodegradable nanoparticles for delivery of quercetin. Colloids Surf, B 80(2):184–192

365. Siddiqui AJ, Jahan S, Ashraf SA, Alreshidi M, Ashraf MS, Patel M, Snoussi M, Singh R, Adnan M (2020) Current status and strategic possibilities on potential use of combinational drug therapy against COVID-19 caused by SARS-CoV-2. J Biomole Struct Dyn 1–14

366. Shyr ZA, Gorshkov K, Chen CZ, Zheng W (2020) Drug discovery strategies for SARS-CoV-2. J Pharmacol Exp Ther 375(1):127–138

367. Kalil AC, Patterson TF, Mehta AK, Tomashek KM, Wolfe CR, Ghazaryan V, Marconi VC, Ruiz-Palacios GM, Hsieh L, Kline S (2020) Baricitinib plus remdesivir for hospitalized adults with Covid-19. N Engl J Med

368. Zhao H, Zhu Q, Zhang C, Li J, Wei M, Qin Y, Chen G, Wang K, Yu J, Wu Z (2020) Tocilizumab combined with favipiravir in the treatment of COVID-19: a multicenter trial in a small sample size. Biomed Pharmacother 133:110825

369. Gautret P, Lagier J-C, Parola P, Meddeb L, Mailhe M, Doudier B, Courjon J, Giordanengo V, Vieira VE, Dupont HT (2020) Hydroxychloroquine and azithromycin as a treatment of COVID-19: results of an open-label non-randomized clinical trial. Int J Antimicrob Agents 105949

370. Chen H, Zhang Z, Wang L, Huang Z, Gong F, Li X, Chen Y (2020) First clinical study using HCV protease inhibitor danoprevir to treat naive and experienced COVID-19 patients. MedRxiv

371. Peng H, Gong T, Huang X, Sun X, Luo H, Wang W, Luo J, Luo B, Chen Y, Wang X (2020) A synergistic role of convalescent plasma and mesenchymal stem cells in the treatment of severely ill COVID-19 patients: a clinical case report. Stem Cell Res Therapy 11(1):1–6

372. Colunga Biancatelli RML, Berrill M, Catravas JD, Marik PE (2020) Quercetin and vitamin C: an experimental, synergistic therapy for the prevention and treatment of SARS-CoV-2 related disease (COVID-19). Front Immunol 11:1451

Chapter 3
Structure of SARS-CoV-2 Proteins

Shokouh Rezaei and Yahya Sefidbakht

3.1 Introduction

Coronaviruses (CoVs) belong to the Coronaviridae family, which is classified into four lineages namely α, β, γ, and δ [1]. Among the lineages, α coronaviruses and β coronaviruses lead to respiratory disease in humans [2]. So far, SARS-CoV-2, MERS-CoV, and novel Coronavirus 2019 (nCoV-2019), named SARS-CoV-2, are identified as important members of lineage β coronaviruses [3]. Due to the risk of previous β-coronaviruses and SARS-CoV-2 outbreaks, it is important to develop an effective and safe vaccine against these viruses that cause infectious diseases [4]. In recent years, structural biology has made it possible to study the complete structure of the virus and the three-dimensional structure of the virus proteins [5]. In fact, investigation of accurate structural information, provides an essential insight into the determinant of viral epitopes and solves some of the challenging problems of virus vaccine production [5, 6]. Therefore, investigation of SARS-CoV-2 structure [7] and structural vaccinology [5, 8] can help design an efficient vaccine to fight SARS-CoV-2.

3.2 Morphology (Size, Structure, and Shape) of SARS-CoV-2

The SARS-CoV-2 is mostly pleomorphic and has a spherical or elliptical shape, with its diameter altering between nearly 60–140 nm [9, 10]. The SARS-CoV-2 has a positive-sense and single-strand RNA genome containing 29891 nucleotides (~30 kb) with 5′-cap structure and 3′-poly-A tail (Fig. 3.1), which encodes 9860

S. Rezaei · Y. Sefidbakht (✉)
Protein Research Center, Shahid Beheshti University, Tehran, Iran
e-mail: y_sefidbakht@sbu.ac.ir

© The Author(s), under exclusive license to Springer Nature Singapore Pte Ltd. 2021
M. Rahmandoust and S.-O. Ranaei-Siadat (eds.), *COVID-19*,
https://doi.org/10.1007/978-981-16-3108-5_3

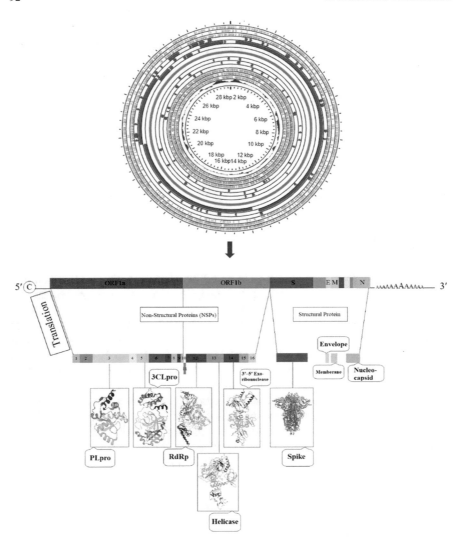

Fig. 3.1 Circular and linear representation of the whole-genome of SARS-CoV-2

amino acids [10, 11]. The SARS-CoV-2 genome is organized as 5′–3′ (Fig. 3.1B) and contains replicase ORF1ab, S, ORF3a, E, M, ORF6, ORF7a, ORF7b, ORF8, N, and ORF10 [12]. The four main structural proteins (spike (S), membrane (M), envelope (E), and nucleocapsid (N) proteins) are encoded within the 3′ end of the viral genome (Fig. 3.1) [13]. In addition, there is another class of proteins, namely, accessory proteins (ORF3, ORF6, ORF7, ORF8, ORF9, and ORF10), which are also encoded by SARS-CoV-2 genome [14]. It seems that these proteins might not be essential for viral replication or structure of virus, but play a role in the viral pathogenicity through modulating the host interferon signaling pathways [15].

3.2.1   Structural Proteins

3.2.1.1   Spike(S) Glycoprotein

Spike glycoprotein is a homotrimer and class I viral fusion protein with multiple glycosylation sites. This protein contains 1273 amino acids and identifies as glycoprotein having multiple domains [16]. As shown in Fig. 3.2, the surface of the spike glycoprotein consists of several (22) potential N-linked glycosylation sites per monomer [17]. The S proteins of coronavirus contain three domains; including (i) an extracellular domain (EC), (ii) a transmembrane anchor domain, and (iii) a short intracellular tail [18, 19]. The EC consists of receptor-binding subunit (S1) and a membrane-fusion subunit (S2), which are two functional and noncovalently associated subunits [18]. The S1 subunit has two independent domains, an N-terminal domain (S1-NTD) and receptor-binding domain (RBD) [18, 20].

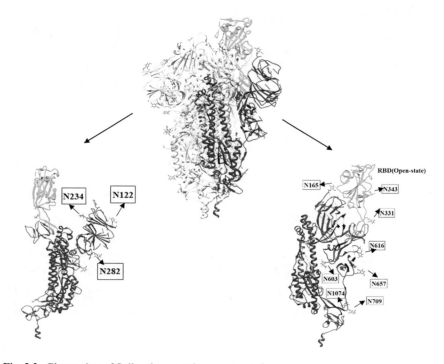

Fig. 3.2   Glycan sites of Spike glycoprotein are presented

The SARS-CoV-2 RBD (residues Arg319–Phe541), a significant domain of whole spike protein, plays an essential role in receptor recognition and binding [21]. To interact with host ACE2, RBD uses hinge-like conformational movements that temporarily hide or expose the determinants of receptor binding [22]. These movements create two states, which are called "down" and "up" conformations [23] (Fig. 3.3A). Down conformation of RBD is related to the unexposed state, and up conformation is related to the exposed state, which is less stable and binds to ACE2 [24]. RBD contains short connecting helices, loops, and a twisted five-stranded antiparallel β sheet; include β1, β2, β3, β4, and β7 [25]. Also, RBD structure consists of an extended insertion, which is called RBM and contains the short β5 and β6 strands, α4 and α5 helices, and loops. In fact, RBM (residues Asn437–Tyr508) has most of the connecting residues of SARS-CoV-2 RBD that bind to host ACE2 [26, 27] (Fig. 3.3B, C). On the other hand, there are nine cysteine residues in the RBD, eight of which contribute to forming four pairs of disulfide bonds. Three of these

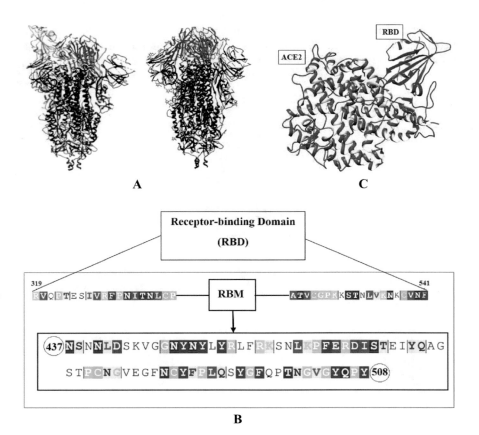

Fig. 3.3 **A** Receptor-Binding Motif (RBM) of SARS-CoV-2 RBD, **B** Up state (left/pink) and Down state (right/pink) of RBD, and **C** Interaction between RBD and ACE2

Table 3.1 A summary of structure of SARS-CoV-2 spike protein

Number	Spike protein	Amino acid
1	S1 subunit	14–685
2	N-terminal domain	14–305
3	receptor-binding domain (RBD)	319–541
4	receptor-binding motif (RBM)	437–508
5	Cleavage site	685–686
6	Cleavage site	815–816
7	S2 subunit	686–1273
8	Fusion Peptide (FP)	788–806
9	HR1	912–984
10	HR2	1163–1213

four pairs including Cys336–Cys361, Cys379–Cys432, and Cys391–Cys525, boost stabilizing of the β sheet structure, and the Cys480–Cys488 pair binds the loops in the distal end of the RBM [27].

In addition, four additional amino acid residues (–PRRA–) or twelve nucleotides were inserted at the interface of S1/S2 subunits of SARS-CoV-2 spike, which suggested insertions affect the efficiency of cleavage of S protein [28]. The presence of a leading proline (P) residue creates the turn, which affects the cleavability of the S1/S2 junction and impacts the accessibility of the cleavage loop with the active site of the protease [29, 30]. Also, it seems that the leading proline residue enables the addition of O-linked glycans to neighboring residues [30]. The S2 subunit consists of one or more fusion peptides (FP), a second proteolytic site (S2′), and two conserved heptad repeats (HRs) [31]. The FP consists of a short segment of conserved residues of the viral family and contains mainly hydrophobic amino acids, including glycine (G) or alanine (A) [32]. After S protein cut at the cleavage sites, the fusion peptide, plays a key role in the fusion of viral membrane into the membrane of host cells and mediates SARS-CoV-2 entry into the host cells [33]. HR1 and HR2 create the six-helical bundle, which is important for the viral fusion [34]. Table 3.1 summarizes some of the important parts of the SARS-CoV-2 spike structure.

3.2.1.2   Effect of Spike Protein Mutations on Viral Fusion

Typically, genetic changes affecting the RBD in the S1 subunit are significant because of the importance of this domain in receptor binding, but studies have shown that there are other key mutations in other S protein domains. For example, a spike mutation is identified, namely, D614G that shows a high frequency [35]. D614 is pocketed adjacent to the fusion peptide (FP) that is the functional fusogenic element of S protein, which is near the predicted cleavage sites [36]. Therefore, studies suggest that the G614 variant might have changed the conformation of Spike protein, affecting the dynamics of the spatially proximal fusion peptide, and subsequently, G614 in comparison with D614 exhibit different fusion ability and it causes higher transmission of virus [36, 37]. It seems that unlike D416G mutation, D839Y (located in FP) mutation can directly change the FP and could also have shaped this motif towards a better-fitted fusion of SARS-CoV-2 with membrane of human cells [36, 38].

3.2.1.3 Envelope (E) Protein

The envelope (E) protein is identified as the smallest (8–12 KDa) of all the structural proteins [39]. The E protein consists of three domains; include a short hydrophilic N-terminus domain (7–12 amino acids), hydrophobic transmembrane domain (25 amino acids), and long hydrophilic C-terminal region [40]. The E protein forms a homopentameric (Fig. 3.4A) cation channel [41, 42]. This protein contains 35 α-helices and 40 loops (Fig. 3.4A) that can affect ion channel activity and help pathogenesis by SARS-CoV-2 in this way [43]. The pore of the channel contains hydrophobic residues such as Asn15, Leu18, Leu21, Val25, Leu28, Ala32, and Thr35 (Fig. 3.4A), which indicate N/C-terminal of the E protein, where Asn15 and Lys28 are the first pore-contacting residues (Fig. 3.4B) [42]. In addition, UCSF Chimera [44] was used to indicate the electrostatics surface of N/C-terminals of the E protein (Fig. 3.4C).

3.2.1.4 Nucleocapsid (N) Protein

The nucleocapsid (N) phosphoprotein consists of an N-terminal (NTD) and a C-terminal (CTD) domain. All of N protein domains indicate the RNA binding affinity [45]. In contrast, the CTD binds the membrane (M) protein, which creates the physical junction between the envelope and positive-RNA [46]. Investigation of the structure of the SARS-CoV-2 N protein indicated that both of the NTD and CTD of SARS-CoV-2 N protein are rich in β-strands, whereas the CTD show several short helices [47]. Particularly, the structure of N protein indicated the right hand-like fold and a β-sheet core with an extended central loop in which the core region consists of a five-stranded U-shaped right-handed antiparallel β-sheet platform (β4–β2–β3–β1–β5) and flanked by two short α-helices [48].

3.2.1.5 Membrane (M) Protein

The Membrane (M) protein consists of three TM domains with C-terminal inside (long) and N-terminal (short) outside [49]. The M protein plays a main role in maintaining the shape of the virus envelope, which is performed through interacting with S, M, N proteins [49]. The details of the protein structure are available in UniProt (https://covid-19.uniprot.org/uniprotkb/P0DTC5).

Fig. 3.4 Structure of SARS-CoV-2 E protein. **A** Pentamer structure contains five chains (**A–E**) and hydrophobic residues of channel pore are indicated. **B** Asn15 and Lys28 are identified as the first pore-contacting residues. **C** Surface electrostatics of the N/C-terminal E protein is shown

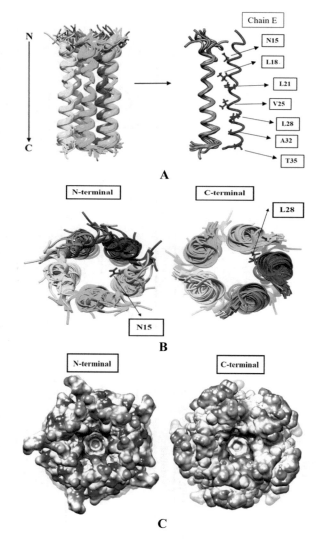

3.2.1.6 Physicochemical Properties of Structural Proteins

The ProtParam [50] is used to investigate several physicochemical properties of SARS-CoV-2 structural proteins (Table 3.2). These properties contain molecular (M) weight, theoretical pI, number of negative and positive charged residues, the formula of proteins, the total number of atoms, instability index, aliphatic index, grand average of hydropathicity (GRAVY), and estimated half-life. The importance of studying the physicochemical properties of SARS-CoV-2 proteins is due to the fact that these properties are valuable for drug design and vaccine development [51,

Table 3.2 Physicochemical properties of structural proteins are shown

	S protein	N protein	E protein	M protein
Amino acid	1273	419	75	222
M weight	141178.47	45625.70	8365.04	25146.62
Theoretical pI	6.24	10.07	8.57	9.51
Total number of (Asp + Glu)	110	36	3	13
Total number of (Arg + Lys)	103	60	5	21
Formula	$C_{6336}H_{9770}N_{1656}O_{1894}S_{54}$	$C_{1971}H_{3137}N_{607}O_{629}S_7$	$C_{390}H_{625}N_{91}O_{103}S_4$	$C_{1165}H_{1823}N_{303}O_{301}S_8$
Instability index	33.01 (stable)	55.09 (unstable)	38.68 (stable)	39.14 (stable)
Aliphatic index	84.67	52.53	144.00	120.86
Grand average of hydropathicity (GRAVY)	−0.079	−0.971	1.128	0.446

Table 3.3 Features of secondary structure of structural proteins

	Alpha helix (%)	Extended strand (%)	Beta turn (%)	Random coil (%)
S protein	28.59	23.25	3.38	44.78
E protein	44.00	26.67	9.33	20.00
N protein	18.62	15.99	7.16	58.23
M protein	34.68	21.17	6.76	37.39

52]. The SOPMA server [53] was used for the secondary structure prediction of the structural proteins. This server presented the conformational information of the α-helices, β-strands, turns, and random coils (Table 3.3).

3.2.2 Accessory Proteins

Another class of SARS-CoV-2 proteins has been identified as accessory proteins. Although very little information is available about the activity of these proteins, in this chapter, the most important properties of the six accessory proteins (ORF3a, ORF6, ORF7a, ORF7b, ORF8, and ORF10) are represented.

3.2.2.1 ORF6

ORF6 (amino acid 61) is known as a membrane-associated interferon antagonist protein [54]. ORF6 interacts with the NSP8 and it can enhance RNA polymerase action [55]. In addition, it has been suggested that ORF6 and ORF8 can inhibit the interferon (type I) signaling pathway [55].

3.2.2.2 ORF7a

ORF7a (amino acid 121) with S, M, and E protein has a critical role in viral assembly, so this protein is important for the viral replication [54, 56]. It seems that ORF7a causes the activation of pro-inflammatory cytokines and chemokines, and previous studies indicated, ORF7a of SARS-CoV with E protein, activates apoptosis [54, 57]. Overall, there is very little information to support a role of ORF7a or ORF7b in the SARS-CoV-2 replication cycle.

3.2.2.3 ORF8

ORF8 (amino acid 121) is identified as a unique accessory protein in SARS-CoV-2 and also, this protein is indicated to cause structural alterations that can affect the ability of the virus outspread [58]. SARS-CoV-2 ORF8 interacts with MHC (class I) molecules and decreases their surface expression on various cells [59].

3.2.2.4 ORF10

ORF10 (amino acid 38) has been known as a unique protein because this accessory protein has 11 cytotoxic T-cell epitopes (with nine amino acids length) [55, 60].

3.2.3 Non-structural Protein

The third class of SARS-CoV-2 proteins includes Non-Structural Proteins (NSPs). The long polyprotein of SARS-CoV-2 encode by ORF1ab and it is processed into 16 NSP by two proteases; including PLpro (NSP3) and 3CLpro (NSP5) [61]. ORF1b (3′ half of ORF1ab) encodes several critical enzymes for viral RNA replication and viral protein translation; including NSP12, NSP13, NSP14, NSP15, and NSP16 [61–63]. NSPs (ORF1a) such as NSP7, NSP8, and NSP10 are identified as a cofactor for NSPs with enzymatic activity (Table 3.4) [64]. Herein, identity and similarity were computed with EMBOSS Needle from alignments of the NSPs sequence of SARS-CoV-2 and SARS-CoV (Table 3.5).

ORF1a creates a large polyprotein that cleaved into 11 NSPs; including NSP1–NSP11 (66), and ribosomal slippage at an RNA pseudoknot structure and slippery sequence at the end of ORF1a, which occasionally led to a frameshift and translation of a joint ORF1a and ORF1b polyprotein (67).

Programmed -1 ribosomal frameshifting (-1 PRF) is one of the important mechanisms during translation of the SARS-CoV-2 RNA genome, which is used for the expression of open reading frames (ORFs) (68). Especially -1 PRF mechanism was shown to be required to translate ORF1ab, which determined that cis-acting elements

Table 3.4 Classification of NSPs that have enzymatic activity

NSPs	Enzymatic activity	Co-factor	Source	PDB code
NSP3	PLpro		SARS-CoV2	6M2Q
NSP5	3CLpro		SARS-CoV2	6M2Q
NSP9	RNA-replicase		SARS-CoV2	6WXD
NSP12	RNA-dependent RNA polymerase (RdRp)	NSP7, NSP8	SARS-CoV2	6M71
NSP13	Helicase	NSP10	SARS-CoV2	6XEZ
NSP14	3′–5′ Exo-ribonuclease, ExoN; Guanine-N7 methyltransferase, N7-MTase	NSP10	SARS-CoV2	–
NSP15	Endo-ribonuclease		SARS-CoV2	6W01
NSP16	2′-O-ribose methyltransferase	NSP10	SARS-CoV2	6W4H

Table 3.5 Identity and similarity between NSPs of SARS-CoV-2 and SARS-CoV are shown

Protein name	Identity (%)	Similarity (%)
NSP1	84.4	91.1
NSP2	68.3	82.9
NSP3	76.0	86.5
NSP4	80.0	90.8
NSP5 (M^{pro})	95.5	98.1
NSP6	87.2	94.8
NSP7	93.2	94.3
NSP8	95.1	96.6
NSP9	90.1	90.9
NSP10	88.8	90.8
NSP12 (RdRp)	93.9	95.8
NSP13 (Helicase)	99.8	100
NSP14 (G-N7-MTase)	95.1	98.7
NSP15 (NendoU)	88.7	95.7
NSP16 (2′-O-MTase)	80.8	84.9

in the mRNA direct elongating ribosomes to shift reading frame using 1 base in the 5′ direction (68). This mechanism is essential for the synthesis of viral RNA-dependent RNA polymerase (RdRp) and downstream viral non-structural proteins in which these proteins with their enzymatic functions play main roles in the capping of RNA, RNA modification and processing, and RNA proof-reading (69).

3.2.3.1 Non-structural Protein 1 (NSP1)

Among the NSPs translated by ORF1a, Non-Structural protein 1 (Nsp1) has a significant role and also, it is identified as a leader protein since this protein is the first production by the N-terminal region of the viral genome [65]. The NSP1 is the main virulence factor of coronaviruses, which binds to the 40S ribosomal subunit and inhibits host gene expression [65]. Comparative structural studies between SARS-CoV-2 and SARS-CoV NSP exhibited that both NSP1 have similar α/β-folds, which contains a six-stranded β-barrel and a long α1-helix coating one opening of the barrel; also, in SARS-CoV-2 nsp1 (153–179 aa), a short C-terminal structure was identified that contains two helices, which may contribute an essential effect in host protein synthesis inhibition and inhibit the type I interferon response [66, 67]. Currently, V121D substitution was known as a key mutation that can destabilize NSP-1 and inactivate the host type-1 Interferon-induced antiviral system [68].

3.2.3.2 Non-structural Protein 3 (NSP3)

NSP3 (1945 aa) has been identified as the largest of the NSPs, which is a 217 kDa [69] protein. This NSP is a multi-domain protein that has several catalytic activities, and plays an essential role during the formation of virus replication complex [70, 71]. ADP-ribose has been known as a macrodomain of NSP3 [72]. In addition, PLpro is one of the two known coronavirus proteases produced by NSP3 and is important for the efficient cleaves between NSP1-NSP2, NSP2-NSP3, and NSP3-NSP4 to release NSP1, NSP2, and NSP3 from the viral polypeptide [73]. Biophysical and structural studies of PLpro have indicated this domain of NSP3 has other functions including (a) hydrolyzing ubiquitin chains that are important for inflammatory responses, and (b) eliminating interferon stimulated gene 15 (ISG15) modifications from proteins that cause reversing antiviral responses [71]. Overall, the results of the amino acid sequence alignment revealed that the NSP3 of SARS-CoV-2 shares 76.0% identity and 86.5% similarity with the NSP3 of SARS-CoV. Especially, investigation of both PLpro shows an 82.9% sequence identity [69, 74]. Interestingly, SARS-CoV-2 PLpro like SARS-CoV PLpro has several key residues (such as Tyr269 and Gln270) that play a critical role in binding to small molecules (Fig. 3.5) [74, 75]. In addition, structural studies have indicated the PLpro structure is similar to ubiquitin-specific proteases (USPs) in humans, which seems like an open hand and contains four domains; including ubiquitin-like modifier (UBL), thumb, palm, and fingers [69, 76].

Fig. 3.5 Structure comparison. Two key residues in PLpro of both CoVs are represented

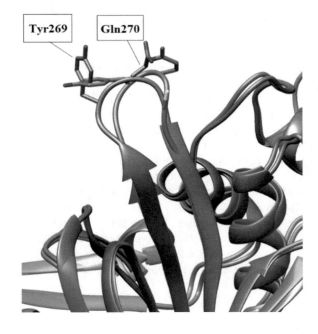

3.2.3.3 Non-structural Protein 5 (NSP5)

NSP5 (amino acids 306) was identified as Main protease (M^{pro}) or 3-chymotrypsin-like cysteine protease (3CLpro), which is one of the SARS-CoV-2 enzymes that plays an essential role in polyprotein processing [77–79]. M^{pro} is active in a homodimer form and each protomer of enzyme consists of three domains [80, 81]; including domain I (residues 8–101) and domain II (residues 102–184) that contain an antiparallel β-barrel structure, and domain III (residues 201–303) that has five α-helices arranged into a largely antiparallel globular cluster (Fig. 3.6) [78, 81]. Domain III is linked to domain II through a long loop region (residues 185–200) [81]. In addition, M has a catalytic dyad (His41 and Cys145 residues) that is located in a cleft between domains I and II [82].

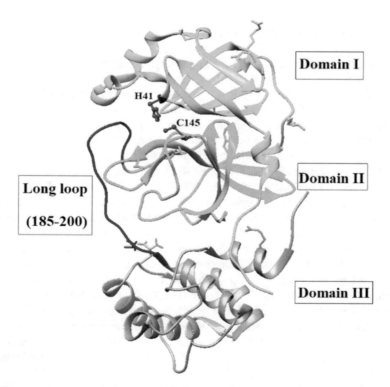

Fig. 3.6 Structure of the SARS-CoV-2 Mpro; catalytic dyad residues (H41 and C145) are indicated by magenta color

3.2.3.4   NSP3/NSP4/NSP6 Complex

The co-expression of three NSPs including Nsp3, Nsp4, and Nsp6 induces double-membrane vesicles (DMVs), which the virus uses to evade detection by host innate immune sensors and to translate its accessory proteins [83].

3.2.3.5   NSP11

SARS-CoV-2 has a short protein namely NSP11 that has only 13 amino acids [77]. In fact, the cleavage of polyprotein 1a by 3CLpro (NSP5) at the NSP10–11 junction leads to produce NSP11 protein [62].

3.2.3.6   NSP7/NSP8/NSP12 Complex

Non-Structural Protein 12 (Nsp12) is a large enzyme that consists of two conserved domains; including the NiRAN and the polymerase domains [84, 85]. The form of NiRAN domain is determined by an $\alpha + \beta$ fold that consists of eight α helices and a five-stranded β-sheet [85]. It seems the core protein consists of a single chain (nearly 900 amino acids) that is identified with minimal activity [86]. With the binding of other essential subunits to the core protein, polymerase activity is increased [86, 87]. In fact, RNA-dependent RNA polymerase (RdRp) of SARS-CoV-2 identified as NSP12, catalyzes the synthesis of viral RNA and thus plays a central role in the replication and transcription cycle of COVID-19 virus [88]. For the RdRp activity, several cofactors are required that contain NSP/NSP8, and additional NSP8 [87]. As shown in Fig. 3.7, the binding site for the second additional NSP8 is different. Generally, the shape of the RdRp looks like a closed right hand that consists of the finger subdomain (residues 398–581, 628–687), palm subdomain (residues 582–627, 688–815), and thumb subdomain (residues 816–919) [89, 90]. For the structural stability, RdRp requires two Zn ions [86], one of the Zn ions is bound to residues His295, Cys301, Cys306, and Cys310 (in N-terminal), and the other Zn ion is bound to residues Cys487, His642, Cys645, Cys646 (in finger subdomain) [87, 91]. Interestingly, the outer surface of nsp12 has a mainly negative electrostatic potential while the RNA template and NTP binding sites contain positive electrostatic potential [87]. Also, the other sites such as binding of NSP7/8 complex, additional NSP8 site, and the template exit site are neutral [86, 87].

The nidovirus-unique domain can be divided into two different regions including the NiRAN (residues 117–250) and an Interface region (residues 251–398) [87]. The interface region is known as a protein-interaction junction that contacts with the NiRAN, the fingers domain, and the additional NSP8 [87]. Although the exact role of NiRAN is not known, structural analysis has suggested that the NiRAN domain indicates structural properties of kinase-like folds, so this domain can be suggested as a target for kinase inhibitors [85].

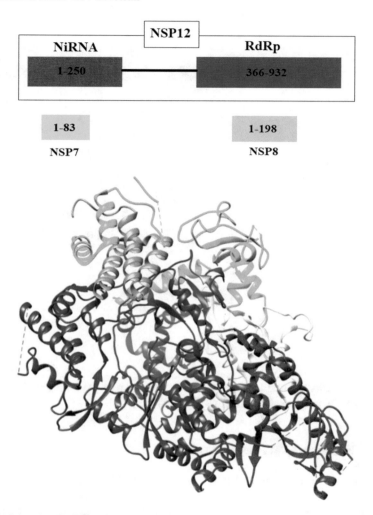

Fig. 3.7 Structure of NSP12/NSP8/NSP7 complex is shown

3.2.3.7 Non-structural Protein 13 (NSP13)

Non-structural protein 13 (NSP13) has two activities including RNA helicase and nucleoside triphosphate hydrolase (NTPase), which is dependent on the special divalent metallic ions [92]. The helicase can open double-stranded RNA and double-stranded DNA using a 5′-ss tail along the polarity of 5′ to 3′ [93]. The Nsp13 (helicase) of SARS-CoV-2 shares a 99.8% sequence identity with NSP13 (helicase) of SARS-CoV (with only one single residue difference) [94]. This enzyme also can hydrolyze all deoxyribonucleotide and ribonucleotide triphosphates [95]. Helicase

Fig. 3.8 Structure of
SARS-CoV-2 Nsp13
(helicase) is indicated

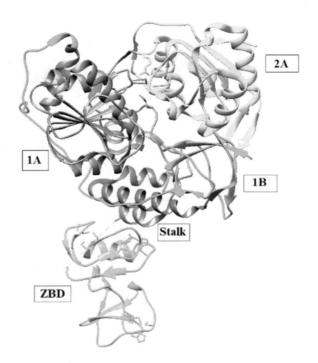

is composed of five domains, namely, 1A, 2A, 1B domain, N-terminal Zinc binding
domain (ZBD), and the stalk domain (Fig. 3.8) [85].

3.2.3.8   Non-structural Protein (Nsp15)

NSP15 protein is an endonuclease from SARS-CoV-2 that plays an important
role in the proofreading of viral RNA [96]. This protein Nsp15 is identified as
a nidoviral RNA uridylate-specific endoribonuclease and possesses a C-terminal
catalytic domain, which is specific for uridine acting on ssRNA and dsRNA [97].

3.2.3.9   Complex of Nsp10/Nsp14/Nsp16

The Nsp14 and nsp16 play a role in the methylation of the cap on the guanine of
the GTP and the C2′ hydroxyl group of the following nucleotide, respectively. These
NSPs are S-adenosylmethionine (SAM)-dependent methyltransferases (MTases) and
known as essential factors for the viral lifecycle. The NSP10 increases enzymatic
activities of both nsp14 and nsp16, which is a key cofactor for their correct function
[98, 99]. Particularly, NSP14 is a bi-functional protein that consists of 3′-to-5′ exoribonuclease (ExoN) and guanine-N7-methyltransferase (N7-MTase) domains [100].

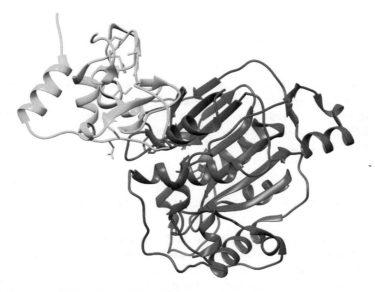

Fig. 3.9 Structure of SARS-CoV-2 NSP16/NSP10 complex (Color pattern: NSP16 (Magenta) and NSP10 (Cyan))

The N7-MTase activity depends on the integrity of the N-terminal ExoN domain, and the flexibility of the whole NSP14 is modulated by a hinge region linking the two domains [101].

The structure of SARS-CoV-2 nsp10 comprises residues 19–133, which contains a positively charged and hydrophobic surface that interacts with a hydrophobic pocket and a negatively charged surface from nsp16 (Fig. 3.9) [102]. The structure of nsp10 of SARS-CoV-2 contains a central antiparallel pair of β-strands surrounded on one side by a crossover large loop. In addition, there is a helical domain with loops that form two zinc fingers [102]. The nsp16 structure has all 298 amino acids that formed through a β-sheet with the canonical 3-2-1-4-5-7-6 arrangement, in which β7 is the only antiparallel strand. This β-sheet is sandwiched by loops and α-helices [102].

3.3 Disorder Intrinsically Region (DIR)

The Intrinsically disordered proteins (IDPs) and intrinsically disordered protein regions (IDPRs) are identified as functional proteins and protein regions without unique structures, and usually, play critical roles in various biological processes [103, 104]. It seems that IDRs of a virus's structure can affect an RNA virus's adaptive capacity [105, 106]. In addition, it was identified that high IDR fraction in some of the virus proteomes contribute to their wider host range, interactions between viral protein and host cell, host tropism, and cross-species transmission compared with other viruses [107–109]. It has been predicted that IDRs may show weak or even completely non-immune responses, thus during the antigen selection step, IDRs should be considered [110].

It seems the SARS-CoV-2 proteome has a significant content of ordered proteins [54]. While several proteins such as Nucleocapsid, Nsp8, ORF6, and cleavage sites in replicase 1ab polyprotein are identified to be highly disordered (Fig. 3.10) [54, 111]. Since the disordered region in proteins has consequences for the structured and unstructured biology of SARS-CoV-2, therefore, the investigation of these regions is of specific importance.

3.4   Importance of Virus Structure on Vaccine Development

Historically, it has been determined that high rate transmission and easy spread of some viruses such as coronaviruses can cause an epidemic risk [112]. Therefore, with the fast increase in infectious cases and its high mortality rate due to epidemics, the world needs an urgent, efficient, and safe vaccine [113–115]. Currently, COVID-19 caused by SARS-CoV-2 is identified as an epidemic, and according to previous studies, it seems that investigating the structure of the virus and key proteins that have been considered as candidates for the vaccine development process are significant [116]. Structural studies indicated that among all proteins of SARS-CoV-2 (structural and Non-structural), spike glycoprotein can be the major antigen for the vaccine design because it can induce neutralizing antibodies and protective immunity [117, 118]. There are many reasons for selecting spike protein as an antigen: it does not belong to the high-disorder proteins [54]; it plays an important role in binding to the host cell receptor [119]; this protein is a virus surface protein that facilitates the neutralization process [120]; in addition, spike is a large protein and contains several domains such as RBD that was known to be one of the key antigens of the virus [7, 121].

3.5   Substitution Mutations

In this study, mutations with a high frequency of SARS-CoV-2 proteins are collected in Table 3.6. The introduction of these mutations is important because these substitutions can affect the structure and function of proteins [122]. Importantly, if the viral protein is the target of the drug or vaccine development, its mutations may affect the stability and function of the protein and lead to disruption of the interaction between target and drug/antibody, such as mutations of receptor-binding domain (RBD) of spike protein [27]. It seems that RBD mutations change the pathogenicity of the virus (UK variants) [123], which these substitutions probably alter binding affinity between protein and drug/antibody. Therefore, the investigation of mutations in virus proteins is significant [122].

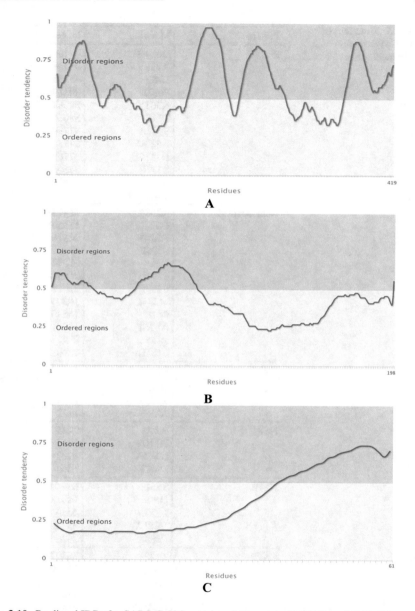

Fig. 3.10 Predicted IDRs for SARS-CoV-2 proteins; **A** N protein, **B** NSP8, and **C** ORF6

Table 3.6 Mutations of SARS-CoV-2 protein with high frequency are collected

Protein	Mutation	Frequency
Spike	D614G	404007
	A222V	95338
	L18F	43355
	P681H	40812
	N501Y	39067
	T716I	38065
	S982A	37780
	A570D	37777
	D1118H	37763
	S477N	22801
	N439K	9861
	W1214G	7141
Nucleocapsid	R203K	154733
	G204R	153506
	A220V	94576
	S235F	38826
	D3L	37374
	S194L	25853
	P199L	22012
	M234I	19304
	P67S	16599
	A376T	11401
	P365S	8799
	P151L	7220
Envelope	P71L	1063
Membrane	D3G	1725
	T175M	1428
	A2S	1172
NSP1	R24C	2225
	H110Y	1845
	D144A	1122
	D48G	1060
NSP2	T85I	63760
	I120F	16828
	L550F	4659
	V381A	3374
	P585S	2628
NSP3	T183I	38397
	I1412T	37570
	A890D	37371
	I1683T	9382
	M1788I	7117
	S543P	6791
	H295Y	5822
	T428I	5103

(continued)

Table 3.6 (continued)

Protein	Mutation	Frequency
NSP4	M324I	11831
	F17L	3801
	F308Y	1891
	A380V	1582
	A231V	1489
NSP5	L89F	19780
	G15S	8082
	P108S	7558
	K90R	6763
	G71S	3075
NSP6	L37F	24736
	A54S	4253
	V149F	3393
	K270R	3371
	M86I	3258
NSP7	L142F	6,307
	S25L	4,305
	M75I	2,640
	M3I	1116
NSP8	T145I	2685
	Q24R	2090
NSP9	M101I	7163
	I65V	1115
NSP10	T58I	608
	T12I	476
	T102I	465
	N85D	461
	A32V	420
NSP12	P323L	402962
	V776L	12022
	A185S	11608
	V720I	7176
	A423V	6911
NSP13	K460R	15875
	E261D	11793
	K218R	11527
	H290Y	9882
	A598S	7302
NSP14	N129D	16868
	P43L	5614
	M501I	4443
NSP15	T33I	6409
	V127F	2851
	R206S	2450
	A81V	1876
	D132Y	1481

(continued)

Table 3.6 (continued)

Protein	Mutation	Frequency
NSP16	R216C	15799
	R287I	5245
	K160R	1966
ORF 3a	Q57H	95091
	G172V	16365
	T223I	6886
	G251V	6230
	V202L	5760
	Q38R	5366
	G172R	5322
ORF6	I33T	1349
ORF 7a	T14I	2884
ORF 7b	S5L	4216
	H37L	1215
	C41F	975
ORF8	Y73C	37639
	R52I	37422
	S24L	25244
	L84S	8324
	A65V	5299
ORF 10	V30L	94745

3.6 Conclusion

In this chapter, we have tried to examine all the proteins of the SARS-CoV-2 and evaluate them based on the classification of the proteins (structural, non-structural, and accessory proteins). According to previous studies, the structure of all proteins has been described in detail. It is important to investigate the structure of proteins because this information can be used in pharmacology in order to design drugs and vaccines based on structure. Substitution mutations in viral proteins (with high frequency) were also introduced, as these mutations can greatly assist in the selection of protein as a therapeutic target; because mutations can change the structure, function, and binding affinity of a protein relative to other molecules (proteins and small molecule).

References

1. He G, Sun Z, Zhao Y, Zhang S, Chen H, Zhao Z, Yang G, Zhou Q (2020) B-coronavirus infectious diseases: recommended strategies for the prevention and control of transmission. Int J Clin Exp Pathol 13:1060–1065
2. Madjid M, Safavi-Naeini P, Solomon SD, Vardeny O (2020) Potential effects of coronaviruses on the cardiovascular system: a review. JAMA Cardiol 5:831–840. https://doi.org/10.1001/

jamacardio.2020.1286
3. Liu Z, Xiao X, Wei X, Li J, Yang J, Tan H, Zhu J, Zhang Q, Wu J, Liu L (2020) Composition and divergence of coronavirus spike proteins and host ACE2 receptors predict potential intermediate hosts of SARS-CoV-2. J Med Virol 92:595–601. https://doi.org/10.1002/jmv.25726
4. Ong E, Wong MU, Huffman A, He Y (2020) COVID-19 coronavirus vaccine design using reverse vaccinology and machine learning. Front Immunol 11:1581. https://doi.org/10.3389/fimmu.2020.01581
5. Anasir MI, Poh CL (2019) Structural vaccinology for viral vaccine design. Front Microbiol 10:1–11. https://doi.org/10.3389/fmicb.2019.00738
6. Xiang S-H (2014) Recent advances on the use of structural biology for the design of novel envelope immunogens of HIV-1. Curr HIV Res 11:464–472. https://doi.org/10.2174/157016 2x113116660053
7. Zhang J, Zeng H, Gu J, Li H, Zheng L, Zou Q (2020) Progress and prospects on vaccine development against sars-cov-2. Vaccines 8:1–12. https://doi.org/10.3390/vaccines8020153
8. Dormitzer PR, Grandi G, Rappuoli R (2012) Structural vaccinology starts to deliver. Nat Rev Microbiol 10:807–813. https://doi.org/10.1038/nrmicro2893
9. Zhu N, Zhang D, Wang W, Li X, Yang B, Song J, Zhao X, Huang B, Shi W, Lu R, Niu P, Zhan F, Ma X, Wang D, Xu W, Wu G, Gao GF, Tan W (2019) A novel coronavirus from patients with pneumonia in China. N Engl J Med 382(2020):727–733. https://doi.org/10.1056/nejmoa 2001017
10. Cascella M, Rajnik M, Cuomo A, Dulebohn SC, Di Napoli R (2020) Features, evaluation, and treatment of coronavirus. StatPearls 1–22
11. Ortega, JT, Serrano ML, Pujol FH, Rangel HR, Biology S (2020) Original article: role of changes in SARS-COV-2 spike protein in the interaction with the human ACE2 receptor. Excli J 410–417
12. Velazquez-Salinas L, Zarate S, Eberl S, Gladue DP, Novella I, Borca MV (2020) Positive selection of ORF1ab, ORF3a, and ORF8 genes drives the early evolutionary trends of SARS-CoV-2 during the 2020 COVID-19 pandemic, front. Microbiol 11 (2020). https://doi.org/10.3389/fmicb.2020.550674
13. Yoshimoto FK (2020) The proteins of severe acute respiratory syndrome coronavirus-2 (SARS CoV-2 or n-COV19), the cause of COVID-19. Protein J 39:198–216. https://doi.org/10.1007/s10930-020-09901-4
14. Li Q, Kang CB (2020) Progress in developing inhibitors of sars-cov-2 3c-like protease. Microorganisms. 8:1–18. https://doi.org/10.3390/microorganisms8081250
15. Michel CJ, Mayer C, Poch O, Thompson JD (2020) Characterization of accessory genes in coronavirus genomes. Virol J 17:1–13. https://doi.org/10.1186/s12985-020-01402-1
16. Mohammad S, Salauddin A, Barua R (2020) Since January 2020 Elsevier has created a COVID-19 resource centre with free information in English and Mandarin on the novel coronavirus COVID- 19. The COVID-19 resource centre is hosted on Elsevier Connect, the company's public news and information
17. Ke Z, Oton J, Qu K, Cortese M, Zila V, McKeane L, Nakane T, Zivanov J, Neufeldt CJ, Cerikan B, Lu JM, Peukes J, Xiong X, Kräusslich HG, Scheres SHW, Bartenschlager R, Briggs JAG (2020) Structures and distributions of SARS-CoV-2 spike proteins on intact virions. Nature 588 (2020) https://doi.org/10.1038/s41586-020-2665-2
18. Huang Y, Yang C, Feng X, Xu W, Wen Liu S (2020) Structural and functional properties of SARS-CoV-2 spike protein: potential antivirus drug development for COVID-19. Acta Pharmacol Sin 41:1141–1149 (2020). https://doi.org/10.1038/s41401-020-0485-4
19. Shang J, Wan Y, Luo C, Ye G, Geng Q, Auerbach A, Li F (2020) Cell entry mechanisms of SARS-CoV-2. Proc Natl Acad Sci USA 117. https://doi.org/10.1073/pnas.2003138117
20. Yan R, Zhang Y, Li Y, Xia L, Guo Y, Zhou Q (2020) Structural basis for the recognition of SARS-CoV-2 by full-length human ACE2. Science (80–367):1444–1448 (2020). https://doi.org/10.1126/science.abb2762

21. Wang Q, Zhang Y, Wu L, Niu S, Song C, Zhang Z, Lu G, Qiao C, Hu Y, Yuen KY, Wang Q, Zhou H, Yan J, Qi J (2020) Structural and functional basis of SARS-CoV-2 entry by using human ACE2. Cell 181:894–904. https://doi.org/10.1016/j.cell.2020.03.045

22. Pierri CL (2020) SARS-CoV-2 spike protein: flexibility as a new target for fighting infection. Signal Transduct Target Ther 5:4–6. https://doi.org/10.1038/s41392-020-00369-3

23. Henderson R, Edwards RJ, Mansouri K, Janowska K, Stalls V, Gobeil SMC, Kopp M, Li D, Parks R, Hsu AL, Borgnia MJ, Haynes BF, Acharya P (2020) Controlling the SARS-CoV-2 spike glycoprotein conformation. Nat Struct Mol Biol 27:925–933. https://doi.org/10.1038/s41594-020-0479-4

24. Wrapp D, Wang N, Corbett KS, Goldsmith KA, Hsieh C, Abiona O, Graham BS, Mclellan JS (2020) Cryo-EM structure of the 2019-nCoV spike in the prefusion conformation 1263:1260–1263

25. Lan J, Ge J, Yu J, Shan S, Zhou H, Fan S, Zhang Q, Shi X, Wang Q, Zhang L, Wang X (2020) Structure of the SARS-CoV-2 spike receptor-binding domain bound to the ACE2 receptor. Nature 581:215–220. https://doi.org/10.1038/s41586-020-2180-5

26. Shang J, Ye G, Shi K, Wan Y, Luo C, Aihara H, Geng Q, Auerbach A, Li F (2020) Structural basis of receptor recognition by SARS-CoV-2. Nature 581:221–224. https://doi.org/10.1038/s41586-020-2179-y

27. Rezaei S, Sefidbakht Y, Uskoković V (2020) Comparative molecular dynamics study of the receptor-binding domains in SARS-CoV-2 and SARS-CoV and the effects of mutations on the binding affinity. J Biomol Struct Dyn 0:1–20. https://doi.org/10.1080/07391102.2020.1860829

28. Wang Q, Qiu Y, Li JY, Zhou ZJ, Liao CH, Ge XY (2020) A unique protease cleavage site predicted in the spike protein of the novel pneumonia coronavirus (2019-nCoV) potentially related to viral transmissibility. Virol. Sin. 35:337–339. https://doi.org/10.1007/s12250-020-00212-7

29. Liu Z, Zheng H, Lin H, Li M, Yuan R, Peng J, Xiong Q, Sun J, Li B, Wu J, Yi L, Peng X, Zhang H, Zhang W, Hulswit RJG, Loman N, Rambaut A, Ke C, Bowden TA, Pybus OG, Lu J (2020) Identification of common deletions in the spike protein of SARS-CoV-2. J Virol 1–9 (2020). https://doi.org/10.1128/jvi.00790-20

30. Andersen KG, Rambaut A, Lipkin WI, Holmes EC, Garry RF (2020) The proximal origin of SARS-CoV-2. Nat Med 26:450–452. https://doi.org/10.1038/s41591-020-0820-9

31. Huang Y, Yang C, Feng Xu X, Xu W, Wen Liu S (2020) Structural and functional properties of SARS-CoV-2 spike protein: potential antivirus drug development for COVID-19, Acta Pharmacol Sin (2020). https://doi.org/10.1038/s41401-020-0485-4

32. Millet JK, Whittaker GR (2018) Physiological and molecular triggers for SARS-CoV membrane fusion and entry into host cells. Virology 517:3–8. https://doi.org/10.1016/j.virol.2017.12.015

33. Duan L, Zheng Q, Zhang H, Niu Y, Lou Y, Wang H (2020) The SARS-CoV-2 spike glycoprotein biosynthesis, structure, function, and antigenicity: implications for the design of spike-based vaccine immunogens. Front Immunol 11:1–12. https://doi.org/10.3389/fimmu.2020.576622

34. Xia S, Zhu Y, Liu M, Lan Q, Xu W, Wu Y, Ying T, Liu S, Shi Z, Jiang S, Lu L (2020) Fusion mechanism of 2019-nCoV and fusion inhibitors targeting HR1 domain in spike protein. Cell Mol Immunol 17:765–767. https://doi.org/10.1038/s41423-020-0374-2

35. Zhang L, Jackson C, Mou H, Ojha A, Rangarajan E, Izard T, Farzan M, Choe H (2020) The D614G mutation in the SARS-CoV-2 spike protein reduces S1 shedding and increases infectivity. BioRxiv Prepr Serv Biol. https://doi.org/10.1101/2020.06.12.148726

36. Borges V, Isidro J, Cortes-Martins H, Duarte S, Vieira L, Leite R, Gordo I, Caetano CP, Nunes B, Sá R, Oliveira A, Guiomar R, Gomes JP (2020) On the track of the D839Y mutation in the SARS-CoV-2 Spike fusion peptide: emergence and geotemporal spread of a highly prevalent variant in Portugal. MedRxiv. 20171884 (2020)

37. Chen C-Y, Chou Y-C, Hsueh Y-P (2020) SARS-CoV-2 D614 and G614 spike variants impair neuronal synapses and exhibit differential fusion ability. BioRxiv

38. Korber B, Fischer W, Gnanakaran S, Yoon H, Theiler J, Abfalterer W, Foley B, Giorgi E, Bhattacharya T, Parker M, Partridge D, Evans C, Freeman T, de Silva T, LaBranche C, Montefiori D (2020) Spike mutation pipeline reveals the emergence of a more transmissible form of SARS-CoV-2. https://doi.org/10.1101/2020.04.29.069054
39. Schoeman D, Fielding BC (2019) Coronavirus envelope protein: current knowledge. Virol J. 16:1–22. https://doi.org/10.1186/s12985-019-1182-0
40. Sarkar M, Saha S (2020) Structural insight into the role of novel SARSCoV-2 E protein: a potential target for vaccine development and other therapeutic strategies. PLoS ONE 15:1–25. https://doi.org/10.1371/journal.pone.0237300
41. Singh Tomar PP, Arkin IT (2020) SARS-CoV-2 E protein is a potential ion channel that can be inhibited by Gliclazide and Memantine. Biochem Biophys Res Commun 530:10–14. https://doi.org/10.1016/j.bbrc.2020.05.206
42. Mandala VS, McKay MJ, Shcherbakov AA, Dregni AJ, Kolocouris A, Hong M (2020) Structure and drug binding of the SARS-CoV-2 envelope protein transmembrane domain in lipid bilayers. Nat Struct Mol Biol 27 (2020). https://doi.org/10.1038/s41594-020-00536-8
43. Gupta MK, Vemula S, Donde R, Gouda G, Behera L, Vadde R (2020) In-silico approaches to detect inhibitors of the human severe acute respiratory syndrome coronavirus envelope protein ion channel. J Biomol Struct Dyn 1–11. https://doi.org/10.1080/07391102.2020.175 1300
44. Pettersen EF, Goddard TD, Huang CC, Couch GS, Greenblatt DM, Meng EC, Ferrin TE (2004) UCSF chimera—a visualization system for exploratory research and analysis. J Comput Chem 25:1605–1612. https://doi.org/10.1002/jcc.20084
45. Kang S, Yang M, Hong Z, Zhang L, Huang Z, Chen X, He S, Zhou Z, Zhou Z, Chen Q, Yan Y, Zhang C, Shan H, Chen S (2020) Crystal structure of SARS-CoV-2 nucleocapsid protein RNA binding domain reveals potential unique drug targeting sites. Acta Pharm Sin B 10:1228–1238. https://doi.org/10.1016/j.apsb.2020.04.009
46. Khan MT, Zeb MT, Ahsan H, Ahmed A, Ali A, Akhtar K, Malik SI, Cui Z, Ali S, Khan AS, Ahmad M, Wei DQ, Irfan M (2020) SARS-CoV-2 nucleocapsid and Nsp3 binding: an in silico study. Arch Microbiol. https://doi.org/10.1007/s00203-020-01998-6
47. Zeng W, Liu G, Ma H, Zhao D, Yang Y, Liu M (2020) Since January 2020 Elsevier has created a COVID-19 resource centre with free information in English and Mandarin on the novel coronavirus COVID-19. The COVID-19 resource centre is hosted on Elsevier Connect, the company's public news and information
48. Dinesh DC, Chalupska D, Silhan J, Veverka V, Boura E (2020) Structural basis of RNA recognition by the SARS-CoV-2 nucleocapsid phosphoprotein. https://doi.org/10.1101/2020. 04.02.022194
49. Prajapat M, Sarma P, Shekhar N, Avti P, Sinha S, Kaur H, Kumar S, Bhattacharyya A, Kumar H, Bansal S, Medhi B (2020) Drug targets for corona virus: a systematic review. Indian J Pharmacol 52:56–65. https://doi.org/10.4103/ijp.IJP_115_20
50. Gasteiger E, Hoogland C, Gattiker A, Duvaud S, Wilkins MR, Appel RD, Bairoch A (2005) The proteomics protocols handbook. Proteomics Protoc Handb 571–608. https://doi.org/10. 1385/1592598900
51. Scheller C, Krebs F, Minkner R, Astner I, Gil-Moles M, Wätzig H (2020) Physicochemical properties of SARS-CoV-2 for drug targeting, virus inactivation and attenuation, vaccine formulation and quality control. Electrophoresis 41:1137–1151. https://doi.org/10.1002/elps. 202000121
52. Almansour I, Alhagri M, Alfares R, Alshehri M, Bakhashwain R, Maarouf A (2019) IRAM: Virus capsid database and analysis resource. Database 2019:1–7. https://doi.org/10.1093/dat abase/baz079
53. Geourjon C, Deléage G (1995) Sopma: Significant improvements in protein secondary structure prediction by consensus prediction from multiple alignments. Bioinformatics 11:681–684. https://doi.org/10.1093/bioinformatics/11.6.681

54. Giri R, Bhardwaj T, Shegane M, Gehi BR, Kumar P, Gadhave K, Oldfield CJ, Uversky VN (2020) Understanding COVID-19 via comparative analysis of dark proteomes of SARS-CoV-2, human SARS and bat SARS-like coronaviruses. Springer International Publishing. https://doi.org/10.1007/s00018-020-03603-x

55. Hassan SS, Choudhury PP, Uversky VN, Dayhoff GW, Aljabali AAA, Uhal BD, Lundstrom K, Rezaei N, Seyran M, Pizzol D, Adadi P, Lal A, Soares A, Abd El-Aziz TM, Kandimalla R, Tambuwala M, Azad GK, Sherchan SP, Baetas-da-Cruz W, Takayama K, Serrano-Aroca A, Chauhan G, Palu G, Brufsky AM (2020) Variability of accessory proteins rules the SARS-CoV-2 pathogenicity, BioRxiv. https://doi.org/10.1101/2020.11.06.372227

56. Holland LRA, Kaelin EA, Maqsood R, Estifanos B, Wu LI, Varsani A, Halden RU, Hogue BG, Scotch M, Lim ES (2020) An 81 base-pair deletion in SARS-CoV-2 ORF7a identified from sentinel surveillance in Arizona. MedRxiv 2–4 (2020). https://doi.org/10.1101/2020.04.17.20069641

57. Ostaszewski M, Mazein A, Gillespie ME, Kuperstein I, Niarakis A, Hermjakob H, Pico AR, Willighagen EL, Evelo CT, Hasenauer J, Schreiber F, Dräger A, Demir E, Wolkenhauer O, Furlong LI, Barillot E, Dopazo J, Orta-Resendiz A, Messina F, Valencia A, Funahashi A, Kitano H, Auffray C, Balling R, Schneider R (2020) COVID-19 disease map, building a computational repository of SARS-CoV-2 virus-host interaction mechanisms. Sci Data 7:8–11. https://doi.org/10.1038/s41597-020-0477-8

58. Pereira F (2020) Evolutionary dynamics of the SARS-CoV-2 ORF8 accessory gene. Infect Genet Evol 85: https://doi.org/10.1016/j.meegid.2020.104525

59. Hassan SS, Ghosh S, Attrish D, Choudhury PP, Seyran M, Pizzol D, Adadi P, Abd El-Aziz TM, Soares A, Kandimalla R, Lundstrom K, Tambuwala M, Aljabali AAA, Lal A, Azad GK, Uversky VN, Sherchan SP, Baetas-Da-Cruz W, Uhal BD, Rezaei N, Brufsky AM (2020) A unique view of SARS-CoV-2 through the lens of ORF8 protein. BioRxiv. https://doi.org/10.1101/2020.08.25.267328

60. Hassan SS, Attrish D, Ghosh S, Choudhury PP, Uversky VN, Uhal BD, Lundstrom K, Rezaei N, Aljabali AAA, Seyran M, Pizzol D, Adadi P, El-Aziz TMA, Soares A, Kandimalla R, Tambuwala M, Lal A, Azad GK, Sherchan SP, Baetas-Da-Cruz W, Palù G, Brufsky AM (2020) Notable sequence homology of the ORF10 protein introspects the architecture of SARS-COV-2. BioRxiv 1–13. https://doi.org/10.1101/2020.09.06.284976

61. Sun Y, Abriola L, Surovtseva YV, Lindenbach BD, Guo JU (2020) Restriction of SARS-CoV-2 Replication by targeting programmed—1 ribosomal frameshifting in vitro. BioRxiv

62. Chan JFW, Kok KH, Zhu Z, Chu H, To KKW, Yuan S, Yuen KY (2020) Genomic characterization of the 2019 novel human-pathogenic coronavirus isolated from a patient with atypical pneumonia after visiting Wuhan. Emerg Microb Infect 9:221–236. https://doi.org/10.1080/22221751.2020.1719902

63. Gasmalbari E, Abbadi OS (2020) Non-structural proteins of SARS-CoV-2 as potential sources for vaccine synthesis, 1–7

64. Qiu Y, Xu K (2020) Functional studies of the coronavirus nonstructural proteins. STEMedicine 1: https://doi.org/10.37175/stemedicine.v1i2.39

65. Thoms M, Buschauer R, Ameismeier M, Koepke L, Denk T, Hirschenberger M, Kratzat H, Hayn M, MacKens-Kiani T, Cheng J, Straub JH, Stürzel CM, Fröhlich T, Berninghausen O, Becker T, Kirchhoff F, Sparrer KMJ, Beckmann R (2020) Structural basis for translational shutdown and immune evasion by the Nsp1 protein of SARS-CoV-2. Science (80–369):1249–1256. https://doi.org/10.1126/science.abc8665

66. Min YQ, Mo Q, Wang J, Deng F, Wang H, Ning YJ (2020) SARS-CoV-2 nsp1: bioinformatics, potential structural and functional features, and implications for drug/vaccine designs. Front Microbiol 11:1–12. https://doi.org/10.3389/fmicb.2020.587317

67. Nomburg J, Meyerson M, DeCaprio JA (2020) Noncanonical junctions in subgenomic RNAs of SARS-CoV-2 lead to variant open reading frames. BioRxiv

68. Novel Mutations in NSP1 and PLPro of SARS-CoV-2 NIB-1 Genome Mount for Effective Therapeutics (2020)

69. Armstrong L, Lange SM, de Cesare V, Matthews SP, Nirujogi RS, Cole I, Hope A, Cunningham F, Toth R, Mukherjee R, Bojkova D, Gruber F, Gray D, Wyatt PG, Cinatl J, Dikic I, Davies P, Kulathu Y (2020) Characterization of protease activity of Nsp3 from SARS-CoV-2 and its in vitro inhibition by nanobodies, BioRxiv

70. Debnath P, Debnath B, Bhaumik S, Debnath S (2020) In silico identification of potential inhibitors of ADP-ribose phosphatase of SARS-CoV-2 nsP3 by combining E-pharmacophore- and receptor-based virtual screening of database. Chem Select 5:9388–9398. https://doi.org/10.1002/slct.202001419

71. Klemm T, Ebert G, Calleja DJ, Allison CC, Richardson LW, Bernardini JP, Lu BG, Kuchel NW, Grohmann C, Shibata Y, Gan ZY, Cooney JP, Doerflinger M, Au AE, Blackmore TR, Heden van Noort GJ, Geurink PP, Ovaa H, Newman J, Riboldi-Tunnicliffe A, Czabotar PE, Mitchell JP, Feltham R, Lechtenberg BC, Lowes KN, Dewson G, Pellegrini M, Lessene G, Komander D (2020) Mechanism and inhibition of the papain-like protease, PLpro, of SARS-CoV-2. EMBO J 39:1–17. https://doi.org/10.15252/embj.2020106275

72. Frick DN, Virdi RS, Vuksanovic N, Dahal N, Silvaggi NR (2020) Molecular basis for ADP-ribose binding to the Mac1 domain of SARS-CoV-2 nsp3. Biochemistry 59:2608–2615. https://doi.org/10.1021/acs.biochem.0c00309

73. McClain CB, Vabret N (2020) SARS-CoV-2: the many pros of targeting PLpro. Signal Transduct Target Ther 5:1–2. https://doi.org/10.1038/s41392-020-00335-z

74. Ibrahim TM, Ismail MI, Bauer MR, Bekhit AA, Boeckler FM (2020) Supporting SARS-CoV-2 papain-like protease drug discovery: in silico methods and benchmarking. Front Chem 8:1–17. https://doi.org/10.3389/fchem.2020.592289

75. Ratia K, Pegan S, Takayama J, Sleeman K, Coughlin M, Baliji S, Chaudhuri R, Fu W, Prab-hakar BS, Johnson ME, Baker SC, Ghosh AK, Mesecar AD (2008) A noncovalent class of papain-like protease/deubiquitinase inhibitors blocks SARS virus replication. Proc Natl Acad Sci USA. 105:16119–16124. https://doi.org/10.1073/pnas.0805240105

76. Bosken YK, Cholko T, Lou Y-C, Wu K-P, Chang CA (2020) Insights into dynamics of inhibitor and ubiquitin-like protein binding in SARS-CoV-2 papain-like protease. Front Mol Biosci 7:1–14. https://doi.org/10.3389/fmolb.2020.00174

77. Chen Y, Liu Q, Guo D (2020) Emerging coronaviruses: genome structure, replication, and pathogenesis. J Med Virol 92:418–423. https://doi.org/10.1002/jmv.25681

78. Goyal B, Goyal D (2020) Targeting the dimerization of the main protease of coronaviruses: a potential broad-spectrum therapeutic strategy. ACS Comb Sci 22:297–305. https://doi.org/10.1021/acscombsci.0c00058

79. Lee J, Worrall LJ, Vuckovic M, Rosell FI, Gentile F, Ton AT, Caveney NA, Ban F, Cherkasov A, Paetzel M, Strynadka NCJ (2020) Crystallographic structure of wild-type SARS-CoV-2 main protease acyl-enzyme intermediate with physiological C-terminal autoprocessing site. Nat Commun 11. https://doi.org/10.1038/s41467-020-19662-4

80. Zhang L, Lin D, Sun X, Curth U, Drosten C, Sauerhering L, Becker S, Rox K, Hilgenfeld R (2020) Crystal structure of SARS-CoV-2 main protease provides a basis for design of improved a-ketoamide inhibitors. Science 80–368:409–412. https://doi.org/10.1126/science.abb3405

81. Jin Z, Du X, Xu Y, Deng Y, Liu M, Zhao Y, Zhang B, Li X, Zhang L, Peng C, Duan Y, Yu J, Wang L, Yang K, Liu F, Jiang R, Yang X, You T, Liu X, Yang X, Bai F, Liu H, Liu X, Guddat LW, Xu W, Xiao G, Qin C, Shi Z, Jiang H, Rao Z, Yang H (2020) Structure of Mpro from SARS-CoV-2 and discovery of its inhibitors. Nature 582:289–293. https://doi.org/10.1038/s41586-020-2223-y

82. Dash JJ, Purohit P, Muya JT, Meher BR (2020) Drug repurposing of allophenylnorstatine containing HIV-protease inhibitors against SARS-CoV-2 Mpro: insights from molecular dynamics simulations and binding free energy estimations (2020). https://doi.org/10.26434/chemrxiv.12402545

83. Santerre M, Arjona SP, Allen CN, Shcherbik N, Sawaya BE (2020) Why do SARS-CoV-2 NSPs rush to the ER? J Neurol. https://doi.org/10.1007/s00415-020-10197-8

84. Zhang WF, Stephen P, Stephen P, Thériault JF, Wang R, Lin SX (2020) Novel coronavirus polymerase and nucleotidyl-transferase structures: potential to target new outbreaks. J Phys Chem Lett 11:4430–4435. https://doi.org/10.1021/acs.jpclett.0c00571

85. Romano M, Ruggiero A, Squeglia F, Maga G, Berisio R (2020) A structural view of SARS-CoV-2 RNA replication machinery: RNA synthesis, proofreading and final capping, cells 9. https://doi.org/10.3390/cells9051267

86. Ahmad J, Ikram S, Ahmad F, Rehman IU, Mushtaq M (2020) SARS-CoV-2 RNA dependent RNA polymerase (RdRp)—a drug repurposing study. Heliyon 6: https://doi.org/10.1016/j.heliyon.2020.e04502

87. Kirchdoerfer RN, Ward AB (2019) Structure of the SARS-CoV nsp12 polymerase bound to nsp7 and nsp8 co-factors. Nat Commun 10:1–9. https://doi.org/10.1038/s41467-019-10280-3

88. Gao Y, Yan L, Huang Y, Liu F, Zhao Y, Cao L, Wang T, Sun Q, Ming Z, Zhang L, Ge J, Zheng L, Zhang Y, Wang H, Zhu Y, Zhu C, Hu T, Hua T, Zhang B, Yang X, Li J, Yang H, Liu Z, Xu W, Guddat LW, Wang Q, Lou Z, Rao Z (2020) Structure of the RNA-dependent RNA polymerase from COVID-19 virus. Science 80–368:779–782. https://doi.org/10.1126/science.abb7498

89. Picarazzi F, Vicenti I, Saladini F, Zazzi M, Mori M (2020) Targeting the RdRp of emerging RNA viruses: The Structure-based drug design challenge. Molecules 25. https://doi.org/10.3390/molecules25235695

90. Mcdonald SM (2013) RNA synthetic mechanisms employed by diverse families of RNA viruses. Wiley Interdiscip Rev RNA 4:351–367. https://doi.org/10.1002/wrna.1164

91. Pormohammad A, Monych NK, Turner RJ, Pormohammad A (2021) Zinc and SARS-CoV-2: a molecular modeling study of Zn interactions with RNA-dependent RNA-polymerase and 3C-like proteinase enzymes. Int J Mol Med 47:326–334. https://doi.org/10.3892/ijmm.2020.4790

92. Shu T, Huang M, Wu D, Ren Y, Zhang X, Han Y, Mu J, Wang R, Qiu Y, Zhang DY, Zhou X (2020) SARS-coronavirus-2 Nsp13 possesses NTPase and RNA helicase activities that can be inhibited by bismuth salts. Virol Sin 35:321–329. https://doi.org/10.1007/s12250-020-00242-1

93. Jang KJ, Jeong S, Kang DY, Sp N, Yang YM, Kim DE (2020) A high ATP concentration enhances the cooperative translocation of the SARS coronavirus helicase nsP13 in the unwinding of duplex RNA. Sci Rep 10:1–13. https://doi.org/10.1038/s41598-020-61432-1

94. White MA, Lin W, Cheng X (2020) Discovery of COVID-19 inhibitors targeting the SARS-CoV2 Nsp13 helicase. BioRxiv. https://doi.org/10.1101/2020.08.09.243246

95. White MA, Lin W, Cheng X (2020) Discovery of COVID-19 inhibitors targeting the SARS-CoV-2 Nsp13 helicase. J Phys Chem Lett 11:9144–9151. https://doi.org/10.1021/acs.jpclett.0c02421

96. Sada M, Saraya T, Ishii H, Okayama K, Hayashi Y, Tsugawa T, Nishina A, Murakami K, Kuroda M, Ryo A, Kimura H (2020) Detailed molecular interactions of favipiravir with SARS-CoV-2, SARS-CoV, MERS-CoV, and influenza virus polymerases in silico. Microorganisms 8:1–9. https://doi.org/10.3390/microorganisms8101610

97. Sinha SK, Prasad SK, Islam MA, Gurav SS, Patil RB, Al Faris NA, Aldayel TS, Al Kehayez NM, Wabaidur SM, Shakya A (2020) Identification of bioactive compounds from Glycyrrhiza glabra as possible inhibitor of SARS-CoV-2 spike glycoprotein and non-structural protein-15: a pharmacoinformatics study. J Biomol Struct Dyn 1–15

98. Decroly E, Imbert I, Coutard B, Bouvet M, Selisko B, Alvarez K, Gorbalenya AE, Snijder EJ, Canard B (2008) Coronavirus nonstructural protein 16 Is a Cap-0 binding enzyme possessing (Nucleoside-2′O)-methyltransferase activity. J Virol 82:8071–8084. https://doi.org/10.1128/jvi.00407-08

99. Li J, Guo M, Tian X, Liu C, Wang X, Yang X, Wu P, Xiao Z, Qu Y, Yin Y, Fu J, Zhu Z, Liu Z, Peng C, Zhu T, Liang Q (2020) Virus-host interactome and proteomic survey of PMBCs from COVID-19 patients reveal potential virulence factors influencing SARS-CoV-2 pathogenesis. https://doi.org/10.1101/2020.03.31.019216

100. Mers-cov R, Ogando NS, Zevenhoven-dobbe JC, Van Der Meer Y, Bredenbeek PJ (2020) The enzymatic activity of the nsp14 exoribonuclease is critical 94:1–24

101. Ogando NS, Zevenhoven-Dobbe JC, van der Meer Y, Bredenbeek PJ, Posthuma CC, Snijder EJ (2020) The enzymatic activity of the nsp14 exoribonuclease is critical for replication of MERS-CoV and SARS-CoV-2. J Virol 94. https://doi.org/10.1128/jvi.01246-20

102. Rosas-Lemus M, Minasov G, Shuvalova L, Inniss N, Kiryukhina O, Wiersum G, Kim Y, Jedrzejczak R, Maltseva N, Endres M, Jaroszewski L, Godzik A, Joachimiak A, Satchell K (2020) The crystal structure of nsp10-nsp16 heterodimer from SARS-CoV-2 in complex with S-adenosylmethionine. BioRxiv Prepr Serv Biol 1–22. https://doi.org/10.1101/2020.04.17. 047498

103. Uversky VN (2019) Intrinsically disordered proteins and their "Mysterious" (meta)physics. Front Phys 7:8–23. https://doi.org/10.3389/fphy.2019.00010

104. Vacic V, Markwick PRL, Oldfield CJ, Zhao X, Haynes C, Uversky VN, Iakoucheva LM (2012) Disease-associated mutations disrupt functionally important regions of intrinsic protein disorder. PLoS Comput Biol 8. https://doi.org/10.1371/journal.pcbi.1002709

105. Charon J, Barra A, Walter J, Millot P, Hébrard E, Moury B, Michon T (2018) First experimental assessment of protein intrinsic disorder involvement in an RNA virus natural adaptive process. Mol Biol Evol 35:38–49. https://doi.org/10.1093/molbev/msx249

106. Walter J, Charon J, Hu Y, Lachat J, Leger T, Lafforgue G, Barra A, Michon T (2019) Comparative analysis of mutational robustness of the intrinsically disordered viral protein VPg and of its interactor eIF4E. PLoS ONE 14:1–13. https://doi.org/10.1371/journal.pone.0211725

107. Mozzi A, Forni D, Cagliani R, Clerici M, Pozzoli U, Sironi M (2020) Intrinsically disordered regions are abundant in simplexvirus proteomes and display signatures of positive selection. Virus Evol 6:1–12. https://doi.org/10.1093/ve/veaa028

108. Barik S (2020) Genus-specific pattern of intrinsically disordered central regions in the nucleocapsid protein of coronaviruses. Comput Struct Biotechnol J 18:1884–1890. https://doi.org/10.1016/j.csbj.2020.07.005

109. Sen S, Dey A, Bandhyopadhyay S, Uversky VN (2012) Understanding structural malleability of the SARS-CoV-2 proteins and their relation to the comorbidities SARS-CoV-2 774, pp 1–17

110. Macraild CA, Richards JS, Anders RF, Norton RS (2016) Antibody recognition of disordered antigens. Structure 24:148–157. https://doi.org/10.1016/j.str.2015.10.028

111. Giri R, Bhardwaj T, Shegane M, Gehi B, Kumar P, Gadhave K, Oldfield C, Uversky V (2020) When darkness becomes a ray of light in the dark times: understanding the COVID-19 via the comparative analysis of the dark proteomes of SARS-CoV-2, human SARS and bat SARS-like coronaviruses, 1–63 (2020). https://doi.org/10.1101/2020.03.13.990598

112. Yang Y, Peng F, Wang R, Guan K, Jiang T, Xu G, Sun J, Chang C (2020) The deadly coronaviruses: the 2003 SARS pandemic and the 2020 novel coronavirus epidemic in china, the company's public news and information. J Autoimmun 109:

113. Gillim-ross L, Subbarao K (2006) Emerging respiratory viruses: challenges and vaccine. Strategies 19:614–636. https://doi.org/10.1128/CMR.00005-06

114. Lee HY, Nyon MP, Strych U (2016) Vaccine Dev Against Middle East Respir Syndr 3:80–86. https://doi.org/10.1007/s40475-016-0084-0

115. Zhang Y, Zeng G, Pan H, Li C, Hu Y, Chu K, Han W, Chen Z, Tang R, Yin W, Chen X (n.d.) Articles safety, tolerability, and immunogenicity of an inactivated SARS-CoV-2 vaccine in healthy adults aged 18–59 years. Lancet Infect Dis. https://doi.org/10.1016/s1473-3099(20)30843-4

116. Sheikhshahrokh A, Ranjbar R, Saeidi E, Sa F (2020) Frontier therapeutics and vaccine strategies for SARS-CoV-2 (COVID-19). A Rev 49:18–29

117. Khalaj-Hedayati A (2020) Review article protective immunity against SARS subunit vaccine candidates based on spike protein : lessons for coronavirus vaccine development

118. Jackson LA, Anderson EJ, Rouphael NG, Roberts PC, Makhene M, Coler RN, McCullough MP, Chappell JD, Denison MR, Stevens LJ, Pruijssers AJ, McDermott A, Flach B, Doria-Rose NA, Corbett KS, Morabito KM, O'Dell S, Schmidt SD, Swanson PA, Padilla M, Mascola

JR, Neuzil KM, Bennett H, Sun W, Peters E, Makowski M, Albert J, Cross K, Buchanan W, Pikaart-Tautges R, Ledgerwood JE, Graham BS, Beigel JH (2020) An mRNA vaccine against SARS-CoV-2—preliminary report. N Engl J Med. https://doi.org/10.1056/nejmoa2022483

119. Benton DJ, Wrobel AG, Xu P, Roustan C, Martin SR, Rosenthal PB, Skehel JJ, Gamblin SJ (2020) Receptor binding and priming of the spike protein of SARS-CoV-2 for membrane fusion. Nature 588 (2020). https://doi.org/10.1038/s41586-020-2772-0

120. Spike P, Huo J, Zhao Y, Ren J, Fry EE, Owens RJ, Stuart DI, Huo J, Zhao Y, Ren J, Zhou D, Duyvesteyn HME, Ginn HM (2020) Article neutralization of SARS-CoV-2 by destruction of the ll neutralization of SARS-CoV-2 by destruction of the prefusion spike. Cell Host Microbe 28:445–454. https://doi.org/10.1016/j.chom.2020.06.010

121. To L, T.H.E. Editor (2020) A novel receptor-binding domain (RBD)-based mRNA vaccine against SARS-CoV-2, 2–5 (2020). https://doi.org/10.1038/s41422-020-0387-5

122. Wu S, Tian C, Liu P, Guo D, Zheng W, Huang X, Zhang Y, Liu L (2020) Effects of SARS-CoV-2 mutations on protein structures and intraviral protein–protein interactions. J Med Virol. https://doi.org/10.1002/jmv.26597

123. Kirby T (2021) New variant of SARS-CoV-2 in UK causes surge of COVID-19. Lancet Respir. Med 9:e20–e21. https://doi.org/10.1016/s2213-2600(21)00005-9

Chapter 4
The Main Protease of SARS COV-2 and Its Specific Inhibitors

Abdulrahman Ghassemlou, Yahya Sefidbakht, and Moones Rahmandoust

4.1 Introduction

The recent outbreak of coronavirus infection in China caused by the SARS-CoV-2 virus (COVID_19) has become a matter of serious concern to the world community, This virus belongs to the Coronaviruses family which caused two epidemics before, the severe acute respiratory syndrome coronavirus (SARS-CoV) and the middle east respiratory syndrome coronavirus (MERS-CoV) [1]. This virus possesses a positive-sense single-stranded RNA genome and causes disease in fishes, birds and mammals [2, 3].

Sequence analysis of SARS-CoV-2 isolates indicates that the 30 kb genome produces five major open reading frames (ORFs), containing a 5′ frameshifted polyprotein (ORF1a/ORF1ab) and four canonical 3′ structural proteins, namely, the spike (S), envelope (E), membrane (M) and nucleocapsid (N) proteins, which are common to all coronaviruses [4]. ORF1a and ORF1b encode two overlapping polyproteins, pp1a and pp1ab, which are auto-proteolytically processed into 16 Non-Structural Proteins (NSP1–NSP16) followed by the replicase–transcriptase complex formation in the host cell. The replicase–transcriptase complex consists of multiple enzymes, including the papain-like protease (NSP3), 3-chymotrypsin-like protease (NSP5), the NSP7–NSP8 primase complex, the primary RNA-dependent RNA polymerase (NSP12), a helicase–triphosphatase (NSP13), an exoribonuclease (NSP14), an endonuclease (NSP15), N7-methyltransferases (N7-MTase) (NSP10) and 2′O-methyltransferases) NSP16 (Fig. 4.1) [5–8]. Pp1a and pp1b undergo proteolytic processing by papain-like protease (PLpro) and 3-chymotrypsin-like protease (3CLpro, also known as the main protease or Mpro) which is essential for viral replication and transcription [8, 9].

A. Ghassemlou · Y. Sefidbakht (✉) · M. Rahmandoust
Protein Research Center, Shahid Beheshti University, Tehran, Iran
e-mail: y_sefidbakht@sbu.ac.ir

© The Author(s), under exclusive license to Springer Nature Singapore Pte Ltd. 2021   121
M. Rahmandoust and S.-O. Ranaei-Siadat (eds.), *COVID-19*,
https://doi.org/10.1007/978-981-16-3108-5_4

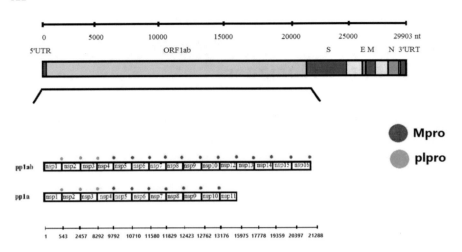

Fig. 4.1 Organization of the RNA genome of SARS-CoV-2 with selected genes and schematic representation of polyprotein cleavage sites of SARS-CoV-2. As shown in Figure, pp1a cleaves at 3 distinct sites, the Mpro cleaves at 11 distinct sites

4.2 Proteolytic Process

Various studies suggested that the proteolytic process is a multi-step mechanism. After the cysteine side chain proton is abstracted by the imidazole of histidine, the yield thiolate nucleophile interacts with the amide bond of the substrate. The *N*-terminal peptide product is unleashed by proton abstraction from histidine before the thioester is hydrolyzed to release the C-terminal product and restore the catalytic dyad [10, 11]. Mpro (NSP5) auto cleaves itself between NSP4 and NSP6 [12, 13], before processing the overlapping polyproteins pp1a and pp1ab at 11 cleavage sites (Fig. 4.1) [14–16]. Mpro mainly identifies substrate residues ranging from *P*4 to *P*1' [17]. Frist site recognition beyond *P*1' is not conserved. Specificity is mostly determined by *P*1, *P*2 and *P*1', which show the highest degree of conservation amongst the cleavage sites [16]. In *P*1, glutamine is highly conserved in all polyprotein cleavage sites of SARS-CoV, MERS-CoV and SARS-CoV-2. In *P*2, hydrophobic residues are tolerated with clear precedence for leucine. *P*1' tolerates small amino acids like alanine or serine [18–23].

In all polyprotein cleavage sites processed by Mpro for three CoVs (SARS-CoV, MERS-CoV and SARS-CoV-2), substrate recognition profiles are very similar to each other. The preference for glutamine in P1 must be considered for inhibitor design [24].

4.3 Why Mpro Is a Potent Target for Drug Design?

The Mpro particularly cleaves polypeptide sequences after a glutamine residue, positioning the Mpro as a good drug target because, until now, no human host-cell proteases are known with this substrate specificity. The catalytic dyad of Mpro (cysteine and histidine) is located in its active center. In comparison with other cysteine and serine proteases, SARS-CoV-2 M^{pro} includes a buried water molecule that occupies this place in the active site, which this differences in the active site of SARS-COV-2 Mpro make it more specified to drug design [5].

Also, there is information about protein–protein interaction networks, which reveals the role of Mpro in blocking the inflammatory and interferon pathway in humans by inhibiting the nuclear transport of epigenetic regulatory proteins. These properties indicate the extremely important role of Mpro in the virus life cycle and stimulating an immune response and makes this enzyme one of the most appropriate targets for drug design [25]. Due to the role of Mpro as a key enzyme in the viral replication, its inhibition can stop the production of infectious viral particles and thus alleviate disease symptoms [26–29].

There are other targets for the development of antiviral drugs, such as an Angiotensin-converting enzyme II (ACE2) entry receptor, RNA-dependent RNA polymerase (RdRp) and papain-like protease (PLpro). But inhibitors that targeting ACE2 (aiming to block coronavirus host interactions, which is critical for virus survival) did not advance clinically due to their side effects. Also, RdRp inhibitors appeared to be not very specific and showed low potency, likewise to ACE2 inhibitors some side effects have been seen in patients which used RdRp inhibitors [30]. On the other hand, enzymes such as RdRp cannot fully function without a prior proteolytic release process which depends on Mpro [18, 31]. The papain-like protease recognizes the C-terminal sequence of ubiquitin. Therefore, it seems that substrate-derived inhibitors of PLpro maybe also inhibit host cells [5].

On the other hand, the Mpro of SARS-CoV-2 has very high sequence similarity with the Mpro of SARS-CoV, so if the active site of Mpro of SARS-CoV and SARS-CoV-2 share a high structural similarity, then reported inhibitors of SARS-CoV Mpro can be a considerable option for drug design against the new virus [32].

4.4 Mpro Properties

The homology models of SARS-CoV-2 Mpro demonstrated that close structural similarity to other coronaviral Mpro amino acid sequence alignments reveals ~96% identity with the previous SARS-CoV Mpro (Fig. 4.2) [5]. The alignment was generated using ESPript 3.0 [33].

The Mpro's weight is almost 33.8 kDa, and it is denied as a homodimer with the two protomers connected at right angles to each other (Fig. 4.3) [34]. The Mpro's total mass is 33792.690 a.m.u., and its total charge is −4 e [35]. Each protomer contains

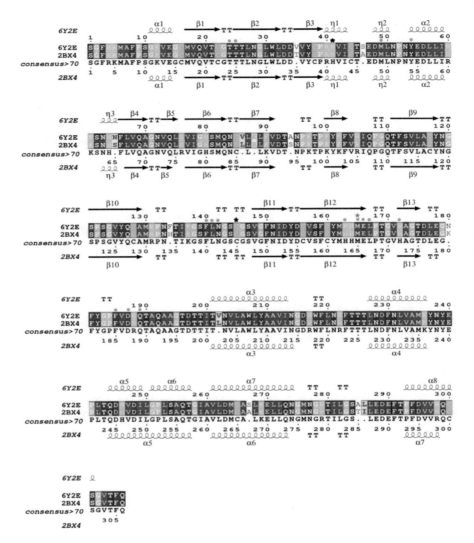

Fig. 4.2 Alignment of sequences and structures of crystallized main proteases of SARS-CoV-2 (PDB: 6Y2E), SARS-CoV (PDB: 2BX4) and MERS-CoV (PDB: 5C3N) are shown. Domains I, II and III comprise residues 8–101, 102–184 and 201–306, respectively. The catalytic dyads are indicated by asterisks, also possible substrate attachment sites are marked with green circles

306 amino acid residues with a 4682 total number of atoms and consists of three domains [7]. In SARS-CoV and SARS-CoV-2, domains I and II contain residues 8–101 and 102–184, respectively, and include an antiparallel β-barrel which is similar to trypsin-like serine proteases [34]. Domain II and Domain III are connected to each other (residues via a longer loop region. Domain III is known by its cluster

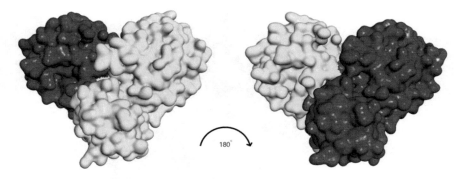

Fig. 4.3 Mpro homodimer is visualized by Discovery studio visualizer (Color pattern: protomer A (red) and protomer B (yellow))

of five α-helices, which plays a role in regulating the dimerization of the Mpro (Fig. 4.4) [36], predominantly through a salt-bridge interaction between Glu290 of one protomer and Arg4 of the other. The tight dimer formed by SARS-CoV-2 Mpro has a contact interface (~1394 Å) mostly between the domain II (molecule A) and the NH2-terminal residues ("N-finger") of molecule B, with the two molecules oriented vertically to one another. The N-finger of each of the two protomers interacts with Glu166 of the other one and as a result, helps to shape the S1 pocket of the substrate-binding site. To provide this interaction site, the N-finger is compressed between domains II and III of the parent monomer and domain II of the other monomer [33, 35–41]. It characterizes that the Mpro monomer is basically inactive so the interface in dimerization can be a suitable option for inhibiting Mpro and drug discovery [10, 27].

The oxyanion hole contains backbone amides or positive-charged amino acids, which is directly corresponded to the Mpro activity and substrate binding. In ligand-bound Mpro, structures of the oxyanion hole consist of a loop (residues 140–145), negatively charged residues Glu166, positively charged residues His172, His163, His41 and remains in an active conformation (Fig. 4.5) [36, 42].

The Pi-stacking (also called π–π stacking) interaction (Phe140/His163) is found in the oxyanion hole. Formation of hydrogen bond and salt bridge due to interaction between Glu166 with water as well as His172 at domain II cause further stability of the oxyanion hole. The oxyanion loop consists of residues 137–145, which is less well ordered and the side chains of Glu166 and Phe140 cannot be fit well because of poor density in the apo state structure. The salt bridge and Pi-stacking interactions between Glu166/His172 and Phe140/His163 are broken, resulting in rearrangements in this region and further collapses of the oxyanion hole (Fig. 4.6) [36]. The N-finger plays a vital role in the formation of Mpro. The active site and auto cleavage activity of Gly2 have interactions with Gly143 in the oxyanion loop in the neighboring protomer, stabilizing the active site and dimer in the active conformation [36].

His163 forms hydrogen bonds with water molecules in our structure, which is not seen in the ligand-bound structures. Another water molecular is found near

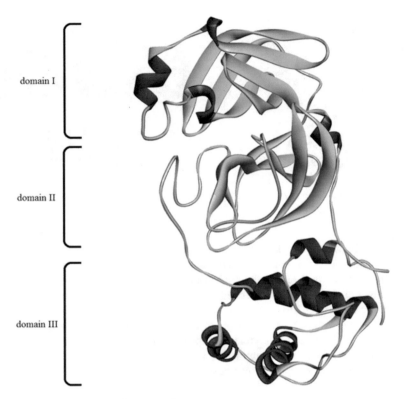

domain I

domain II

domain III

Fig. 4.4 The domain I (residues 8–101) is comprised of three small α-helices and six β-strands. The domain II (residues 102–184) consists of six β-strands. The domain III is composed of five α-helices, which are closely related to the proteolytic activity

the Cys145-His41 catalytic dyad in the active site, acting as a bridge for proton transfer. probably These water molecules influence the negative-charged oxygen of the substrate or inhibitor, which makes rational drug design more difficult (Fig. 4.6) [36].

The analysis and comparisons of Mpro in different states indicate that the substrate-binding site and the active site are more flexible in the apo state than that in the ligand-bound structures. For drug discovery campaigns, the water molecules Embedded in the oxyanion hole and the related interactions should be considered. The ligand-bound structure has no water molecule, while the oxyanion hole in the apo state structure has two water molecules, in the same region. The water molecules, which are found near His163 and His41 in the involved pocket, stabilize the positive-charged His residues and increasing the steric hindrance that may affect the catalytic efficiency of the enzyme [36].

In the SARS-CoV, there is a polar interaction between the two domains III involving a 2.60 Å hydrophobic contact between the side chain of Ile286 and Thr285

Fig. 4.5 Structure of Mpro binding site with 11b in an active conformation. Hydrogen bonds are shown as a black thin line

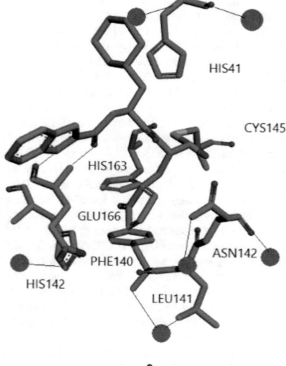

Fig. 4.6 Structure of Mpro in the apo state. Water 1 and 2 are shown in red spheres. Hydrogen bonds are indicated by thin black lines

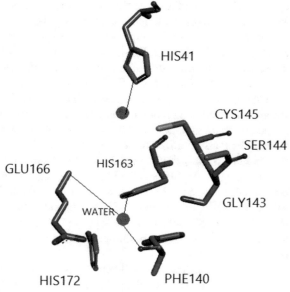

supports hydrogen bond between the side-chain hydroxyl groups of residues Thr285 of each protomer, these properties have not seen in SARS-CoV-2. In SARS-CoV-2, the Thr is substituted by Ala and the isoleucine by Leu. It was previously indicated that the replacing of Ser284 by Thr285 and Ile286 by alanine residues in SARS-CoV Mpro leads to an increase in the catalytic activity of the protease by a factor of 3.6, concurrent with a slightly closer packing of the two domains III of the dimer against one another [37, 43]. This was accompanied by changes in Mpro dynamics that transmit the effect of the mutation on the catalytic center. Thr285 → Ala285 substitution observed in the Mpro of SARS-CoV-2, which leads to approach the two domains III to each other more closely (the distance between the Ca atoms of residues 285 in molecules A and B is 6.77 Å in SARS-CoV Mpro and 5.21 Å in SARS-CoV-2 Mpro, and the distance between the centers of mass of the two domains III shrinks from 33.4 Å to 32.1 Å). However, the catalytic efficiency of SARS-CoV-2 Mpro is indicated slightly higher. Further, the dissociation constant of dimerization for both Mpro is the same (~2.5 mM) [37]. It seems that SARS-COV-2 Mpro has slightly higher catalytic efficiency compared to SARS-COV [37].

The mutations Ser284Ala, Thr285Ala and Ile286Ala in SARS-CoV Mpro results in increasing catalytic activity [43]. Two similar mutations including Thr285Ala and Ile286Leu are observed in SARS-CoV-2 Mpro, which tells us why higher activity was observed for SARS-CoV-2 Mpro compared to SARS-CoV Mpro [7, 37]. It seems that Thr285Ala and Ile286Leu lead to bring domain III of both protomers closer to each other and these mutations can produce a significant change in the dimerization process [5, 32].

The substrate-binding site is the deep cleft between domain I and II which lined by hydrophobic residues with the catalytic site present in the center of the cleft [44]. Mpro features unique catalytic residues (Cys145 and His41) dyad and surrounded by other residues which help to substrate specificity. The N-terminal peptide of Mpro prefers Thr-Ser-Ala-Val-Leu- Gln as positions P6 to P1, which interact with the substrate-binding site amino acids [26, 45, 46]. Despite the S46A mutation, the active sites of SARS-CoV and SARS-CoV-2 Mpro are highly conserved [47, 48].

In the S1 subsite, the imidazole side chain of conserved histidine interacts with the carboxamide side-chain of P1; this interaction is generally accepted to be conducive of specificity for glutamine residue at P1 (Fig. 4.7) [14, 44, 49]. The side chain of Leu-P2 and Thr-P4 is stabilized by deep hydrophobic S4 and S2 subsites, respectively. Ser-P5 and Thr-P6 interact with Pro168 and Ala191 of the Mpro by van der Waals(VDW) interactions [26, 46, 50]. On the C-terminal side of the substrate, the P1 is occupied by a small residue such as Ala, Ser and Gly which directly interact with the S1' subsite by VDW interactions [45]. A long side-chain leucine is accommodated by the hydrophilic S2' subsite of the Mpro [45]. These structural attributes of the Mpro are essential for substrate binding and have a crucial role in the activity of Mpro (Fig. 4.7) [49].

The catalytic mechanism of Mpro starts with the deprotonation of the thiol group of Cys145 followed by a nucleophilic attack of resulting sulfur anion on the substrate carbonyl carbon [10, 51]. This step leads to the release of a peptide product with an amide terminus, and in concert His41 residue is restored to its deprotonated form [10,

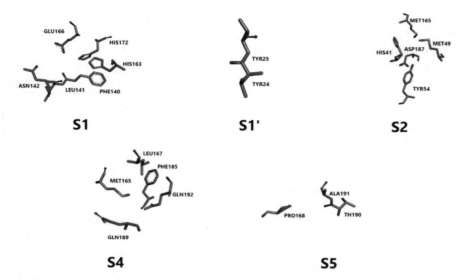

Fig. 4.7 Substrate binding subsites (S1, S2, S4, S5, S1′) of Mpro from SARS CoV-2

38]. The last step is the hydrolysis of the resultant thioester to release a carboxylic acid which then results in the regeneration of the protease [10, 38, 52]. This general acid-base mechanism which processes the hydrolysis of the substrate defines the catalytic action of Mpro [49].

4.5   Drug Design

Understanding the mechanism of action at the atomic level may provide insights for more rational drug design [53]. Computational methods are commonly used for structure-based drug discovery (SBDD) and ligand-based drug discovery (LBDD) [54]. LBDD is a technique for searching and designing new drugs based on experimental information and structural information of known compounds [55, 56]. On the other hand, SBDD is a method based on the tertiary structural information of the target protein [57]. Molecular dynamics (MD) simulations, which analyzed the dynamics of biopolymers in solution at the atomic level, is a typical SBDD method used to predict the interaction between proteins and inhibitors [58]. MD simulations can be applied to clarify the binding mechanism between proteins and ligands at the molecular level, which is highly useful for rational drug design [58, 59].

Fortunately, many complex structures of SARS-CoV Mpro and inhibitors are available in the Protein Data Bank. Therefore, by modelling the complex structure of SARS-CoV-2 Mpro and its inhibitors and using the information on the known structures of SARS-CoV-Mpro, it is possible to analyze the characteristics of functional groups needed for the molecular detection of ligands through SARS-CoV-2 Mpro. His41and Gln189 are adjacent to the HBD sphere, and Gly143, Ser144, Cys145 and Glu166 are adjacent to the HBA sphere. The side chain of His41 is located where the lone pair of nitrogen atoms on the imidazole ring can contact the donor sphere. In addition, the carbonyl oxygen in the side chain of Gln189 is posited near the donor sphere. These residues probably create hydrogen bonds with the HBD located on the donor sphere. On the other hand, the HBA sphere is posited near the main chain of three significant residues (Gly143, Ser144, and Cys145). The HBA sphere has a high affinity to the backbone NH Group. The backbone of Glu166 is also posited near the HBA sphere, which authorizes the NH group on the Glu166 backbone to bind with the HBA sphere. The distance between His41, Gly143, Met145, Glu166, Gln189, and each pharmacophore sphere is 3.58 Å, 3.16 Å, 3.12 Å, 3.37 Å, 1.72 Å, respectively (Fig. 4.8) [54].

His41, Gly143, Ser144, Cys145, Glu166, and Gln189 of SARS-CoV Mpro were located near these pharmacophore spheres. Since these amino acids are conserved in SARS-CoV-2 M^{pro} (Fig. 4.9) [54], it seems SARS-CoV M^{pro} inhibitors are located at similar positions in SARS-CoV-2 M^{pro} and thus have the potential to inhibit SARSCoV-2 Mpro. Also, the three-dimensional structure of SARS-CoV Mpro and SARS-CoV-2 Mpro is almost conserved, and the amino acid sequence identity value shows 96%. The pharmacophores do not contain non-conserved amino acid residues. Thus, inhibitors that are matched with SARS-CoV Mpro pharmacophore may have the same potential to inhibit SARSCoV-2 Mpro [54].

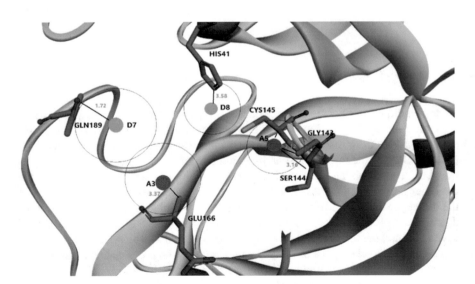

Fig. 4.8 The amino acid residues around the chemical groups which are defined as the pharmacophore

Fig. 4.9 Alignment of SARS-CoV and SARS-CoV-2 Mpro X-ray structure. Sphere model indicates residues that are not conserved between both sequences

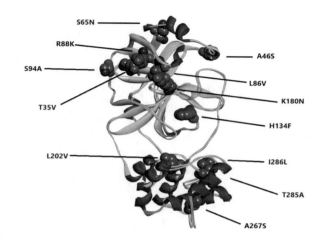

4.5.1 Investigation of Interactions Between SARS–CoV–2 Mpro and the Inhibitors

The interaction with His41 was maintained with a high probability in all MD results (78, 92, and 94%) [54]. There are two types of Interactions with His41: Pi-stacking and hydrogen bonding. During simulation, His41 hydrogen bonds to Gly143 and Cys145 were observed with a probability of over 50% [54]. Interaction of His41, Gly143, Met165, and Glu166 was observed in all MD simulations [54]. The main chains of Gly143 and Glu166 and the side chains of His41 were involved in the interaction, and Met165 creates VDW interaction with the inhibitors. It seems that Gly143 (Acceptor) and Met145 (Acceptor) involved in the same pharmacophore point. Among pharmacophore interaction, His41-Donor and Glu166-Acceptor are highly stable during MD simulation for all ligands. The main chains of Ser144, Gly143 and Cys145 also interact with each inhibitor this suggests that the interaction with these amino acid residues may not be affected by side-chain mutations unless the dynamics of each chain or binding site shape are changed. Interactions with His41 were affirmed as hydrogen bonding and Pi-stacking. In the hydrogen bond, NH in the imidazole ring of His41 works as HBD. Also, the imidazole ring of His41 forms Pi-stacking with each inhibitor. The results of pharmacophore modeling and the MD simulations suggested that His41 works as HBD. In contrast, the HBD pharmacophore sphere is located near His41. Therefore, HBA functional group has the potential to interact with His41. MD simulations also suggested that aromatic functional groups have a high affinity to His41. In each MD simulation, Gly143, Ser144, Cys145, Glu166, and Gln189 interact with functional groups defined as pharmacophore of peptide-like inhibitors [54]. Therefore, interactions with these residues are important for binding to SARS-CoV-2 Mpro. In the MD simulation results, all ligand has one or two water bridges. Therefore, maybe that water bridges are involved

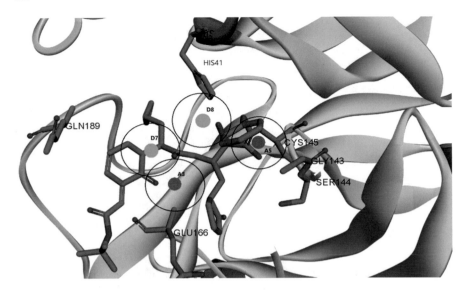

Fig. 4.10 Alignment of α-ketoamide inhibitors and pharmacophore models. SARS-CoV-2 Mpro with α-ketoamide inhibitors (PDBID: 6Y2G) was aligned for 6LU7 and pharmacophore model

in Mpro and ligand complex structures to stabilize the structure; functional groups of ligands can be extended to the space occupied by these waters [54].

A study on SARS-CoV-2 Mpro with α-ketoamide inhibitors shows that one hydroxyl group and two carbonyl groups of α-ketoamide inhibitors are matched the pharmacophore model (Fig. 4.10) [54]. Comparing the structures of Gln189 in Figs. 4.8 and 4.10, the conformations of the side chains are different. It suggested that the side-chain conformation of Gln189 flexibly changes depending on the binding inhibitor [54].

Kneller et al. identified the room temperature X-ray structure of the ligand-free Mpro and compared it with the low temperature one of ligand-free Mpro and N3 inhibitor covalently bound to Mpro. They found that the active site of Mpro had flexible conformation and the conformational change was induced by ligand binding [60]. Teruhisa et al. [61] found five representative drug binding sites on Mpro and named them as "sites 1-5" (Fig. 4.11) [61]. This study investigated the access of drugs to these binding sites, on the fluctuating surface of Mpro by analysis MD trajectories analysis of dimeric Mpro with several HIV inhibitors, including indinavir, lopinavir, darunavir, ritonavir, saquinavir, nelfinavir, and tipranavir. The results showed that most of the contacts were located at the predicted binding sites. The frequent contacts with the active site (site 1) were achieved for indinavir, nelfinavir, tipranavir and ritonavir; site 1 contains catalytic amino acids [61].

Adjacent to the active site, a mostly observed site existed at the border of the chains, indicated as "site 4", except lopinavir all other ligands visited this site frequently. The contact frequency to site 2 was high for lopinavir, ritonavir and saquinavir, modest

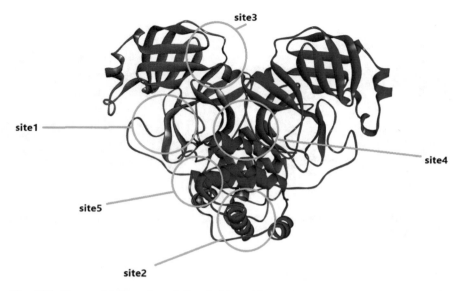

Fig. 4.11 Structural information of dimeric Mpro. The catalytic dyad, His41 and Cys145 are represented by the space-filling model. The five possible ligand binding sites were predicted

for darunavir, indinavir and nelfinavir, and weak for tipranavir. In domain III, another shallow site "site 5" was observed for lopinavir, nelfinavir, and tipranavir (Fig. 4.11) [61].

The contact frequency to site 3 was generally low, except for lopinavir. These values of contact strength should be analyzed carefully since they could observe only a few unbinding events, and they did not accurately reflect the quantified values in equilibrium [61].

Understanding the structural dynamic processes can guide researchers for rational drug design. The analysis of B′-factor profiles from PDB structures can be utilized for this aim [62]. B′-factor analysis can be used to differentiate between protein binding sites and crystal packing. Contacts to estimate protein-ligand binding affinities, fluctuations in B′-factors, show an enhancement or weakening of molecular interactions on an atomic resolution level.

The binding of reversible ligands to their targets usually leads to tighten of the protein scaffold and show itself in a decrease in the B'-factor which approximately correlate to the binding strength of the ligand [62]. Here we used (https://bandit.uni-mainz.de) [62] to check B′-factor and ΔB′-factor of SARS-COV-2 and SARS-COV Mpro, which provide a deeper insight about structural based drug design against SARS-COV-2 (Fig. 4.12).

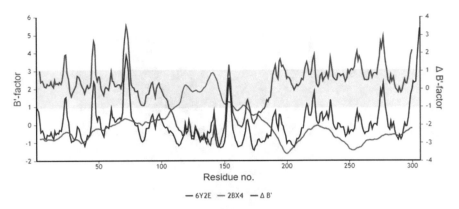

Fig. 4.12 Representation of B′-factor analysis of SARS-COV Mpro (2XB4) and SARS-COV-2 (6Y2E) Mpro

Table 4.1 Proposed compounds to inhibition of Mpro [49]

Drug	Description
Flavonoids	Group of polyphenolic plant metabolites with a general structure consisting of two phenyl rings and a heterocyclic ring
Peptides	Compounds which sport the chemical structure typical of a peptide classifying them as peptide inhibitors
Terpenes	Constitute a large hydrocarbon class constructed from five-carbon isoprene units which are combined in a great variety of skeletons
Quinolines	These compounds have a characteristic structure of a benzene ring fused with a pyridine making it a double-ringed heterocyclic aromatic organic compound
Nucleoside and nucleotide analogues	A popularly identified drug with potential antiviral activity against several RNA viruses
Protease inhibitor	Protease inhibitors (PIs) are a class of antiviral drugs that are widely used to treat HIV/AIDS and hepatitis C
Phenalene	Polycyclic aromatic hydrocarbons
Antibiotic derivatives	Compounds which destroy or slow down the growth of bacteria
Indoles	Compounds classified as indoles are characterized with a benzene ring fused with a pyrrole ring

4.6 Inhibitors

As shown in Table 4.1, different types of compounds for inhibition of SARS-COV-2 Mpro are identified, where the corresponding IC50 values for inhibition of the SARS-CoV Mpro and the MERS CoV Mpro should be considered as 0.90 ± 0.29 mM and 0.58 ± 0.22 mM, respectively [37, 63].

4.6.1 First-Line SARS-COV-2 Mpro Inhibitors

4.6.1.1 Ebselen and N3

Ebselen and N3 illustrated the strongest antiviral effects at a specific concentration (10 μM) treatment in SARS-CoV-2-infected Vero cells. Ebselen and N3 displayed inhibition against SARS-CoV-2 with individual half-maximal effective concentration (EC50) values of 4.67 μM and 16.77 μM, respectively. The dose–response curves suggest that these compounds are probably able to penetrate the cellular membrane to connect their targets. The structure of SARS-COV-2 Mpro with N3 is available on RCSB with PDB ID:6LU7 [7].

Ebselen is an organoselenium compound with anti-oxidant, anti-inflammatory and cytoprotective properties. This compound has been investigated for the treatment of multiple diseases, including hearing loss and bipolar disorders [64–66]. Ebselen known for its low cytotoxicity (the median lethal dose in rats is >4,600 mg kg^{-1}, when taken orally) [7], and its safety in humans has been evaluated in a number of clinical trials. These data strongly suggest the clinical potential of Ebselen for the treatment of coronaviruses [65–67] (Fig. 4.13).

Fig. 4.13 Left: N3 (*Source* RCSB) and Right: Ebselen (*Source* PubChem)

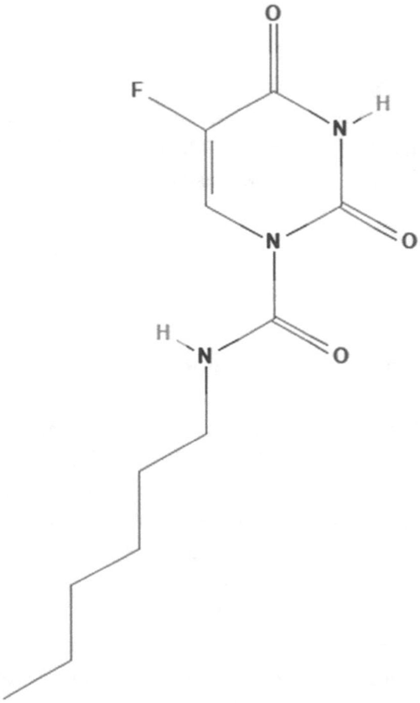

Fig. 4.14 Carmofur (*Source* PubChem)

4.6.1.2 Carmofur

In the previous study [68] clinical isolation of SARS-CoV-2 was proliferated in Vero E6 cells, and Vero E6 cells were from ATCC with authentication [68]. The results show that Carmofur inhibits viral replication in cells (EC50 = 24.30 μM), and it can be a suitable candidate to design new antiviral for COVID-19 [68]. Carmofur EC50 is 24.30 μM, and its IC50 is 1.82 μM. [68]. Coordinates and structure factors for SARS-CoV-2 Mpro in complex with Carmofur have been deposited in Protein Data Bank (PDB code 7BUY) [68] (Fig. 4.14).

4.6.1.3 Lopinavir and Ritonavir

Lopinavir is an HIV-1 protease inhibitor, which is combined with ritonavir to increase its plasma half-life. Also, Lopinavir is an inhibitor of SARS-CoV's main protease [69]. In another study, 199 patients with SARS-CoV-2 infection were investigated [70]. 99 patients were assigned to the lopinavir–ritonavir group, and 100 patients were considered as the standard care group. Mortality at 28 days was similar in both groups (19.2% vs. 25.0%; difference, −5.8% points). The percentages of patients with detectable viral RNA at various time points were similar. In a modified intention-to-treat analysis, lopinavir–ritonavir led to a median time to clinical improvement that was shorter by one day than that observed with standard care (hazard ratio, 1.39) [70]. Lopinavir–ritonavir group shows more Gastrointestinal adverse events, but serious adverse events were more common in the standard-care group [70]. However, there are some reports about the beneficial role of lopinavir/ritonavir as a treatment of COVID-19 [71]. A retrospective study including 120 patients shows that early administration of ritonavir-lopinavir could shorten the time of virus shedding [72]. A controlled study on 47 patients with COVID-19 infection showed that a combination of ritonavir-lopinavir and adjuvant drugs can significantly decrease the number of days for virus clearance compared to adjuvant drugs alone [73]. Another set of studies [74] investigated ritonavir-lopinavir's or ritonavir-lopinavir combined with arbidol effectiveness against COVID-19, and the results show no evidence proved that ritonavir, lopinavir or ritonavir-lopinavir combined with arbidol can shorten the disease course [74]. In another retrospective cohort study, 50 patients were investigated and they were divided into the ritonavir-lopinavir group and results indicated that viral clearance is faster in the arbidol group [74].

In another study, the effect of a triple combination of lopinavir–ritonavir, interferon beta-1b, and ribavirin against COVID19 was investigated [75]. Between February and March 2020, 127 patients were investigated; 86 patients were randomly assigned to the combination group and 41 patients were considered as the control group. The median time from symptom onset to start of study treatment was five days. The combination group had a significantly shorter median time from the start of study treatment to negative nasopharyngeal swab (seven days) than the control group (12 days; hazard ratio 4·37, p = 0·0010). One patient in the control group stopped using lopinavir–ritonavir because of biochemical hepatitis. No one died during the study. It seems that early triple antiviral therapy was safe and more effective compared to lopinavir–ritonavir alone and its shortening the duration of viral shedding and hospital stay and alleviating symptoms in patients with mild to moderate COVID-19 [75] (Fig. 4.15).

4.6.1.4   Atazanavir

One study showed that [76] atazanavir docks in the active site of SARS-CoV-2 Mpro with greater strength than lopinavir. They confirmed that atazanavir inhibits SARS-CoV-2 replication, alone or in combination with ritonavir (ritonavir) in Vero cells and a human pulmonary epithelial cell line. Atazanavir/ritonavir also impaired virus-induced enhancement of interleukin 6 (IL-6) and tumor necrosis factor-alpha (TNF-α) levels. Together, their data strongly suggest that atazanavir and atazanavir/ritonavir should be considered among the candidate repurposed drugs undergoing clinical trials in the fight against COVID-19 [76] (Fig. 4.16).

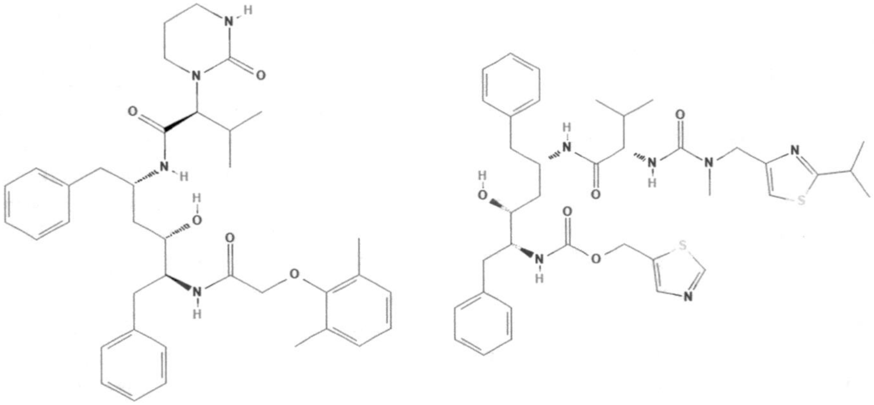

Fig. 4.15 Left: Lopinavir, and Right: Ritonavir (*Source* PubChem)

Fig. 4.16 Atazanavir
(*Source* PubChem)

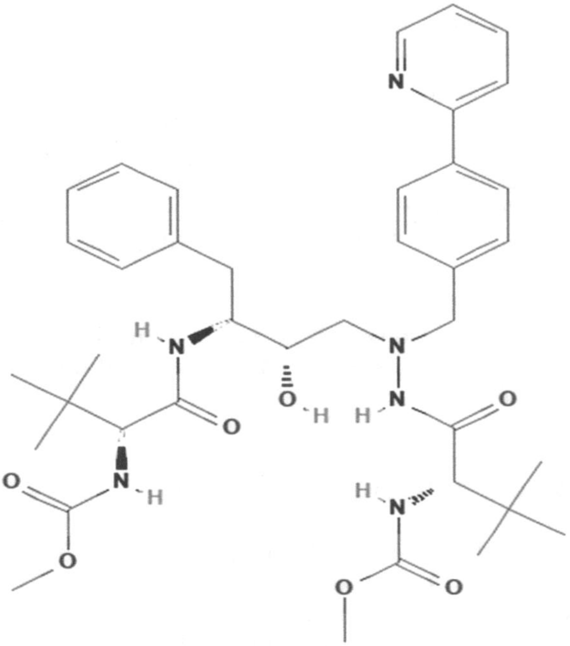

4.6.1.5 Nelfinavir

It is an HIV-1 protease inhibitor [77]. In one study, researchers found that nelfinavir is most probably a multi-target agent. Therefore, its antiviral activity was performed and repeated three times in duplicates in Vero E6 cells. The SARS-CoV-2 virus

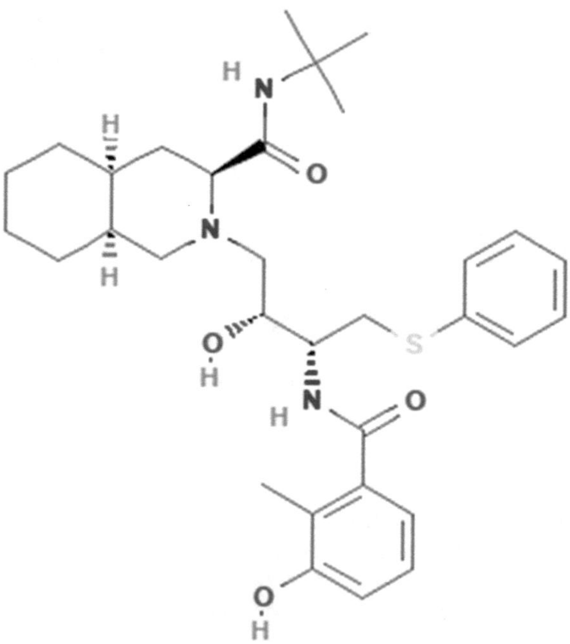

Fig. 4.17 Nelfinavir (*Source PubChem*)

was isolated from a clinical isolate of SARS-CoV-2 infected patients [77]. The half-maximal effective concentration (EC50) of nelfinavir mesylate against the SARS-CoV-2 was determined to be 2.89 ± 0.65 μM [77]. In other study, the effective concentrations for 50% and 90% inhibition (EC50 and EC90) of nelfinavir were 1.13 μM and 1.76 μM, respectively [78]. Hirofumi Ohashi and co-workers found that the combining Nelfinavir/Cepharanthine can be a good option for the treatment of COVID-19 [79] (Fig. 4.17).

4.6.2   New Synthetic Compounds

4.6.2.1   11a and 11b Compounds

Dai and coworkers designed two Mpro inhibitors **11a** and **11b**, which showed perfect anti-COVID-19 activity. The structure–function relationship showed that the aldehyde group of the two compounds can covalently link to the Mpro Cys145 residue, and the IC50 for 11a is 0.053 ± 0.005 μM and this number for 11b is 0.040 ± 0.002 μM [42] (Fig. 4.18).

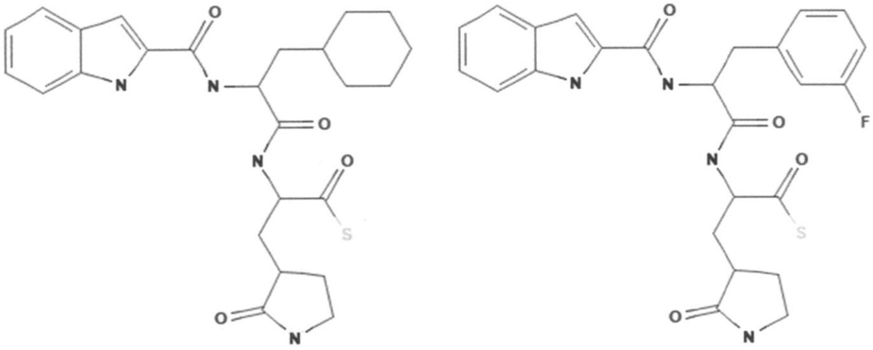

Fig. 4.18 Left: 11a (left) and Right: 11b (*Source* PubChem)

4.6.2.2   11r Compound

Zhang and co-workers reported the complex structure of SARS-COV-2 Mpro and **11r**, and they found that **11r** showed excellent inhibitory activity and potent anti-COVID-19 activity. **11r** could be used as a lead compound to develop potent inhibitors of COVID-19 Mpro, and the IC50, for **11r** calculated 0.18 ± 0.02 mM [37] (Fig. 4.19).

Fig. 4.19 11r (*Source* PubChem)

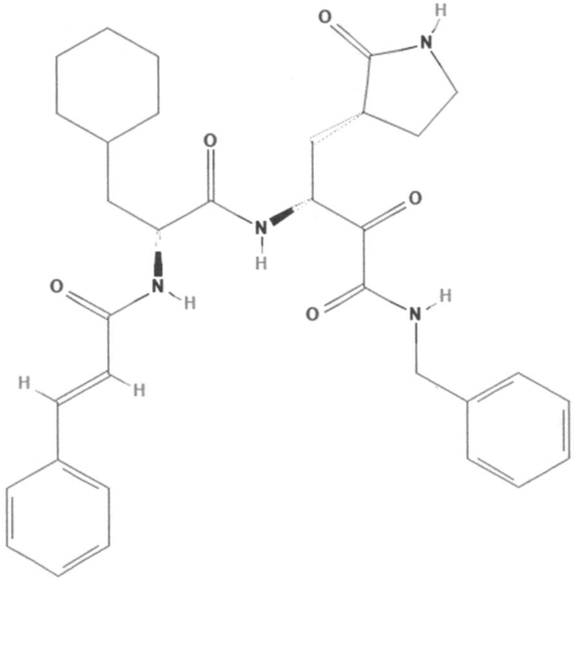

4.6.3 Natural Products Derived from Chinese Traditional Medicines

Su and co-workers investigated the inhibition of COVID-19 Mpro by natural products derived from Chinese traditional medicines. They found that baicalein and baicalin showed non-covalent, non-peptidomimetic inhibition of SARS-COV-2 Mpro, and shows efficient antiviral activities in both in vitro and in a cell-based system. The in vitro study results and favorable safety data from clinical trials showed that baicalein has great potential to become a candidate for a much-needed anti-coronaviral drug [80].

4.7 Conclusion

The SARS-COV-2 virus infected people across the planet; due to the high range of mutations, probably this virus stays longer in the societies, and unfortunately, the vaccines have not been accessible for a significant percentage of people so far, due to their price and geographical aspects and transportations. On the other hand, even after the advances in the global vaccination, there is still a considerable demand for the development of antiviral drugs. SARS-COV-2 Mpro differs significantly from human proteases and has a high sequence similarity to the two previous coronaviruses Mpro. This chapter provided some information about Mpro structure and its attributes and also presented and argued about several known compounds against this enzyme, which can give a wider view to researchers. Due to text and previous research's, using synthetic compounds such as **11a,11b** and **11r** must be especially considered, and also combination drug strategy seems to be a good option for fighting against the disease. It seems that SARS-CoV-2 is not the last human coronavirus; the previous experiences about SARS-COV disease indicated the long-term nature of drug discovery projects, so the virus changes and their impacts on mechanisms and its attributes must be constantly monitored to shorten the drug development time and increase the readiness of the scientific community to fight new coronaviruses' diseases.

References

1. Gorbalenya AE, Baker SC, Baric RS, de Groot RJ, Drosten C, Gulyaeva AA, Haagmans BL, Lauber C, Leontovich AM, Neuman BW, Penzar D, Perlman S, Poon LLM, Samborskiy DV, Sidorov IA, Sola I, Ziebuhr J, C.S.G. of the I.C. on T. of Viruses (2020) The species severe acute respiratory syndrome-related coronavirus: classifying 2019-nCoV and naming it SARS-CoV-2. Nat Microbiol 5:536–544. https://doi.org/10.1038/s41564-020-0695-z
2. Stawicki S, Jeanmonod R, Miller A, Paladino L, Gaieski D, Yaffee A, De Wulf A, Grover J, Papadimos T, Bloem C, Galwankar S, Chauhan V, Firstenberg M, DI Somma S, Jeanmonod

D, Garg S, Tucci V, Anderson H, Fatimah L, Worlton T, Dubhashi S, Glaze K, Sinha S, Opara I, Yellapu V, Kelkar D, El-Menyar A, Krishnan V, Venkataramanaiah S, Leyfman Y, Saoud Al Thani H, Nanayakkara PB, Nanda S, Cioè-Peña E, Sardesai I, Chandra S, Munasinghe A, Dutta V, Dal Ponte S, Izurieta R, Asensio J, Garg M (2020) The 2019–2020 novel coronavirus (severe acute respiratory syndrome coronavirus 2) pandemic: a joint American college of academic international medicine-world academic council of emergency medicine multidisciplinary COVID-19 working group consensus paper. J Glob Infect Dis. https://doi.org/10.4103/jgid.jgid_86_20

3. Khan S, Siddique R, Shereen MA, Ali A, Liu J, Bai Q, Bashir N, Xue M (2020) Emergence of a novel coronavirus, severe acute respiratory syndrome coronavirus 2: biology and therapeutic options. J Clin Microbiol. https://doi.org/10.1128/JCM.00187-20

4. Michel CJ, Mayer C, Poch O, Thompson JD (2020) Characterization of accessory genes in coronavirus genomes. Virol J 17:1–13. https://doi.org/10.1186/s12985-020-01402-1

5. Ullrich S, Nitsche C (2020) The SARS-CoV-2 main protease as drug target. Bioorganic Med Chem Lett 30: https://doi.org/10.1016/j.bmcl.2020.127377

6. Helmy YA, Fawzy M, Elaswad A, Sobieh A, Kenney SP, Shehata AA (2020) The COVID-19 pandemic: a comprehensive review of taxonomy, genetics, epidemiology, diagnosis, treatment, and control. J Clin Med 9. https://doi.org/10.3390/jcm9041225

7. Jin Z, Du X, Xu Y, Deng Y, Liu M, Zhao Y, Zhang B, Li X, Zhang L, Peng C, Duan Y, Yu J, Wang L, Yang K, Liu F, Jiang R, Yang X, You T, Liu X, Yang X, Bai F, Liu H, Liu X, Guddat LW, Xu W, Xiao G, Qin C, Shi Z, Jiang H, Rao Z, Yang H (2020) Structure of Mpro from SARS-CoV-2 and discovery of its inhibitors. Nature 582:289–293. https://doi.org/10.1038/s41586-020-2223-y

8. Gordon DE, Jang GM, Bouhaddou M, Xu J, Obernier K, White KM, O'Meara MJ, Rezelj VV, Guo JZ, Swaney DL, Tummino TA, Hüttenhain R, Kaake RM, Richards AL, Tutuncuoglu B, Foussard H, Batra J, Haas K, Modak M, Kim M, Haas P, Polacco BJ, Braberg H, Fabius JM, Eckhardt M, Soucheray M, Bennett MJ, Cakir M, McGregor MJ, Li Q, Meyer B, Roesch F, Vallet T, Mac Kain A, Miorin L, Moreno E, Naing ZZC, Zhou Y, Peng S, Shi Y, Zhang Z, Shen W, Kirby IT, Melnyk JE, Chorba JS, Lou K, Dai SA, Barrio-Hernandez I, Memon D, Hernandez-Armenta C, Lyu J, Mathy CJP, Perica T, Pilla KB, Ganesan SJ, Saltzberg DJ, Rakesh R, Liu X, Rosenthal SB, Calviello L, Venkataramanan S, Liboy-Lugo J, Lin Y, Huang XP, Liu YF, Wankowicz SA, Bohn M, Safari M, Ugur FS, Koh C, Savar NS, Tran QD, Shengjuler D, Fletcher SJ, O'Neal MC, Cai Y, Chang JCJ, Broadhurst DJ, Klippsten S, Sharp PP, Wenzell NA, Kuzuoglu-Ozturk D, Wang HY, Trenker R, Young JM, Cavero DA, Hiatt J, Roth TL, Rathore U, Subramanian A, Noack J, Hubert M, Stroud RM, Frankel AD, Rosenberg OS, Verba KA, Agard DA, Ott M, Emerman M, Jura N, von Zastrow M, Verdin E, Ashworth A, Schwartz O, d'Enfert C, Mukherjee S, Jacobson M, Malik HS, Fujimori DG, Ideker T, Craik CS, Floor SN, Fraser JS, Gross JD, Sali A, Roth BL, Ruggero D, Taunton J, Kortemme T, Beltrao P, Vignuzzi M, García-Sastre A, Shokat KM, Shoichet BK, Krogan NJ (2020) A SARS-CoV-2 protein interaction map reveals targets for drug repurposing. Nature 583. https://doi.org/10.1038/s41586-020-2286-9

9. Andrianov AM, Kornoushenko YV, Karpenko AD, Bosko IP, Tuzikov AV (2020) Computational discovery of small drug-like compounds as potential inhibitors of SARS-CoV-2 main protease. J Biomol Struct Dyn. https://doi.org/10.1080/07391102.2020.1792989

10. Pillaiyar T, Manickam M, Namasivayam V, Hayashi Y, Jung SH (2016) An overview of severe acute respiratory syndrome-coronavirus (SARS-CoV) 3CL protease inhibitors: peptidomimetics and small molecule chemotherapy. J Med Chem. https://doi.org/10.1021/acs.jmedchem.5b01461

11. Huang C, Wei P, Fan K, Liu Y, Lai L (2004) 3C-like proteinase from SARS coronavirus catalyzes substrate hydrolysis by a general base mechanism. Biochemistry. https://doi.org/10.1021/bi036022q

12. Muramatsu T, Kim YT, Nishii W, Terada T, Shirouzu M, Yokoyama S (2013) Autoprocessing mechanism of severe acute respiratory syndrome coronavirus 3C-like protease (SARS-CoV 3CLpro) from its polyproteins. FEBS J. https://doi.org/10.1111/febs.12222

13. Xia B, Kang X (2011) Activation and maturation of SARS-CoV main protease. Protein Cell. https://doi.org/10.1007/s13238-011-1034-1
14. Ziebuhr J, Snijder EJ, Gorbalenya AE (2000) Virus-encoded proteinases and proteolytic processing in the Nidovirales. J Gen Virol. https://doi.org/10.1099/0022-1317-81-4-853
15. Du QS, Wang SQ, Zhu Y, Wei DQ, Guo H, Sirois S, Chou KC (2004) Polyprotein cleavage mechanism of SARS CoV M pro and chemical modification of the octapeptide. Peptides. https://doi.org/10.1016/j.peptides.2004.06.018
16. Hegyi A, Ziebuhr J (2002) Conservation of substrate specificities among coronavirus main proteases. J Gen Virol. https://doi.org/10.1099/0022-1317-83-3-595
17. Muramatsu T, Takemoto C, Kim YT, Wang H, Nishii W, Terada T, Shirouzu M, Yokoyama S (2016) SARS-CoV 3CL protease cleaves its C-terminal autoprocessing site by novel subsite cooperativity. Proc Natl Acad Sci USA. https://doi.org/10.1073/pnas.1601327113
18. Thiel V, Ivanov KA, Putics Á, Hertzig T, Schelle B, Bayer S, Weißbrich B, Snijder EJ, Rabenau H, Doerr HW, Gorbalenya AE, Ziebuhr J (2003) Mechanisms and enzymes involved in SARS coronavirus genome expression. J Gen Virol. https://doi.org/10.1099/vir.0.19424-0
19. Zhang L, Lin D, Kusov Y, Nian Y, Ma Q, Wang J, Von Brunn A, Leyssen P, Lanko K, Neyts J, De Wilde A, Snijder EJ, Liu H, Hilgenfeld R (2020) α-ketoamides as broad-spectrum inhibitors of coronavirus and enterovirus replication: structure-based design, synthesis, and activity assessment. J Med Chem. https://doi.org/10.1021/acs.jmedchem.9b01828
20. Fan K, Wei P, Feng Q, Chen S, Huang C, Ma L, Lai B, Pei J, Liu Y, Chen J, Lai L (2004) Biosynthesis, purification, and substrate specificity of severe acute respiratory syndrome coronavirus 3C-like proteinase. J Biol Chem. https://doi.org/10.1074/jbc.M310875200
21. Rut W, Groborz K, Zhang L, Sun X, Zmudzinski M, Pawlik B, Wang X, Jochmans D, Neyts J, Młynarski W, Hilgenfeld R, Drag M (2021) SARS-CoV-2 Mpro inhibitors and activity-based probes for patient-sample imaging. Nat Chem Biol 17:222–228. https://doi.org/10.1038/s41589-020-00689-z
22. Fan K, Ma L, Han X, Liang H, Wei P, Liu Y, Lai L (2005) The substrate specificity of SARS coronavirus 3C-like proteinase. Biochem Biophys Res Commun. https://doi.org/10.1016/j.bbrc.2005.02.061
23. Hilgenfeld R (2014) From SARS to MERS: crystallographic studies on coronaviral proteases enable antiviral drug design. FEBS J. https://doi.org/10.1111/febs.12936
24. Rut W, Groborz K, Zhang L, Sun X, Zmudzinski M, Pawlik B, Młynarski W, Hilgenfeld R, Drag M (2020) Substrate specificity profiling of SARS-CoV-2 main protease enables design of activity-based probes for patient-sample imaging BioRxiv. https://doi.org/10.1101/2020.03.07.981928
25. Koulgi S, Jani V, Uppuladinne M, Sonavane U, Nath AK, Darbari H, Joshi R (2020) Drug repurposing studies targeting SARS-CoV-2: an ensemble docking approach on drug target 3C-like protease (3CLpro). J Biomol Struct Dyn. https://doi.org/10.1080/07391102.2020.1792344
26. Anand K, Ziebuhr J, Wadhwani P, Mesters JR, Hilgenfeld R (2003) Coronavirus main proteinase (3CLpro) structure: basis for design of anti-SARS drugs. Science 80. https://doi.org/10.1126/science.1085658
27. Grum-Tokars V, Ratia K, Begaye A, Baker SC, Mesecar AD (2008) Evaluating the 3C-like protease activity of SARS-coronavirus: recommendations for standardized assays for drug discovery. Virus Res. https://doi.org/10.1016/j.virusres.2007.02.015
28. Hilgenfeld R, Peiris M (2013) From SARS to MERS: 10 years of research on highly pathogenic human coronaviruses. Antiviral Res. https://doi.org/10.1016/j.antiviral.2013.08.015
29. Liang P-H (2006) Characterization and Inhibition of SARS-coronavirus main protease. Curr Top Med Chem. https://doi.org/10.2174/156802606776287090
30. Ton AT, Gentile F, Hsing M, Ban F, Cherkasov A (2020) Rapid identification of potential inhibitors of SARS-CoV-2 main protease by deep docking of 1.3 billion compounds. Mol Inform. https://doi.org/10.1002/minf.202000028
31. Poduri R, Joshi G, Jagadeesh G (2020) Drugs targeting various stages of the SARS-CoV-2 life cycle: exploring promising drugs for the treatment of Covid-19. Cell Signal. https://doi.org/10.1016/j.cellsig.2020.109721

32. Gahlawat A, Kumar N, Kumar R, Sandhu H, Singh IP, Singh S, Sjöstedt A, Garg P (2020) Structure-based virtual screening to discover potential lead molecules for the SARS-CoV-2 main protease. J Chem Inf Model. https://doi.org/10.1021/acs.jcim.0c00546

33. Gouet P, Courcelle E, Stuart DI, Mètoz FM (1999) ESPript: analysis of multiple sequence alignments in postscript. Bioinformatics 15:305–308. https://doi.org/10.1093/bioinformatics/15.4.305

34. Yang H, Yang M, Ding Y, Liu Y, Lou Z, Zhou Z, Sun L, Mo L, Ye S, Pang H, Gao GF, Anand K, Bartlam M, Hilgenfeld R, Rao Z (2003) The crystal structures of severe acute respiratory syndrome virus main protease and its complex with an inhibitor. Proc Natl Acad Sci USA. https://doi.org/10.1073/pnas.1835675100

35. Cardoso WB, Mendanha SA (2021) Molecular dynamics simulation of docking structures of SARS-CoV-2 main protease and HIV protease inhibitors. J Mol Struct. https://doi.org/10.1016/j.molstruc.2020.129143

36. Zhou X, Zhong F, Lin C, Hu X, Zhang Y, Xiong B, Yin X, Fu J, He W, Duan J, Fu Y, Zhou H, McCormick PJ, Wang Q, Li J, Zhang J (2020) Structure of SARS-CoV-2 main protease in the apo state. Sci China Life Sci. https://doi.org/10.1007/s11427-020-1791-3

37. Zhang L, Lin D, Sun X, Curth U, Drosten C, Sauerhering L, Becker S, Rox K, Hilgenfeld R (2020) Crystal structure of SARS-CoV-2 main protease provides a basis for design of improved a-ketoamide inhibitors. Science 80:368, 409–412. https://doi.org/10.1126/science.abb3405

38. Hsu MF, Kuo CJ, Chang KT, Chang HC, Chou CC, Ko TP, Shr HL, Chang GG, Wang AHJ, Liang PH (2005) Mechanism of the maturation process of SARS-CoV 3CL protease. J Biol Chem. https://doi.org/10.1074/jbc.M502577200

39. Chou CY, Chang HC, Hsu WC, Lin TZ, Lin CH, Chang GG (2004) Quaternary structure of the severe acute respiratory syndrome (SARS) coronavirus main protease. Biochemistry. https://doi.org/10.1021/bi0490237

40. Tahir ul Qamar M, Alqahtani SM, Alamri MA, Chen LL (2020) Structural basis of SARS-CoV-2 3CLpro and anti-COVID-19 drug discovery from medicinal plants. J Pharm Anal. https://doi.org/10.1016/j.jpha.2020.03.009

41. Shi J, Song J (2006) The catalysis of the SARS 3C-like protease is under extensive regulation by its extra domain. FEBS J. https://doi.org/10.1111/j.1742-4658.2006.05130.x

42. Dai W, Zhang B, Jiang XM, Su H, Li J, Zhao Y, Xie X, Jin Z, Peng J, Liu F, Li C, Li Y, Bai F, Wang H, Cheng X, Cen X, Hu S, Yang X, Wang J, Liu X, Xiao G, Jiang H, Rao Z, Zhang LK, Xu Y, Yang H, Liu H (2020) Structure-based design of antiviral drug candidates targeting the SARS-CoV-2 main protease. Science 80. https://doi.org/10.1126/science.abb4489

43. Lim L, Shi J, Mu Y, Song J (2014) Dynamically-driven enhancement of the catalytic machinery of the SARS 3C-like protease by the S284-T285-I286/A mutations on the extra domain. PLoS ONE. https://doi.org/10.1371/journal.pone.0101941

44. Anand K, Palm GJ, Mesters JR, Siddell SG, Ziebuhr J, Hilgenfeld R (2002) Structure of coronavirus main proteinase reveals combination of a chymotrypsin fold with an extra α-helical domain. EMBO J. https://doi.org/10.1093/emboj/cdf327

45. Xue X, Yu H, Yang H, Xue F, Wu Z, Shen W, Li J, Zhou Z, Ding Y, Zhao Q, Zhang XC, Liao M, Bartlam M, Rao Z (2008) Structures of two coronavirus main proteases: implications for substrate binding and antiviral drug design. J Virol. https://doi.org/10.1128/jvi.02114-07

46. Yang H, Xie W, Xue X, Yang K, Ma J, Liang W, Zhao Q, Zhou Z, Pei D, Ziebuhr J, Hilgenfeld R, Kwok YY, Wong L, Gao G, Chen S, Chen Z, Ma D, Bartlam M, Rao Z (2005) Design of wide-spectrum inhibitors targeting coronavirus main proteases. PLoS Biol. https://doi.org/10.1371/journal.pbio.0030324

47. Morse JS, Lalonde T, Xu S, Liu WR (2020) Learning from the past: possible urgent prevention and treatment options for severe acute respiratory infections caused by 2019-nCoV. ChemBioChem. https://doi.org/10.1002/cbic.202000047

48. Bzówka M, Mitusińska K, Raczyńska A, Samol A, Tuszyński JA, Góra A (2020) Structural and evolutionary analysis indicate that the sars-COV-2 mpro is a challenging target for small-molecule inhibitor design. Int J Mol Sci. https://doi.org/10.3390/ijms21093099

49. Singh E, Khan RJ, Jha RK, Amera GM, Jain M, Singh RP, Muthukumaran J, Singh AK (2020) A comprehensive review on promising anti-viral therapeutic candidates identified against main protease from SARS-CoV-2 through various computational methods. J Genet Eng Biotechnol. https://doi.org/10.1186/s43141-020-00085-z

50. Wang F, Chen C, Tan W, Yang K, Yang H (2016) Structure of main protease from human coronavirus NL63: insights for wide spectrum anti-coronavirus drug design. Sci Rep. https://doi.org/10.1038/srep22677

51. Wang H, He S, Deng W, Zhang Y, Li G, Sun J, Zhao W, Guo Y, Yin Z, Li D, Shang L (2020) Comprehensive insights into the catalytic mechanism of middle east respiratory syndrome 3C-like protease and severe acute respiratory syndrome 3C-like protease. ACS Catal. https://doi.org/10.1021/acscatal.0c00110

52. Chou KC, Wei DQ, Zhong WZ (2003) Binding mechanism of coronavirus main proteinase with ligands and its implication to drug design against SARS. Biochem Biophys Res Commun. https://doi.org/10.1016/S0006-291X(03)01342-1

53. Ruigrok RW, Crépin T, Hart DJ, Cusack S (2010) Towards an atomic resolution understanding of the influenza virus replication machinery. Curr Opin Struct Biol. https://doi.org/10.1016/j.sbi.2009.12.007

54. Yoshino R, Yasuo N, Sekijima M (2020) Identification of key interactions between SARS-CoV-2 main protease and inhibitor drug candidates. Sci Rep. https://doi.org/10.1038/s41598-020-69337-9

55. Ou-Yang SS, Lu JY, Kong XQ, Liang ZJ, Luo C, Jiang H (2012) Computational drug discovery. Acta Pharmacol Sin. https://doi.org/10.1038/aps.2012.109

56. Jorgensen WL (2004) The Many roles of computation in drug discovery. Science 80. https://doi.org/10.1126/science.1096361

57. Kuntz ID (1992) Structure-based strategies for drug design and discovery. Science 80. https://doi.org/10.1126/science.257.5073.1078

58. Buch I, Giorgino T, De Fabritiis G (2011) Complete reconstruction of an enzyme-inhibitor binding process by molecular dynamics simulations. Proc Natl Acad Sci USA. https://doi.org/10.1073/pnas.1103547108

59. Yoshino R, Yasuo N, Sekijima M (2019) Molecular dynamics simulation reveals the mechanism by which the influenza cap-dependent endonuclease acquires resistance against Baloxavir marboxil. Sci Rep. https://doi.org/10.1038/s41598-019-53945-1

60. Kneller DW, Phillips G, O'Neill HM, Jedrzejczak R, Stols L, Langan P, Joachimiak A, Coates L, Kovalevsky A (2020) Structural plasticity of SARS-CoV-2 3CL Mpro active site cavity revealed by room temperature X-ray crystallography. Nat Commun. https://doi.org/10.1038/s41467-020-16954-7

61. Komatsu TS, Okimoto N, Koyama YM, Hirano Y, Morimoto G, Ohno Y, Taiji M (2020) Drug binding dynamics of the dimeric SARS-CoV-2 main protease, determined by molecular dynamics simulation. Sci Rep 10:16986. https://doi.org/10.1038/s41598-020-74099-5

62. Barthels F, Schirmeister T, Kersten C (2021) BANΔIT: B'-factor analysis for drug design and structural biology. Mol Inform 40:2000144. https://doi.org/10.1002/minf.202000144

63. Kusov Y, Tan J, Alvarez E, Enjuanes L, Hilgenfeld R (2015) A G-quadruplex-binding macrodomain within the "SARS-unique domain" is essential for the activity of the SARS-coronavirus replication-transcription complex. Virology. https://doi.org/10.1016/j.virol.2015.06.016

64. Singh N, Halliday AC, Thomas JM, Kuznetsova O, Baldwin R, Woon ECY, Aley PK, Antoniadou I, Sharp T, Vasudevan SR, Churchill GC (2013) A safe lithium mimetic for bipolar disorder. Nat Commun. https://doi.org/10.1038/ncomms2320

65. Kil J, Lobarinas E, Spankovich C, Griffiths SK, Antonelli PJ, Lynch ED, Le Prell CG (2017) Safety and efficacy of ebselen for the prevention of noise-induced hearing loss: a randomised, double-blind, placebo-controlled, phase 2 trial. Lancet. https://doi.org/10.1016/S0140-6736(17)31791-9

66. Lynch E, Kil J (2009) Development of ebselen, a glutathione peroxidase mimic, for the prevention and treatment of noise-induced hearing loss. Semin Hear. https://doi.org/10.1055/s-0028-1111106

67. Masaki C, Sharpley AL, Cooper CM, Godlewska BR, Singh N, Vasudevan SR, Harmer CJ, Churchill GC, Sharp T, Rogers RD, Cowen PJ (2016) Effects of the potential lithium-mimetic, ebselen, on impulsivity and emotional processing. Psychopharmacology. https://doi.org/10.1007/s00213-016-4319-5

68. Jin Z, Zhao Y, Sun Y, Zhang B, Wang H, Wu Y, Zhu Y, Zhu C, Hu T, Du X, Duan Y, Yu J, Yang X, Yang X, Yang K, Liu X, Guddat LW, Xiao G, Zhang L, Yang H, Rao Z (2020) Structural basis for the inhibition of SARS-CoV-2 main protease by antineoplastic drug carmofur. Nat Struct Mol Biol. https://doi.org/10.1038/s41594-020-0440-6

69. Horby PW, Mafham M, Bell JL, Linsell L, Staplin N, Emberson J, Palfreeman A, Raw J, Elmahi E, Prudon B, Green C, Carley S, Chadwick D, Davies M, Wise MP, Baillie JK, Chappell LC, Faust SN, Jaki T, Jefferey K, Lim WS, Montgomery A, Rowan K, Juszczak E, Haynes R, Landray MJ (2020) Lopinavir–ritonavir in patients admitted to hospital with COVID-19 (RECOVERY): a randomised, controlled, open-label, platform trial. Lancet. https://doi.org/10.1016/S0140-6736(20)32013-4

70. Cao B, Wang Y, Wen D, Liu W, Wang J, Fan G, Ruan L, Song B, Cai Y, Wei M, Li X, Xia J, Chen N, Xiang J, Yu T, Bai T, Xie X, Zhang L, Li C, Yuan Y, Chen H, Li H, Huang H, Tu S, Gong F, Liu Y, Wei Y, Dong C, Zhou F, Gu X, Xu J, Liu Z, Zhang Y, Li H, Shang L, Wang K, Li K, Zhou X, Dong X, Qu Z, Lu S, Hu X, Ruan S, Luo S, Wu J, Peng L, Cheng F, Pan L, Zou J, Jia C, Wang J, Liu X, Wang S, Wu X, Ge Q, He J, Zhan H, Qiu F, Guo L, Huang C, Jaki T, Hayden FG, Horby PW, Zhang D, Wang C (2020) A trial of lopinavir-ritonavir in adults hospitalized with severe covid-19. N Engl J Med. https://doi.org/10.1056/nejmoa2001282

71. Lim J, Jeon S, Shin HY, Kim MJ, Seong YM, Lee WJ, Choe KW, Kang YM, Lee B, Park SJ (2020) Case of the index patient who caused tertiary transmission of coronavirus disease 2019 in Korea: the application of lopinavir/ritonavir for the treatment of COVID-19 pneumonia monitored by quantitative RT-PCR. J Korean Med Sci. https://doi.org/10.3346/jkms.2020.35.e79

72. Yan D, Liu X-Y, Zhu Y, Huang L, Dan B, Zhang G, Gao Y (2020) Factors associated with prolonged viral shedding and impact of lopinavir/ritonavir treatment in hospitalised non-critically ill patients with SARS-CoV-2 infection. Eur Respir J 56:2000799. https://doi.org/10.1183/13993003.00799-2020

73. Ye XT, Luo YL, Xia SC, Sun QF, Ding JG, Zhou Y, Chen W, Wang XF, Zhang WW, Du WJ, Ruan ZW, Hong L (2020) Clinical efficacy of lopinavir/ritonavir in the treatment of Coronavirus disease 2019. Eur Rev Med Pharmacol Sci. https://doi.org/10.26355/eurrev_202003_20706

74. Wang X, Guan Y (2021) COVID-19 drug repurposing: a review of computational screening methods, clinical trials, and protein interaction assays. Med Res Rev. https://doi.org/10.1002/med.21728

75. Hung IFN, Lung KC, Tso EYK, Liu R, Chung TWH, Chu MY, Ng YY, Lo J, Chan J, Tam AR, Shum HP, Chan V, Wu AKL, Sin KM, Leung WS, Law WL, Lung DC, Sin S, Yeung P, Yip CCY, Zhang RR, Fung AYF, Yan EYW, Leung KH, Ip JD, Chu AWH, Chan WM, Ng ACK, Lee R, Fung K, Yeung A, Wu TC, Chan JWM, Yan WW, Chan WM, Chan JFW, Lie AKW, Tsang OTY, Cheng VCC, Que TL, Lau CS, Chan KH, To KKW, Yuen KY (2020) Triple combination of interferon beta-1b, lopinavir–ritonavir, and ribavirin in the treatment of patients admitted to hospital with COVID-19: an open-label, randomised, phase 2 trial. Lancet. https://doi.org/10.1016/S0140-6736(20)31042-4

76. Fintelman-Rodrigues N, Sacramento CQ, Lima CR, da Silva FS, Ferreira AC, Mattos M, de Freitas CS, Soares VC, da Silva Gomes Dias S, Temerozo JR, Miranda MD, Matos AR, Bozza FA, Carels N, Alves CR, Siqueira MM, Bozza PT, Souza TML (2020) Atazanavir, alone or in combination with ritonavir, inhibits SARS-CoV-2 replication and proinflammatory cytokine production. Antimicrob Agents Chemother. https://doi.org/10.1128/aac.00825-20

77. Xu Z, Yao H, Shen J, Wu N, Xu Y, Lu X, Zhu W, Li LJ (2020) Nelfinavir is active against SARS-CoV-2 in Vero E6 cells. ChemRxiv. https://doi.org/10.26434/chemrxiv.12039888.v1

78. Yamamoto N, Matsuyama S, Hoshino T, Yamamoto N (2020) Nelfinavir inhibits replication of severe acute respiratory syndrome coronavirus 2 in vitro. BioRxiv. https://doi.org/10.1101/2020.04.06.026476

79. Ohashi H, Watashi K, Saso W, Shionoya K, Iwanami S, Hirokawa T, Shirai T, Kanaya S, Ito Y, Kim KS, Nishioka K, Ando S, Ejima K, Koizumi Y, Tanaka T, Aoki S, Kuramochi K, Suzuki T, Maenaka K, Matano T, Muramatsu M, Saijo M, Aihara K, Iwami S, Takeda M, McKeating JA, Wakita T (2020) Multidrug treatment with nelfinavir and cepharanthine against COVID-19. BioRxiv. https://doi.org/10.1101/2020.04.14.039925
80. Su H, Yao S, Zhao W, Li M, Liu J, Shang W, Xie H, Ke C, Gao M, Yu K, Liu H, Shen J, Tang W, Zhang L, Zuo J, Jiang H, Bai F, Wu Y, Ye Y, Xu Y (2020) Discovery of baicalin and baicalein as novel, natural product inhibitors of SARS-CoV-2 3CL protease. BioRxiv. https://doi.org/10.1101/2020.04.13.038687

Chapter 5
Vaccine Development and Immune Responses in COVID-19: Lessons from the Past

Fataneh Fatemi, Zahra Hassani Nejad, and Seyed Ehsan Ranaei Siadat

Graphical Abstract

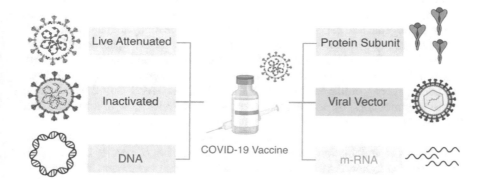

F. Fatemi (✉)
Protein Research Center, Shahid Beheshti University, Tehran, Iran
e-mail: f_fatemi@sbu.ac.ir

Z. Hassani Nejad
Institute of Biochemistry and Biophysics, University of Tehran, Tehran, Iran

S. E. R. Siadat (✉)
Sobhan Recombinant Protein, No. 22, 2nd Noavari St, Pardis Technology Park, 20th Km of
Damavand Road, Tehran, Iran
e-mail: ersiadat@sobhanbiotech.com

5.1 Introduction

Coronavirus Disease 2019 (COVID-19), caused by severe acute respiratory syndrome coronavirus 2 (SARS-CoV-2), was first discovered in Wuhan, China, in late 2019 [1]. As of 24 December 2020, COVID-19 has spread to more than 200 countries and territories and has infected more than 78.7 million people and killed more than 1.7 million people globally and counting [2]. Elderly and those with underlying ailments such as diabetes, hypertension, heart disease, chronic respiratory disease, and cancer are at a higher risk of getting infected and exhibiting symptoms [3]. COVID-19 can be easily transmitted by inhalation of respiratory droplets and human-to-human contact [4]. Hence, most countries have implemented lockdowns and stringent social distancing policies.

Coronaviruses (CoVs) are enveloped, positive-sense single-stranded RNA (ssRNA) viruses that belong to the *Coronaviridae* family [5]. Currently, seven genera of CoVs that can cause infection in humans have been identified. Four of these viruses, including human (h)CoV 229E, hCoV OC43, hCoV NL63, and hCoV HKU1, are all associated with relatively mild symptoms [6]. The remaining three genera of CoVs are highly pathogenic and can sometimes be fatal. The potentially lethal CoVs, severe acute respiratory syndrome (SARS), Middle East respiratory syndrome (MERS), and the newly discovered SARS-CoV-2 have caused or continue to cause severe illness in humans. Most human CoVs cause mild upper respiratory tract infections like common cold, unlike SARS-CoV, MERS-CoV, and SARS-CoV-2 which cause severe pneumonia [7, 8]. However, in comparison to SARS-CoV and MERS-CoV, SARS-CoV-2 has a lower fatality but higher transmission rate [8].

RNA viruses like CoVs evolve by mutations and homologous and non-homologous recombination, which facilitates the crossing of species barriers. It is still unclear as to how SARS-CoV-2 was first transmitted to humans. Its origin can be traced to bats, which is also the original host for other CoV infections in humans [9–11].

The genome and sub-genomes of a typical CoV include at least six open reading frames (ORFs). The first ORFs (ORF1a/b), which is about two-thirds of the whole genome length, encodes 16 NSPs (NSP1-16). The remaining one-third encodes at least four key structural proteins: spike (S) protein, envelope (E) protein, membrane (M) protein, and nucleocapsid (N) protein (Fig. 5.1) [12]. The S protein, which is expressed on the surface of the virus, is required for cell entry and plays a key role in eliciting immune response during disease progression [13]. The trimeric S protein consists of two subunits (S1 and S2), which mediate receptor binding (S1) and membrane fusion (S2). These subunits are further split into different functional domains. The S1 subunit includes a fragment called receptor-binding domain (RBD) that binds to angiotensin-converting enzyme 2 (ACE2), a receptor which is expressed in all organs [14], but primarily found in the lungs [15], brain [16], and gut [17]. After binding, the S protein is cleaved and activated by host transmembrane protease, serine 2 (TMPRSS2), for cell entry [18]. Studies based on the full-length genome phylogenetic analysis show that SARS-CoV-2 shares approximately 79% sequence identity

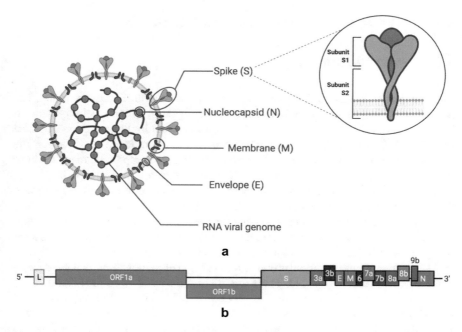

Fig. 5.1 Schematic representation of structure and genome of SARS-CoV-2 **a** SARS-CoV-2 is an enveloped RNA virus with four main structural proteins: spike (S), membrane (M), envelope (E) and nucleocapsid (N) proteins. **b** The single-stranded RNA genome of COVID-19 encodes two large overlapping open reading frames (ORFs) that makeup about two-thirds of the viral genome. These two ORFs encode 16 non-structural proteins (nsp1-nsp16). The 3′-end (the remaining one-third of the viral genome) encodes the four structural proteins (S, M, E and N) and accessory proteins

and has a similar cell entry mechanism and human receptor usage as SARS-CoV [10]. The SARS-CoV-2 genome shares about 50% sequence identity with MERS-CoV [9]. Therefore, previous investigations mostly on SARS-CoV and to some extent on MERS-CoV can provide insights into vaccination strategies for COVID-19.

It is indisputable that the world will not return to normality until an anti-SARS-CoV-2 vaccine is developed. Therefore, a lot of efforts are being put into developing safe and effective vaccines against SARS-CoV-2. Some vaccines are currently in advanced clinical stages, and a few of them have already received emergency authorization in some countries. Here, we summarize the SARS-CoV-2-immune system interaction and provide an overview of previous efforts made on SARS-CoV and MERS-CoV vaccine development. In addition, we will discuss recent efforts made towards COVID-19 vaccine development.

5.2 COVID-19 and Immune System Interaction

5.2.1 *Innate Immunity*

The first line of defense against viral infections is known as the innate immunity. The innate immunity provides an initial, non-specific response, with no memory induced [19]. Due to its novelty, our understanding of the human immune response against SARS-COV-2 is in its infancy and much remains to be understood. However, the virus-host interactions in SARS-CoV-2 are most likely to recapitulate many of those that are seen in previously discovered CoVs due to their resemblance [19].

Upon binding of SARS-CoV-2 and entry to the cells through the ACE2 receptor, the innate immune system gets triggered [13]. The innate immune response is key for targeting and restricting infected cells and for the subsequent activation of the adaptive immune response. The host immune system recognizes the pathogen via pattern-recognition receptors (PRRs) signaling such as toll-like receptors (TLRs), a family of 11 transmembrane receptor proteins that recognize pathogen-associated molecular patterns (PAMPs) [20, 21]. TLRs are important sensor molecules that detect a wide range of microbial pathogens. For RNA viruses like CoVs, the receptors, TLR3 and TLR7, recognize the viral single-stranded RNA (ssRNA) and double-stranded RNA (dsRNA) [22]. After PRR activation, downstream signaling cascade leads to the secretion of proinflammatory cytokines including type I/III interferons (IFNs), interleukin-1 (IL-1), IL-2, IL-6, IL-7, IL-18, and tumor necrosis factor-α (TNF-α), therefore, initiating the defense mechanism against viral infection [19]. Together with the pro-inflammatory cytokines, there will be a strong local inflammatory response that can lead to an influx of neutrophils and other myeloid cells into the lung [19] (Fig. 5.2). This is also seen in SARS-CoV and MERS-CoV infections indicating the importance of cytokine storm and lymphopenia in the COVID-19 pathogenesis [23, 24].

Cytokines induce the infiltration of immune cells to remove infectious viral agents. Most individuals infected with COVID-19 recover from the disease symptoms once the infiltrated immune cells clear the infection. However, dysregulation of proinflammatory cytokines can have detrimental consequences for the host and lead to pathogenesis [19]. Therefore, the activation of the innate immune system must be strictly modulated since excessive activation can cause systemic inflammation and tissue damage [25].

Similar to other CoVs, SARS-CoV-2 has several mechanisms to evade innate immune recognition, such as masking its antigenic epitopes, RNA shielding, and synthesizing viral proteins that hinder anti-viral responses [26, 27]. SARS-like CoVs encodes multiple proteins that antagonize IFN responses and are critical for promoting early viral pathogenesis [28, 29]. Similarly, severe cases of COVID-19 have shown impaired IFN-I/III signaling profile as compared to moderate or healthy cases [30]. Therefore, SARS-CoV-2 promotes prolonged survival via suppressing IFN-I/III signaling, which gives it adequate time to spread inside pneumocytes and alveolar cells [31].

Innate Immunity

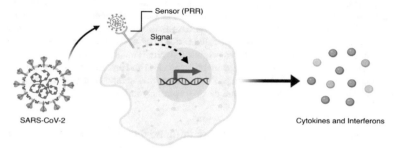

a

Adaptive Immunity

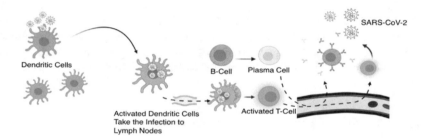

b

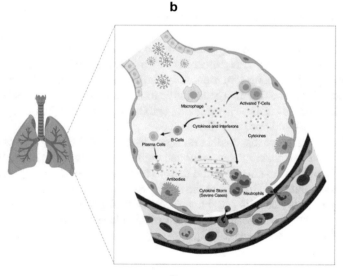

c

◀**Fig. 5.2** Overview of innate immunity (**a**), adaptive immunity (**b**), and cytokine storm (**c**) longs.
a Innate immunity is initiated upon detection of the viral pathogen by the pattern recognition
receptor (PRR). Following viral recognition, the expression of cytokines and interferons is induced.
b Adaptive immunity is triggered when infected dendritic cells travel to lymph nodes to activate the
T and B cells. The T cells and antibodies produced by the B cells attack viruses and virus-infected
cells. **c** After viral infection, immune cells identify the virus and produce cytokines. These cytokines
attract more immune cells which in turn produce more cytokines and lead to a cycle of inflammation
that can eventually damage the lungs and cause respiratory failure

Another strategy to escape the innate immune recognition is the evolution of low
genomic cytosine phosphate guanosine (CpG). Typically, the CpG motifs in genomes
of RNA viruses are targeted and degraded by the zinc-finger antiviral protein (ZAP).
Among the beta-CoVs, SARS-CoV-2 has the most severe CpG deficiency [27, 32].
Another way to protect mRNA is the processing of capping the 5' ends of the viral
RNA. Degradation is decreased by capping since it prevents viral recognition by
cytosolic PPRs. Similar to other CoVs, SARS-CoV-2 uses its own capping machinery
to synthesize 2'-o-methyltransferase caps [33]. These RNA caps are indistinguish-
able from cellular mRNAs caps, thereby they do not get targeted for degradation. All
and all, these mechanisms pave the way for widespread infection of the virus.

5.2.2 Adaptive (Active) Immune Response

5.2.2.1 Cell-Mediated Immune Response

After virus entry and the subsequent activation of the innate immune response, the
adaptive immune response gets triggered to eliminate the virus. Adaptive immunization typically produces long-term immunity due to stimulation of the immune system
by the exposure to an antigen. The adaptive immune system consists of three main
lymphocytes: B cells (antibody-producing cells), CD4+ T cells (helper T cells), and
CD8+ T cells (cytotoxic or killer T cells).

The CD4+ T cells are responsible for regulating CD8+ T cell responses, humoral
immunity, as well as macrophage-mediated antiviral activity. In addition, the CD4+
T cells play a pivotal role in recruiting cells to infection sites. On the other hand, the
CD8 + cells regulate viral infections via lysing the infected cells, secreting cytokines,
and forming memory cells to provide protection if reinfection occurs. Nonetheless,
these viral sensing mechanisms can often over induce the immune response and
eventually lead to tissue damage [25].

5.2.2.2   Antibody-Mediated Immune Response

SARS-CoV-2 triggers a robust B cell response. The activation of B cells triggers antiviral antibody secretion and acts against the virus through a number of different mechanisms, including neutralization, opsonization and activation of complementary proteins [34]. Antibody-mediated immune response plays an essential role against CoV infections. The S protein, specifically the RBD subunit, is the main target for neutralizing antibody (NAb)-mediated inactivation of SARS-CoV-2 by inhibiting it from binding to ACE2 receptors. Therefore, NAbs remain even after infection to prevent the virus from re-infecting the host. In addition, NAbs are responsible for viral clearance during acute phases of the infection and modulate disease progression [35]. Most COVID-19 vaccine efforts focus on eliciting NAbs and CD4+ or CD8+ T cells. Therefore, the main role of a vaccine is to induce both arms of the adaptive immune system and to elicit adequate amounts of T and B cells.

5.3   Vaccine Development

5.3.1   Pre-clinical Studies

Vaccine development is a lengthy process and typically takes many years to develop a safe and effective vaccine. The first step in vaccine development involves basic laboratory research and computational modeling to identify natural and synthetic antigens that can be exploited as a vaccine candidate. Afterwards, the pre-clinical stage is initiated which involves evaluating the safety and the immunogenicity of the vaccine by testing it in various animal models. These studies provide insights into the cellular responses that might be expected in humans. Also, the safest starting dose and methods for administering the vaccine for the next phase of research will be determined through this stage of development.

There are various animal models for preclinical testing of SARS-CoV-2. These animal species have different degrees of susceptibility to SARS-CoV-2 depending on the binding affinity of the virus to the host ACE2 receptor or on host protease activities on the S protein [18]. Cats, ferrets, hamsters, mouse models, and non-human-primate models are all susceptible to SARS-CoV-2 and are usually used as animal models in pre-clinical studies [36]. The non-human primate models, particularly rhesus macaques, exhibit the highest binding affinity for SARS-CoV-2 among the animal models tested [37]. These species show viral shedding from the upper and lower respiratory tract. Nevertheless, their symptoms, clinical signs, and disease severity are different from humans [38, 39].

Mice are also a popular animal model for pathogenesis studies of many viruses. Nonetheless, conventional mice lack appropriate receptors to initiate CoV viral infection. The ACE2 receptor of these organisms does not bind well to the S protein of SARS-CoV-2 [40]. To circumvent this issue, transgenic mice expressing human

ACE2 have been developed. Previously these mice were tested with SARS-CoV and are now being tested with SARS-CoV-2. These mice have been shown to replicate SARS-CoV-2 in the lung and exhibit similar interstitial pneumonia as humans [41]. Another approach involves utilizing mouse-adapted CoV strains that could mimic and induce health conditions identical to human infection [42]. This strategy has been employed for SARS-CoV using conventional mouse strains without the requirement of transgenic mice expressing the human ACE2 receptor and is now being employed for SARS-CoV-2 [43, 44].

5.3.2 Clinical Studies

After assessing the immunogenicity and safety of the vaccine in animal models, progress is made to human clinical studies. The clinical-stage of development consists of at least three phases and progresses sequentially. After the clinical stage, the vaccine proceeds to regulatory approval and licensing. The phase I trial consists of short-term studies in which the vaccine is administered to a small number of healthy adult participants usually between 20 and 80 subjects to assess the safety, reactogenicity and the type of immune response the vaccine may produce. The optimal dose range and the desired route of administration are evaluated. In some cases, the participants are challenged with the pathogens under a controlled environment to find the true effect of the vaccine [45]. The preliminary information on efficacy and immunogenicity is analyzed and if satisfactory results are achieved, the trial advances to the next phase.

Following the completion of phase I trials, the phase II trials are conducted. In phase II, a large-scale (several hundred) of the target population will be tested. These trials are randomized, well-monitored, with a placebo control group. The purpose of this phase of testing is to examine the vaccine's safety, efficacy, and proposed doses.

The final phase in clinical evaluation before product licensing is the phase III trial, in which the vaccine is tested in a larger group of people. The designs of phase II and phase III are alike, but the target population of phase III is much larger. These trials are designed to evaluate the efficacy and safety of the final formulation. The immunogenicity (production of antibodies/cell-mediated immunity) is tested in this stage. After a successful phase III trial, the vaccine will go through licensing, as part of post-marketing surveillance.

In normal circumstances, vaccine development takes approximately 10–15 years [46]. The fastest vaccine development has been for mumps, which took about five years to get approved. The speed with which the COVID-19 vaccine has been developing is faster than conventional vaccines against other diseases and there is an overlapping of clinical trial phases and the whole process is compressed into 12–18 months.

5.4 Vaccine Platform Technologies

A variety of novel vaccine platform technologies have been established over the past decades. These platforms range from inactivated or live attenuated pathogens, protein subunit, nucleic acid-based (RNA or DNA), virus-like particle (VLPs), non-replicating, and replicating viral vectors, with all of them having different advantages and disadvantages (Table 5.1). Most viral vaccines that are currently available for human use are virus or protein-based. The virus-based vaccines employ inactivated or live attenuated viruses. These vaccines are highly effective for contagious diseases. However, their production is time-consuming and complicated. In addition, extensive safety testing is required to ensure that the virus does not revert to its infective state [47]. Protein-based vaccines exhibit more safety and are easier for mass production than whole-virus vaccines. However, their immunogenicity is lower and might need

Table 5.1 Vaccine production platforms and their advantages and limitations

Vaccine platform	Advantage	Disadvantage
Live-attenuated vaccine	High potency; long-lasting immune response; provides 'one shot' immunity	Requires low-temperature storage and transportation (cannot be used in countries with limited access to refrigerators); cannot be used in people with the weak immune system
Inactivated virus vaccine	It offers broad protection; strong immune response; safer than live-attenuated virus vaccine	Low production titer; might require multiple doses or an adjuvant
Viral vector-based vaccine	Induces strong T cell response without requiring an adjuvant; native antigens are preserved	Complicated manufacturing process; cannot be used in people with the suppressed immune system
RNA-based vaccines	Safe and well-tolerated; native antigen expression; low-cost and easy manufacturing	Lower immunogenicity; might require multiple doses; requires low-temperature storage and transportation
DNA-based vaccines	Low-cost and easy manufacturing; highly adaptable to the new pathogen; stable under room temperature	Requires expensive equipment; low immunogenicity; risk of genomic integration
Protein subunit vaccine	Can be used on almost everyone including people with weakened immune systems; does not have the live component of the viral particles; has fewer side-effects	Low immunogenicity; might require booster shots; requirement of adjuvant or conjugate
Virus-like particles (VLPs)	A well-established technology; safe and well-tolerated; native antigen expression; stable under room temperature	Low immunogenicity; might require multiple doses or an adjuvant

multiple doses. Nucleic acid-based vaccines have gotten considerable attention in the new generation vaccines field. One of their advantages is the short time required from the design to clinical trials and their potential for mass-production. But there may be a need for multiple doses to enhance immunity against the virus.

The COVID-19 vaccine development is proceeding at an unprecedented speed. The efforts on vaccine development against COVID-19 started initially in China as soon as the disease was discovered. Since then, many countries have been directing their efforts towards the development of safe and effective vaccines against COVID-19. As of December 22, 2020, the worldwide SARS-CoV-2 vaccine landscape includes 233 vaccine candidates, out of which, 61 are in the clinical stage of development [48].

5.4.1 Live Attenuated Vaccines

Live attenuated and inactivated vaccine technology is one of the most traditional and effective approaches. Live attenuated vaccines provide a robust and long-lasting immune response due to preserving the native antigenic moiety. These vaccines have been commercially available for viruses like influenza, chickenpox, smallpox, polio, measles, and yellow fever virus [49]. Currently, attenuated virus strains are devised via deleting or mutating virulence genes. Thus, eliminating its ability to cause disease in vivo. Deletion of structural E protein [50–53] and non-structural proteins (NSP) [54, 55] has been used to devise vaccines against strains of several CoVs. The E protein is known to trigger inflammasomes and is correlated with intensified inflammation in the lung parenchyma [56]. Therefore, deletion or mutating the E protein can reduce the virulence of CoVs [56]. An alternative approach for the development of attenuated virus is known as codon deoptimization which involves hampering the translation of the viral protein during viral infection [57]. The development of a live attenuated vaccine for CoVs is challenging since they are known to recombine in nature with other CoVs, leading to new pathogenic strains [11]. In addition, live attenuated vaccines can have the potential to return to their virulent state [58]. Therefore, the safety of these vaccines should be thoroughly assessed in animal models before proceeding into the clinical stage. Currently, there is only one COVID-19 live attenuated vaccine in the clinical stage [48]. The COVI-VAC live attenuated vaccine, developed by Codagenix Inc. in collaboration with the serum institute of India is currently in phase I clinical trial (NCT04619628).

5.4.1.1 SARS-CoV and MERS-CoV Live Attenuated Vaccine

A number of effective SARS-CoV and MERS-CoV live attenuated vaccines have been developed and tested in vitro and in vivo. Nonetheless, all of these vaccines have remained in the pre-clinical trials [50–55, 57]. Most CoV live attenuated vaccines are designed through deletion of the E gene. Lamirande et al. developed a recombinant

SARS-CoV (rSARS-CoV) live attenuated vaccine lacking the structural E gene. The rSARS-CoV-Δ E elicited serum-NAbs and completely protected the upper and lower respiratory tract against challenge with homologous (SARS-CoV Urbani) and heterologous (GD03) virus in the Golden Syrian hamster model [50].

The nonstructural protein 16 (nsp16), a conserved 2'O methyltransferase (MTase) that encodes essential functions in immune modulation and infection [54], can be used as an alternative target for attenuation of CoVs. However, targeting the nsp16 gene alone can have the potential for reversion of the virus to its virulence state in aged animal models [59]. To overcome this issue, Menachery and colleagues designed a SARS-CoV vaccine by mutating the nsp16 gene in combination with another conserved attenuating mutation in nsp14, exonuclease (ExoN) activity. They evaluated this vaccine in the mouse model and noticed that combining the 2'O MTase mutation with a second attenuating mutation provides a vaccine strain that offers protection from heterologous virus challenge with no evidence of reversion to virulence state [54]. These results indicate that CoV 2'O MTase in parallel with other conserved attenuating mutations can be a suitable strategy for the production of live attenuated CoV vaccines. The dnsp16 can also serve as a target for live attenuated vaccine development against MERS-CoV. Menachery et al. devised a dnsp16 mutant MERS-CoV strain as an alive attenuated vaccine platform which showed robust protection from challenge with a mouse-adapted MERS-CoV strain [55]. Therefore, the viral 2'O-MTase activity can be a potential universal platform for vaccine development of human and animal CoVs.

5.4.2 Inactivated Vaccines

Inactivated vaccines are made non-infectious through chemical (e.g., formaldehyde) or physical (e.g., heat) methods [60, 61] and are a suitable option given their ability to induce robust immune responses and their feasibility for large-scale production [62]. However, inactivated vaccines' potency is less than that of live attenuated vaccines. Therefore, there may be a need for multiple doses over time to develop ongoing immunity against the disease. Inactivated vaccines have the risk of viral reactivation if improperly inactivated and deformation of immunogenic epitopes during the inactivation process can sometimes weaken the protection that inactivated viruses provide. In addition, inactivated viral vaccine studies have shown that CoVs may induce antibody-dependent enhancement (ADE) effect, recommending more attention when evaluating the safety of vaccines against these viruses [63, 64].

5.4.2.1 SARS-CoV and MERS-CoV Inactivated Vaccine

A few inactivated vaccines have been reported for SARS-CoV. Most of these vaccines induce high levels of specific NAbs in animal models[65-69]. Xiong and colleagues formulated a formaldehyde inactivated SARS-CoV vaccine treated with formaldehyde and supplemented with aluminum hydroxide, $Al(OH)_3$ [65]. Three doses of the vaccine induced specific IgM and IgG antibodies in BALB/c mice on day four and day eight, respectively, with no significant change in $CD4^+$ and $CD8^+$ levels [65]. Luo et al. reported an inactivated SARS-CoV Z-1 vaccine that elicits neutralizing and protective antibody response in rhesus macaques upon challenge. They also examined whether the vaccine could trigger antibody-dependent enhancement (ADE). They found that low levels of antibodies induced by the inactivated SARS-CoV Z-1 vaccine might not induce ADE in rhesus macaques [66]. Takasuka et al. immunized mice with UV-inactivated SARS-CoV with or without an adjuvant (alum) [67]. They noticed that the UV-inactivated SARS-CoV virion induced a high level of humoral immunity and long-term antibody production and memory B cells even in the absence of an adjuvant. However, the serum IgG production was enhanced with the addition of alum to the vaccine [67].

Despite exhibiting potency in eliciting protective antibody responses, some UV and formaldehyde inactivated CoV vaccines that include the N protein are reported to cause eosinophil-related lung pathology in animal models upon SARS-CoV challenge [68, 69]. In addition, adverse effects may arise from SARS-CoV N protein-specific T cells and Th2-skewed cytokine profile [68]. As a result, it is highly crucial to boost the protective S-specific immune response and decrease the potentially pathological anti-N response. In addition, some studies have shown that alum unadjuvanted or adjuvanted inactivated SARS-CoV vaccines provide inadequate protection in mice and induce eosinophilic pro-inflammatory pulmonary response post-viral challenge [68, 70]. Therefore, inactivated vaccines must be thoroughly evaluated prior to clinical studies.

It has been reported that the MERS-CoV vaccine accompanied with alum or MF59 adjuvant can induce NAbs [71]. However, these vaccines can also induce eosinophil-related lung pathology on the challenge with the virus [71]. Inactivated vaccines can still be a suitable platform for CoV vaccine development since the incorporation of suitable inactivation techniques and adjuvants can overcome the bottleneck of lung pathology. Iwata-Yoshikawa and co-workers have shown that UV-inactivated SARS-CoV adjuvanted with TLR agonists could elicit protective antibodies and can reduce eosinophilic responses in the BALB/c mouse model [72]. In addition, inactivated MERS-CoV inactivated with formaldehyde and adjuvanted with a combined alum and unmethylated CpG adjuvant can decrease or prevent pulmonary immunopathology and enhance protective immunity against MERS-CoV in mice post-challenge [73].

5.4.2.2 COVID-19 Inactivated Vaccine

As of now, there are eight vaccine candidates based on inactivated SARS-CoV-2 going through clinical trials. Among these vaccines, the CoronaVacc (also known as PiCoVacc) developed by Sinovac Biotech Ltd in China is the most advanced with published preclinical results [74]. CoronaVacc vaccine is produced in Vero cells and inactivated using β-propiolactone [74]. This vaccine has been tested on non-human primates, rats, and mice, demonstrating potency, safety, and good immunogenicity with vaccine-induced NAbs that neutralize representative strains of SARS-CoV-2 [74]. Data published from pre-clinical trials in macaque and mice models demonstrate that adequate specific IgG response and NAb titer levels were achieved with no notable cytokine changes and ADE in the macaques [74]. In addition, no pathological changes in vaccinated macaques' vital organs were observed after the SARS-CoV-2 challenge [74]. CoronaVacc has completed its phase I/II clinical trial and is currently in phase III of the clinical trial [48]. Results from phase I/II show that the vaccine is well-tolerated and induces NAbs with a seroconversion rate of 90% [75].

The SARS-CoV-2 inactivated vaccine developed by Sinopharm Inc. in collaboration with Wuhan Institute of Biological Products works through propagating WIV04 strain from different COVID-19 patients in Vero cells doubly inactivated with two rounds of β-propiolactone exposure [76]. In their phase I and II studies, different dosage and injection timelines were tested. All the vaccinated patients who received different vaccination regimens had produced NAbs with little to no adverse effects in their phase I/II studies [76]. In addition, Sinopharm Inc. is developing another COVID-19 vaccine called BBIBP-CorV, with $Al(OH)_3$ as an adjuvant [77]. This vaccine has a similar manufacturing process as the other Sinopharm Inc. vaccine except that the HB02 strain is used instead of the WIV04 strain. The BBIBP-CorV has been tested in pre-clinical models and has demonstrated efficacy in non-human primates when immunized with two doses of BBIBP-CorV with no disease enhancement upon SARS-CoV-2 challenge [77]. The results from their I/II clinical trial demonstrate the vaccine's efficacy, tolerability, and good humoral response in all participants 42 days post-immunization in all tested doses [78]. Both of these Sinopharm Inc. inactivated vaccines are currently in Phase III clinical trial.

Three vaccine manufacturing and academic/research institutions in Iran are in the process of developing inactivated COVID-19 vaccines. The alum adjuvanted inactivated vaccine developed by Shifa-Pharmed, a part of the state-owned pharmaceutical conglomerate, has recently started its clinical trial in Iran. In the Phase I clinical trial, they enrolled 56 participants, each receiving two shots of the vaccine within a period of two weeks. This company has not disclosed any of its pre-clinical results, and its Phase I clinical results are to be announced roughly a month after the second shot [79].

5.4.3   Viral Vectored Vaccines

Viral vector vaccines consist of a recombinant virus (often attenuated) in which genes encoding antigens of interest (usually S gene for CoVs) have been cloned using recombinant DNA techniques. The viral vector elicits antigen-specific humoral and cell-mediated immune responses through antigen presentation [80]. The most commonly employed non-replicating vectors are the adenovirus (Ad), measles virus, and vesicular stomatitis virus vectors [81]. These vectors mimic natural viral infection and elicit the production of the target viral proteins inside host cells. The main advantage of vector-based vaccines is their ability to induce both humoral and innate immune responses [82]. In addition, vector-based vaccines provide stronger cellular immune responses compared to the recombinant protein vaccines. One of the main drawbacks of using recombinant viruses is the possibility of genome integration with the host cell genome. Moreover, the manufacturing process of viral vector vaccines is very complex and includes optimizing cellular systems and eliminating contaminants that can potentially impede the efficiency of viral vectors [81].

5.4.3.1 SARS-CoV and MERS-CoV Viral Vectored Vaccines

Some studies have assessed the efficacy of the adenovirus-based SARS-CoV vaccine. Gao et al. and Liu et al. have shown that Ad vector expressing S1 can induce potent NAbs responses in rhesus macaques and rats [83, 84]. However, these experiments were done in vitro and whether these vaccines provide protection against SARS-CoV challenge in vivo is still questionable. The potency of Ad vaccine expressing the SARS-CoV S protein has been compared with the whole inactivated SARS-CoV vaccine by See and co-workers. It was found that these vaccines provide immunity in mice when challenged with SARS-CoV. Nevertheless, the NAb response is stronger in the whole-inactivated virus vaccine than the adenovirus-based vaccine [85].

A number of Ad-based MERS-CoV vaccines have been developed. MERS-CoV vaccines based on human Ad type 5 and type 41 (Ad41) expressing S or S1 protein have demonstrated the induction of NAb in mice [86, 87]. An example is the Ad5-MERS-S vaccine which works with S protein nanoparticles. This vaccine provides protection in hDPP4-transduced mice against viral challenge. In addition, heterologous immunization with Ad5/MERS prime and S protein nanoparticles boost has demonstrated enhanced Th1/Th2 responses than the Ad5- or nanoparticles-alone homologous prime-boost vaccines [88].

5.4.3.2 COVID-19 Viral Vectored Vaccines

Currently, 18 viral vectored COVID-19 vaccines are undergoing clinical trials. Out of these vaccines, the ChAdOx1nCoV-19 and now designated AZD1222 developed by Oxford University in collaboration with AstraZeneca is the most clinically advanced viral-vector based COVID-19 vaccine. This vaccine consists of a replication-deficient chimpanzee Ad vector ChAdOx1, containing the S glycoprotein gene. The AZD1222 is designed via the deletion of E1 and E3 genes. The deletion of E1 inhibits replication and deletion of E3 allows integration of larger genetic cargo into the viral vector [89, 90]. The AZD1222 has demonstrated high NAb levels in 91% of participants following the first dose. Participants receiving booster dose had a high NAb response, therefore, indicating the need for a two-dose regimen to enhance the NAb response [89]. Phase I/II results demonstrate its acceptable safety profile with no serious adverse events with induction of binding and NAb antibodies as well as production of interferon-γ enzyme-linked immunospot responses with higher antibody titers after the second dose of vaccine [89, 91]. In addition, no severe adverse effects were seen during this phase. [89] The interim analysis from the AZD1222 vaccine's phase I/II clinical trial showed that AZD1222 has an efficacy of about 70% [92].

Another viral vector-based vaccine is the Ad5-nCoV co-developed by CanSino Biological Inc. and the Beijing Institution of Biotechnology. The ad5-nCoV is designed similarly to the AZD1222. The full-length S gene along with the plasminogen activator signal peptide gene is cloned into the Ad5 vector missing the E1 and E3 genes. Deletion of E1 inactivates the replication potential of the vaccine, and deletion of E3 allows for the addition of large genes (up to 8 kb) [90]. Phase-III safety studies of this vaccine have shown success with both groups of vaccinated participants developing NAb responses in 47–59% of the volunteers and seroconversion of binding antibodies in 96–97% of them [93].

5.4.4   MRNA Vaccines

Over the past decade, there has been a lot of effort and investment in enabling mRNA to become a promising therapeutic tool for vaccine development and for cancer prophylaxis and therapy [94]. In this approach, the antigen-encoding mRNA is complexed with a carrier that can efficiently deliver it to the cytoplasm of host cells for protein translation and post-translation modification [95]. RNA vaccines utilize lipid- and polymer-based nanoparticles or protamine, for increased efficacy [96]. These vaccines are synthesized in vitro transcription and are non-infectious, making them ideal for rapid and inexpensive production. In addition, unlike DNA vaccines, RNA vaccines are able to synthesize viral proteins without the risk of integration with the host cell genome. Also, DNA vaccines require special devices for administration, whereas RNA vaccines can be administered through various ways including intravenous injection. Nonetheless, RNA vaccines do have their own demerits such as having low immunogenicity and instability concerns.

No RNA vaccine studies for SARS-CoV or MERS-CoV have been previously reported. Nonetheless, eight RNA vaccines for SARS-CoV-2 are currently in the clinical stage with two of them receiving emergency authorization in the US. The first mRNA-based vaccine for COVID-19 to get approved is the BNT162b1 vaccine developed by BioNTech company in collaboration with the Pfizer company. This vaccine exploits a lipid nanoparticle (LNP)-formulated nucleoside-modified mRNA that encodes the RBD of SARS-CoV-2 S protein [97]. The mRNA in this vaccine is modified with single nucleoside incorporations of 1-methylpseudouridine, which reduces the immunogenicity of the mRNA in vivo and enhances its translation [98]. The results from phase I/II showed that the BNT162b1 elicited RBD-binding and NAbs. This vaccine has been reported to trigger the production of T helper type 1 (1_H1)-skewed T cell immune responses with RBD-specific CD8$^+$ and CD4$^+$ T cells and robust release of immune-modulatory cytokines such as IFN$_y$, [97, 99] which is essential for several antiviral responses and inhibits replication of SARS-CoV-2

[99]. On 18, November 2020, it was announced that the BNT162b2 vaccine exhibits more than 95% effectiveness in preventing the disease in participants of 16 years or older [100]. This result was based on examining a total of 178 confirmed COVID-19 cases, out of which 162 cases were in the placebo group and the remaining were in the BNT162b2 group [100]. Also, nine severe COVID-19 cases were in the placebo group with one of them being in the BNT162b2 group. No severe adverse effects were detected among the 43,000 enrolled participants [100].

Another leading mRNA vaccine is the mRNA-1273, co-developed by researchers at the National Institute of Allergy and Infectious Diseases (NIAID) and Moderna (Cambridge, MA). The mRNA-1273 vaccine recently received FDA approval for emergency use. Moderna started its clinical testing just two months after sequence identification of SARS-CoV-2. This vaccine is made out of synthetic mRNA encapsulated in LNPs that encodes for the full-length, pre-fusion stabilized S protein of SARS-CoV-2. Two proline substitutions in the vaccine mRNA at amino acids 986 and 987, located in the central helix of the S2 subunit, keep the protein in its prefusion conformation [101]. The mRNA is also modified to increase the half-life and its translation, as well as to prevent the activation of interferon-associated genes upon cell entry [102].

Their phase I clinical trial report showed that NAbs were detected in 45 participants after two doses of immunization. In addition, antibody titers were higher than convalescent serum after two doses of vaccination in immunized individuals. There were some systemic adverse effects such as headache, fatigue, myalgia, chills, and pain after the second dose of vaccination, particularly with the highest dose. However, no grade 4 adverse effects were reported [102, 103]. Based on these results, they concluded that 100 μg dosage can lead to acceptable immune response. In addition, the vaccine was tested in elderly participants (55 or older) and their results showed that 100 μg doses can lead to higher binding and NAb titers as compared to 25 μg dose, and the adverse effects were moderate in these elderly participants. On 16 November 2020, the results from their phase III trial were reported. Out of 95 participants who had symptomatic COVID-19, five were from the vaccinated group and the remaining participants were from the placebo group. The vaccine efficacy was estimated to be 94.5% without any significant safety concerns [104].

Both Moderna's and Pfizer-BioNTech vaccines have shown efficacy levels near 95%. Unlike the Pfizer vaccine which needs to be kept at −75 °C, Moderna's vaccine does require really cold temperatures. Moderna's vaccine can be kept at about −20 °C and can be kept in a refrigerator for 30 days before it expires. Therefore, Pfizer's vaccine can be more suitable for places with established infrastructure like hospitals.

5.4.5 DNA-Based Vaccine

DNA-based vaccines have a lot of potential due to their ability to induce humoral and cell-mediated immune responses, low-cost manufacturing, and their long shelf life, which makes them suitable for use in endemic areas [105, 106]. DNA vaccines consist of genes or fragments of viral antigens that are transferred to the host cells via DNA plasmid vectors. Once the genetic material is translocated to the host cell's nucleus, the transcription of the gene is triggered. The antigen-presenting cells (APCs) are the main target cells to acquire the genetic material [107]. One advantage of DNA-based vaccines is that the native conformation and post-translational modification will be recapitulated since the antigens are produced in the target cells. One of the disadvantages associated with this type of vaccine is its low immunogenicity and that the DNA molecule must be able to first cross the nuclear membrane barrier and then get transcribed. Therefore, enhancing the efficacy of these vaccines by using an adjuvant or a multiple shot might be required. In addition, a major safety concern is the integration of DNA vaccines with the host DNA, which may cause mutagenesis and oncogenesis [81].

5.4.5.1   SARS-CoV and MERS-CoV DNA Vaccines

A number of DNA-based vaccines for SARS-CoV have been developed [108–111]. All of these vaccines have shown to produce antibody and cell-immune responses. Among the S, E, and N antigens, the S protein-based SARS-CoV DNA vaccine has shown to induce a protective immune response. Yang and colleagues have reported that a DNA vaccine encoding full-length S protein can induce T cell and NAb responses in the mouse model. In addition, alternative forms of the S protein were analyzed by DNA immunization. All of these vaccines were able to elicit strong immune responses mediated by the CD4+ and CD8+ cells. [108]. Furthermore, the expression vector encoding a form of S that includes its transmembrane domain induced NAb production [108]. Prime-boost strategies can be employed to augment the strength of the S protein-based SARS-CoV DNA vaccine. For instance, DNA vaccine augmented with recombinant S protein booster has shown to induce higher NAbs titers than DNA or protein subunit vaccine alone [112]. In addition, combining DNA and whole-inactivated SARS-CoV vaccines can increase the antibody response and induce a better Th1-skewed immune response [113].

A few MERS-CoV DNA vaccines have been developed. Muthmani and co-workers have devised a full-length S protein-based MERS-CoV DNA vaccine, GLS-5300 or INO-4700 that can induce strong cellular immunity and NAbs in mice, macaques, and camels. Vaccinated macaques were immune against MERS-CoV challenge without exhibiting histopathological or radiological evidence of pneumonia [114]. Therefore, a phase I study based on the GLS-5300 vaccine was done to assess its safety and immunogenicity in humans. The vaccine showed robust immunogenicity in 85% of participants after two vaccinations. In addition, the vaccine was well tolerated with no serious adverse events were reported [115].

In addition to full-length S protein DNA vaccines, the S1 subunit can also serve as a suitable target for DNA vaccine development against MERS-CoV. In a study done by Al-Amri and co-workers, the immunogenicity of full-length S-based and S1-based MERS-CoV vaccines was compared by using the same expression vector. It was found that plasmids expressing full-length pS1 immunization elicited a balanced

Th1/Th2 response and higher levels of all IgG isotypes compared to pS vaccination. Based on these results, it can be concluded that vaccines expressing S1-subunit of the MERS-CoV could be a more suitable target than full-length S protein [116].

5.4.5.2   COVID-19 DNA Vaccines

Thus far, five SARS-CoV-2 DNA vaccines are under clinical trials. The most clinically advanced SARS-CoV-2 DNA vaccine is the Invivio's vaccine, INO-4800, which has published results on MERS-CoV and SARS-CoV-2 DNA vaccines. This vaccine employs a plasmid pGX9501 designed to encode the full-length SARS-CoV-2 S protein and is administered intradermally via electroporation of the skin by a device called CELLECTRA [117, 118]. Pre-clinical studies of this vaccine have shown that it can induce NAb that inhibits the binding of SARS-CoV-2 S protein to the ACE2 receptor and elicits Th1-skewed immune responses in animal models [117, 119].

5.4.6   Protein Subunit Vaccine

Protein subunit vaccines are based on recombinant antigenic proteins or synthetic polysaccharides [120]. These recombinant proteins are synthesized in various expression systems, including insect cells and mammalian cells (CHO cells), yeast, or plants [121–123]. They are easy to manufacture and safer compared to some viral vector vaccines and inactivated or live-attenuated virus vaccines that include infectious components of the virus. Protein-subunit vaccines provide a strong immune response targeted towards key parts of the virus without including any virulent components of the virus [120]. Therefore, eliminating concerns of virulence recovery or pre-existing immunity [124]. One of the limitations associated with these vaccines is their low immunogenicity. Thus, an adjuvant or a booster shot may be required to potentiate the vaccine-induced immune response and to enhance the immunomodulatory cytokine response [125].

Among the SARS-CoV-2 structural proteins, the prime target for subunit vaccines is the S protein especially RBD, S1, and S2 as in the case of MERS and SARS vaccines. This is because the S protein is regarded as the most suitable antigen to induce the production of NAbs. The S protein of CoVs, particularly the RBD, is known to induce NAbs and T cell immune responses [126–128]. Therefore, RBD is a promising target for COVID-19 vaccine development, and previous investigations from using RBD-base vaccines for SARS-CoV and MERS-CoV can provide information based on the design of RBD-based vaccines against SARS-CoV-2. The S protein is a dynamic protein and has two conformational states: pre-fusion state and post-fusion state. In order to trigger good quality antibody responses, the antigen must maintain its surface chemistry of the original pre-fusion spike protein [129]. It is known that recombinant S protein vaccines could have improper epitope confirmation unless mammalian cells are utilized for their production [130].

5.4.6.1   SARS-CoV and MERS-CoV Protein Subunit Vaccine

None of the SARS-CoV protein subunit vaccines have proceeded to the clinical phase despite showing potent antibody responses and protective immunity against infection in animal models. Previous studies indicate that vaccines based on the full-length S protein, trimeric S protein, and its antigenic fragments including S1, RBD, and S2 subunit can all provide protection against SARS-CoV. He et al. showed that immunized mice with full-length recombinant S protein or its extracellular domain vaccines develop increased titers of anti-S antibodies with strong NAbs activities against SARS-CoV [131]. Even though full-length S protein vaccines have the ability to induce strong immune responses, some in vitro studies have found that immunization with SARS-CoV full-length S protein can cause antibody-mediated enhancement viral infection, raising safety concerns for the development of these types of vaccines against CoVs [132, 133].

The S protein RBD-based vaccines have shown high-titer NAbs with no apparent adverse effects [134–136]. In addition, RBD-based vaccines can induce S-specific antibodies that can last for almost a year [134] and can induce the production of RBD-specific IFN-γ which induce cellular immune responses in mice [135]. Therefore, the RBD serves as a promising vaccine target for inducing NAbs against viral infection. Guo et al. investigated the immune responses against the S2 domain of SARS-CoV in BALB/c mice and noticed that the S2 domain could elicit specific cellular and humoral responses with little NAb against infection by SARS-CoV [137].

Besides the S protein, the N and M proteins have been also utilized as the target antigen in developing subunit vaccines against SARS-CoV [123, 138]. Liu et al. have shown that the N protein of SARS-CoV is immunogenic in the mouse and macaque models. They devised a recombinant N (rN) protein vaccine formulated with ISA/CpG adjuvants. This vaccine-elicited potent Th1 immune responses and the immunodominant B-and T-cell epitopes of the rN protein were present in both mice and macaques [138]. Zheng and colleagues formulated a plant-expressed SARS-CoV rN protein-based vaccine. This vaccine also induced potent humoral and cellular responses in mice [123]. The N-based subunit vaccines have also shown their efficacy in eliciting specific antibody responses. However, the protective efficacy of non-S protein-based SARS-CoV vaccines is still unclear [139, 140].

Most protein subunit vaccines against MERS-CoV are based on the RBD of the S protein. The MERS-CoV RBD-based vaccines have shown potent immunogenicity and elicited strong neutralizing antibodies, cell-mediated immunity, and protective effect against MERS-CoV challenge [127, 141, 142]. Lan et al. evaluated a recombinant RBD protein vaccine in the rhesus macaque model and noticed robust immunological responses including the production of NAbs after rRBD vaccination. In addition, the rRBD vaccine reduced tissue impairment and clinical side-effects in monkeys [127]. Tai and co-workers have reported that the RBD of MERS-CoV in its native trimeric form can induce strong RBD-specific NAb in mice against challenge [141]. In addition, the recombinant RBD from various MERS-CoV strains can elicit NAbs in animals that cross-neutralize with different human and animal MERS-CoV strains [142]. Therefore, the RBD domain is a promising target for protein subunit vaccines against CoVs.

5.4.6.2 COVID-19 Protein Subunit Vaccine

Similar to SARS-CoV and MERS-CoV, COVID-19 protein subunit vaccines account for most of the vaccines that are currently under development. Most COVID-19 protein subunit vaccines contain full length or portions of the SARS-CoV-2 S protein. Currently, there are 18 COVID-19 subunit vaccines in clinical trials. An example is the NVX-CoV2373 vaccine developed by Novavax. NVX-CoV2373 is a nanoparticle-based immunogenic vaccine based on recombinant expression of the CoV trimeric full-length S protein stabilized in the prefusion conformation [143]. This recombinant protein is optimized in the baculovirus (Sf9) insect cell-expression system. During the pre-clinical studies, it was shown that low-dose NVX-CoV2373 supplemented with the Matrix-M1™ adjuvant induces NAbs and high levels of anti-S protein antibodies which block the hACE2 receptor-binding domain in mice and non-human primate [144]. The vaccine also elicits $CD4^+$ and $CD8^+$ T cells, $CD4^+$ T helper cells and induced the production of antigen-specific germinal center (GC) B cells in the spleen [144]. More importantly, vaccinated non-human primates had little to zero viral shedding in either upper or lower respiratory tracks [145]. In Phase I-II trial, the vaccine-induced binding and NAbs in all participants. In addition, compared to the placebo group and the unadjuvanted 25 μg dose group, both adjuvanted regimens induced higher levels of NAb titers [146]. Currently, the NVX-CoV2373 vaccine is being evaluated in the Phase III trial.

5.5   Other Vaccine Platforms

5.5.1   Virus-like Particles (VLPs)

Virus-like particles also known as VLPs are composed of some key structural viral components that are either admixed or co-expressed in a manner that resembles the conformation of the native virus. VLPs are non-infectious and non-replicating due to the lack of genetic materials [147]. Compared to protein subunit vaccines, VLP vaccines have better immunization responses. Also, the manufacturing process of VLP vaccines is simpler and more convenient than inactivated or attenuated vaccines since the inactivation step is skipped and there is no need for the live virus. Currently available VLP-based vaccines marketed for human use are against human papillomavirus and hepatitis B virus [148].

VLPs for CoVs are formed when the viral protein S, M, and E, with or without N, are co-expressed in eukaryotic cells [149]. The N protein encapsulates the viral genome into virions and does not have an essential role in SARS-CoV-2 VLPs assembly. The S protein present on the surface of the VLPs allows them to bind and fuse into ACE2+ cells similar to the native virus and elicits immune response [150]. Similar to subunit and inactivated viral vaccines, VLPs usually require multiple doses and an adjuvant because of their poor immunogenicity.

5.5.1.1 SARS-CoV and MERS-CoV VLP Vaccines

VLP vaccines have shown satisfactory results in eliciting both humoral and cellular immunity in SARS-CoV and MERS-CoV preclinical studies. Lokugamage et al. devised chimeric VLPs composed of the S protein of SARS-CoV and mouse hepatitis virus E, M and N proteins that elicit the production of NAb responses and reduce SARS-CoV virus shedding in mice lung [151]. In addition, no apparent lung pathology in the chimeric-VLP-treated mice was reported when compared to the negative control mice [151]. Another study done on chimeric Sf9 cell-based VLPs consisting of SARS-CoV full-length S protein along with M1 protein of influenza virus expressed in the baculovirus insect cells has shown that chimeric VLPS can induce NAbs and provide protection against challenge in mice [152]. However, possible adverse effects of CoV VLP vaccines should be examined carefully. For instance, Tseng et al. employed the same chimeric VLPs as Lokugamage et al. and observed pulmonary immunopathology post SARS-CoV challenge [69]. Wang and colleagues have constructed recombinant baculovirus co-expressing the S, E, and M genes of MERS-CoV. The assembled VLPs were able to elicit robust antibody response and Th1-mediated immunity in rhesus macaques and can serve as a promising vaccine candidate [153]. In addition, they devised a chimeric canine parvovirus (CPV) VLP expressing the RBD of MERS-CoV self-assembled into chimeric spherical VLP (sVLP). The sVLP vaccine was shown to induces RBD-specific antibody response and T-cell immunity in mice [154].

5.5.1.2 COVID-19 VLP Vaccines

Currently, only two VLP-based COVID-19 vaccines are in the clinical trial. None of them has publicly reported their vaccine studies until now. One of them is a plant-derived VLP called CoVLP vaccine developed by Medicago, a Canadian pharmaceutical company. The CoVLP vaccine is composed of recombinant S protein expressed as VLPs. Medicago exploits living plants as bioreactors to produce VLPs. A synthetic gene containing a fragment of SARS-CoV-2 genes is transfected to a species of tobacco by means of a bacterial vector. The expressed VLPs can then be purified by various purification techniques. The Phase I results showed that after receiving two doses of Medicago's COVID-19 adjuvanted vaccine, all the participants developed NAbs antibody responses [155]. This vaccine has been tested with two separate adjuvants: GSK's propriety pandemic adjuvant technology and Dyanavax's CpG 1018^{TM}. These adjuvants improved humoral and cellular responses compared to the non-adjuvanted formulation. Medicago's COVID-19 vaccine candidate is currently in Phase II/III clinical trials and the company has estimated to hold a production capacity of 10 million doses a month [156]. The second VLP vaccine is being developed by the Serum Institute of India, which recently entered phase I/II trial in Australia.

5.5.2 Bacillus Calmette Guerin (BCG) Vaccine

Although several COVID-19 vaccines are under clinical trials with some of them being in advanced stages, scientists have been also testing existing licensed vaccines to fight COVID-19. One example is the Bacillus Calmette Guerin (BCG) vaccine, which contains an attenuated strain of the bovine tubercle bacillus *Mycobacterium bovis*, which is an old age vaccine used for the prevention of tuberculosis. Currently, about 100 million children are vaccinated annually worldwide. However, the vaccine does not exhibit satisfactory results for the adult pulmonary tuberculosis but provides a broad protection against other diseases [157]. This BCG vaccine has the ability to exert a potent nonspecific immunity (off-target protection) against viral and bacterial infections [158]. The increase in immunogenicity against pathogens induced by this vaccine is due to heterologous effects on adaptive immunity, such as T cell-mediated cross-reactivity, as well as trained immunity [159].

To see whether BCG-induced immunity could influence the adaptive immune response against SARS-CoV-2, Urbán et al. conducted a study in which the T-cell and B-cell epitopes of the BCG strain Pasteur 1173P2 were compared with T-cell and B-cell epitopes of SARS-CoV-2 to find similar epitopes that might induce adaptive cross-immunity and to explain the protective qualities of BCG vaccination against SARS-CoV-2. They discovered shared MHC-I restricted T-cell epitopes and B-cell epitopes between SARS-CoV-2 and BCG-Pasteur which might induce cross-immunity. Their results suggest that BCG can be a potential preventive immunotherapy against COVID-19 and to enhance innate immunity [160].

BCG vaccine has shown to offer protection against COVID-19 and reduced mortality in countries with a routine BCG vaccination program, even when the BCG vaccination was performed during childhood [161]. For instance, in South Asian regions where the BCG vaccine is administered at birth, delayed or less infection rate and less deaths due to COVID-19 infections have been reported [162]. A few other studies have also suggested that there is a correlation between BCG and the SARS-CoV-2 epidemic [163, 164], while others have disbelieved its relation to COVID-19 mortality and morbidity [165]. Therefore, whether the BCG vaccine is effective against COVID-19 is highly debatable and further elucidation on the use of the BCG vaccine as a preventive therapy against SARS-CoV-2 infection is still required.

5.6 Conclusion

Coronaviruses are expected to continue to cross species barriers and cause severe illness in humans. Undoubtedly COVID-19 will not be the last pandemic of the century due to changes in climate and the increased interactions of humans with animals. Therefore, the development of novel and effective technology platforms is required to expedite vaccine development. With record numbers of human causalities and confirmed cases being reported daily, an anti-SARS-CoV-2 vaccine is imperative. While vaccines against SARS-CoV-2 are being developed at a fast pace, the special nature of this virus and safety concerns make the development of these vaccines very challenging. Some individuals get infected with SARS-CoV-2 while remaining asymptomatic, whereas some exhibit severe illness and succumb to the disease. Given the variability of host immune responses to SARS-CoV-2, there is no guarantee that vaccination could provide uniform long-lasting immunity in whoever gets vaccinated. However, with the increasing accumulation of knowledge about SARS-CoV-2 and massive efforts from the scientific community to develop anti-SARS-CoV-2 vaccines, the COVID-19 pandemic will come to an end eventually.

References

1. Zhu N, Zhang D, Wang W, Li X, Yang B, Song J, Zhao X, Huang B, Shi W, Lu R, Niu P, Zhan F, Ma X, Wang D, Xu W, Wu G, Gao GF, Tan W (2019) A novel coronavirus from patients with pneumonia in China. N Engl J Med 382(2020):727–733. https://doi.org/10.1056/nejmoa 2001017
2. WHO Coronavirus Disease (COVID-19) Dashboard (n.d.)
3. Bale BF, Doneen AL, Vigerust DJ (2020) Microvascular disease confers additional risk to COVID-19 infection. Med Hypotheses 144: https://doi.org/10.1016/j.mehy.2020.109999
4. Medicine TLR (2020) COVID-19 transmission-up in the air. Lancet Respir Med 8:1159. https://doi.org/10.1016/S2213-2600(20)30514-2
5. Perlman S, Netland J (2009) Coronaviruses post-SARS: update on replication and pathogenesis. Nat Rev Microbiol 7:439–450. https://doi.org/10.1038/nrmicro2147
6. van der Hoek L (2007) Human coronaviruses: what do they cause? Antivir Ther 12:651–658
7. Chen G, Wu D, Guo W, Cao Y, Huang D, Wang H, Wang T, Zhang X, Chen H, Yu H, Zhang X, Zhang M, Wu S, Song J, Chen T, Han M, Li S, Luo X, Zhao J, Ning Q (2020) Clinical and immunological features of severe and moderate coronavirus disease 2019. J Clin Invest 130:2620–2629. https://doi.org/10.1172/JCI137244
8. Petrosillo N, Viceconte G, Ergonul O, Ippolito G, Petersen E (2020) COVID-19, SARS and MERS: are they closely related? Clin Microbiol Infect 26:729–734. https://doi.org/10.1016/j.cmi.2020.03.026
9. Lu R, Zhao X, Li J, Niu P, Yang B, Wu H, Wang W, Song H, Huang B, Zhu N, Bi Y, Ma X, Zhan F, Wang L, Hu T, Zhou H, Hu Z, Zhou W, Zhao L, Chen J, Meng Y, Wang J, Lin Y, Yuan J, Xie Z, Ma J, Liu WJ, Wang D, Xu W, Holmes EC, Gao GF, Wu G, Chen W, Shi W, Tan W (2020) Genomic characterisation and epidemiology of 2019 novel coronavirus: implications for virus origins and receptor binding. Lancet (London, England) 395:565–574. https://doi.org/10.1016/S0140-6736(20)30251-8
10. Zhou P, Yang X-L, Wang X-G, Hu B, Zhang L, Zhang W, Si H-R, Zhu Y, Li B, Huang C-L, Chen H-D, Chen J, Luo Y, Guo H, Jiang R-D, Liu M-Q, Chen Y, Shen X-R, Wang X, Zheng

X-S, Zhao K, Chen Q-J, Deng F, Liu L-L, Yan B, Zhan F-X, Wang Y-Y, Xiao G-F, Shi Z-L (2020) A pneumonia outbreak associated with a new coronavirus of probable bat origin. Nature 579:270–273. https://doi.org/10.1038/s41586-020-2012-7

11. Tao Y, Shi M, Chommanard C, Queen K, Zhang J (2017) Surveillance of bat coronaviruses in Kenya identifies relatives of human coronaviruses NL63 and 229E and their recombination history. J Virol 91:e01953–16. https://doi.org/10.1128/JVI.01953-16

12. Chen Y, Liu Q, Guo D (2020) Emerging coronaviruses: genome structure, replication, and pathogenesis. J Med Virol 92:418–423. https://doi.org/10.1002/jmv.25681

13. Hofmann H, Hattermann K, Marzi A, Gramberg T, Geier M, Krumbiegel M, Kuate S, Uberla K, Niedrig M, Pöhlmann S (2004) S protein of severe acute respiratory syndrome-associated coronavirus mediates entry into hepatoma cell lines and is targeted by neutralizing antibodies in infected patients. J Virol 78:6134–6142. https://doi.org/10.1128/JVI.78.12.6134-6142.2004

14. Zhang Y, Geng X, Tan Y, Li Q, Xu C, Xu J, Hao L, Zeng Z, Luo X, Liu F, Wang H (2020) New understanding of the damage of SARS-CoV-2 infection outside the respiratory system. Biomed Pharmacother 127: https://doi.org/10.1016/j.biopha.2020.110195

15. Kuba K, Imai Y, Penninger JM (2006) Angiotensin-converting enzyme 2 in lung diseases. Curr Opin Pharmacol 6:271–276. https://doi.org/10.1016/j.coph.2006.03.001

16. Xia H, Lazartigues E (2008) Angiotensin-converting enzyme 2 in the brain: properties and future directions. J Neurochem 107:1482–1494. https://doi.org/10.1111/j.1471-4159.2008.05723.x

17. Wang J, Zhao S, Liu M, Zhao Z, Xu Y, Wang P, Lin M, Xu Y, Huang B, Zuo X, Chen Z, Bai F, Cui J, Lew AM, Zhao J, Zhang Y, Luo H-B, Zhang Y (2020) ACE2 expression by colonic epithelial cells is associated with viral infection, immunity and energy metabolism. MedRxiv. https://doi.org/10.1101/2020.02.05.20020545

18. Hoffmann M, Kleine-Weber H, Schroeder S, Krüger N, Herrler T, Erichsen S, Schiergens TS, Herrler G, Wu N-H, Nitsche A, Müller MA, Drosten C, Pöhlmann S (2020) SARS-CoV-2 cell entry depends on ACE2 and TMPRSS2 and is blocked by a clinically proven protease inhibitor. Cell 181:271–280.e8. https://doi.org/10.1016/j.cell.2020.02.052

19. Vabret N, Britton GJ, Gruber C, Hegde S, Kim J, Kuksin M, Levantovsky R, Malle L, Moreira A, Park MD, Pia L, Risson E, Saffern M, Salomé B, Esai Selvan M, Spindler MP, Tan J, van der Heide V, Gregory JK, Alexandropoulos K, Bhardwaj N, Brown BD, Greenbaum B, Gümüş ZH, Homann D, Horowitz A, Kamphorst AO, Curotto de Lafaille MA, Mehandru S, Merad M, Samstein RM, Project SIR (2020) Immunology of COVID-19: current state of the science. Immunity 52:910–941. https://doi.org/10.1016/j.immuni.2020.05.002

20. Thompson MR, Kaminski JJ, Kurt-Jones EA, Fitzgerald KA (2011) Pattern recognition receptors and the innate immune response to viral infection. Viruses 3:920–940. https://doi.org/10.3390/v3060920

21. Fitzgerald KA, Kagan JC (2020) Toll-like receptors and the control of immunity. Cell 180:1044–1066. https://doi.org/10.1016/j.cell.2020.02.041

22. Lester SN, Li K (2014) Toll-like receptors in antiviral innate immunity. J Mol Biol 426:1246–1264. https://doi.org/10.1016/j.jmb.2013.11.024

23. Mahallawi WH, Khabour OF, Zhang Q, Makhdoum HM, Suliman BA (2018) MERS-CoV infection in humans is associated with a pro-inflammatory Th1 and Th17 cytokine profile. Cytokine 104:8–13. https://doi.org/10.1016/j.cyto.2018.01.025

24. Wong CK, Lam CWK, Wu AKL, Ip WK, Lee NLS, Chan IHS, Lit LCW, Hui DSC, Chan MHM, Chung SSC, Sung JJY (2004) Plasma inflammatory cytokines and chemokines in severe acute respiratory syndrome. Clin Exp Immunol 136:95–103. https://doi.org/10.1111/j.1365-2249.2004.02415.x

25. Blanco-Melo D, Nilsson-Payant BE, Liu W-C, Uhl S, Hoagland D, Møller R, Jordan TX, Oishi K, Panis M, Sachs D, Wang TT, Schwartz RE, Lim JK, Albrecht RA, tenOever BR (2020) Imbalanced host response to SARS-CoV-2 drives development of COVID-19. Cell 181:1036–1045.e9. https://doi.org/10.1016/j.cell.2020.04.026

26. Walls AC, Tortorici MA, Frenz B, Snijder J, Li W, Rey FA, DiMaio F, Bosch B-J, Veesler D (2016) Glycan shield and epitope masking of a coronavirus spike protein observed by cryo-electron microscopy. Nat Struct Mol Biol 23:899–905. https://doi.org/10.1038/nsmb.3293

27. Amor S, Fernández Blanco L, Baker D (2020) Innate immunity during SARS-CoV-2: evasion strategies and activation trigger hypoxia and vascular damage. Clin Exp Immunol 202:193–209. https://doi.org/10.1111/cei.13523

28. Békés M, Rut W, Kasperkiewicz P, Mulder MPC, Ovaa H, Drag M, Lima CD, Huang TT (2015) SARS hCoV papain-like protease is a unique Lys48 linkage-specific di-distributive deubiquitinating enzyme. Biochem J 468:215–226. https://doi.org/10.1042/BJ20141170

29. Mielech AM, Kilianski A, Baez-Santos YM, Mesecar AD, Baker SC (2014) MERS-CoV papain-like protease has deISGylating and deubiquitinating activities. Virology 450–451:64–70. https://doi.org/10.1016/j.virol.2013.11.040

30. Hadjadj J, Yatim N, Barnabei L, Corneau A, Boussier J, Pere H, Charbit B, Bondet V, Chenevier-Gobeaux C, Breillat P, Carlier N, Gauzit R, Morbieu C, Pene F, Marin N, Roche N, Szwebel T-A, Smith N, Merkling S, Treluyer J-M, Veyer D, Mouthon L, Blanc C, Tharaux P-L, Rozenberg F, Fischer A, Duffy D, Rieux-Laucat F, Kerneis S, Terrier B (2020) Impaired type I interferon activity and exacerbated inflammatory responses in severe Covid-19 patients. MedRxiv. https://doi.org/10.1101/2020.04.19.20068015

31. Park A, Iwasaki A (2020) Type I and type III interferons—induction, signaling, evasion, and application to combat COVID-19. Cell Host Microbe 27:870–878. https://doi.org/10.1016/j.chom.2020.05.008

32. Xia X (2020) Extreme genomic CpG deficiency in SARS-CoV-2 and evasion of host antiviral defense. Mol Biol Evol 37:2699–2705. https://doi.org/10.1093/molbev/msaa094

33. Decroly E, Debarnot C, Ferron F, Bouvet M, Coutard B, Imbert I, Gluais L, Papageorgiou N, Sharff A, Bricogne G, Ortiz-Lombardia M, Lescar J, Canard B (2011) Crystal structure and functional analysis of the SARS-coronavirus RNA cap 2'-O-methyltransferase nsp10/nsp16 complex. PLoS Pathog 7:e1002059–e1002059. https://doi.org/10.1371/journal.ppat.1002059

34. Ganji A, Farahani I, Khansarinejad B, Ghazavi A, Mosayebi G (2020) Increased expression of CD8 marker on T-cells in COVID-19 patients. Blood Cells Mol Dis 83: https://doi.org/10.1016/j.bcmd.2020.102437

35. Weisel F, Shlomchik M (2017) Memory B cells of mice and humans. Annu Rev Immunol 35:255–284. https://doi.org/10.1146/annurev-immunol-041015-055531

36. Muñoz-Fontela C, Dowling WE, Funnell SGP, Gsell P-S, Riveros-Balta AX, Albrecht RA, Andersen H, Baric RS, Carroll MW, Cavaleri M, Qin C, Crozier I, Dallmeier K, de Waal L, de Wit E, Delang L, Dohm E, Duprex WP, Falzarano D, Finch CL, Frieman MB, Graham BS, Gralinski LE, Guilfoyle K, Haagmans BL, Hamilton GA, Hartman AL, Herfst S, Kaptein SJF, Klimstra WB, Knezevic I, Krause PR, Kuhn JH, Le Grand R, Lewis MG, Liu W-C, Maisonnasse P, McElroy AK, Munster V, Oreshkova N, Rasmussen AL, Rocha-Pereira J, Rockx B, Rodríguez E, Rogers TF, Salguero FJ, Schotsaert M, Stittelaar KJ, Thibaut HJ, Tseng C-T, Vergara-Alert J, Beer M, Brasel T, Chan JFW, García-Sastre A, Neyts J, Perlman S, Reed DS, Richt JA, Roy CJ, Segalés J, Vasan SS, Henao-Restrepo AM, Barouch DH (2020) Animal models for COVID-19. Nature 586:509–515. https://doi.org/10.1038/s41586-020-2787-6

37. Wan Y, Shang J, Graham R, Baric RS, Li F (2020) Receptor recognition by the novel coronavirus from Wuhan: an analysis based on decade-long structural studies of SARS coronavirus. J Virol 94:e00127–20. https://doi.org/10.1128/JVI.00127-20

38. Rockx B, Kuiken T, Herfst S, Bestebroer T, Lamers MM, Oude Munnink BB, de Meulder D, van Amerongen G, van den Brand J, Okba NMA, Schipper D, van Run P, Leijten L, Sikkema R, Verschoor E, Verstrepen B, Bogers W, Langermans J, Drosten C, Fentener van Vlissingen M, Fouchier R, de Swart R, Koopmans M, Haagmans BL (2020) Comparative pathogenesis of COVID-19, MERS, and SARS in a nonhuman primate model. Science 80:368. 1012 LP – 1015. https://doi.org/10.1126/science.abb7314

39. Munster VJ, Feldmann F, Williamson BN, van Doremalen N, Pérez-Pérez L, Schulz J, Meade-White K, Okumura A, Callison J, Brumbaugh B, Avanzato VA, Rosenke R, Hanley PW, Saturday G, Scott D, Fischer ER, de Wit E (2020) Respiratory disease in rhesus macaques inoculated with SARS-CoV-2. Nature 585:268–272. https://doi.org/10.1038/s41 586-020-2324-7

40. Vrancken B, Zhao B, Li X, Han X, Liu H, Zhao J, Zhong P, Lin Y, Zai J, Liu M, Smith DM, Dellicour S, Chaillon A (2020) Comparative circulation dynamics of the five main HIV types in China. J Virol 94:e00683–20. https://doi.org/10.1128/JVI.00683-20

41. Bao L, Deng W, Huang B, Gao H, Liu J, Ren L, Wei Q, Yu P, Xu Y, Qi F, Qu Y, Li F, Lv Q, Wang W, Xue J, Gong S, Liu M, Wang G, Wang S, Song Z, Zhao L, Liu P, Zhao L, Ye F, Wang H, Zhou W, Zhu N, Zhen W, Yu H, Zhang X, Guo L, Chen L, Wang C, Wang Y, Wang X, Xiao Y, Sun Q, Liu H, Zhu F, Ma C, Yan L, Yang M, Han J, Xu W, Tan W, Peng X, Jin Q, Wu G, Qin C (2020) The pathogenicity of SARS-CoV-2 in hACE2 transgenic mice. Nature 583:830–833. https://doi.org/10.1038/s41586-020-2312-y

42. Regla-Nava JA, Nieto-Torres JL, Jimenez-Guardeño JM, Fernandez-Delgado R, Fett C, Castaño-Rodríguez C, Perlman S, Enjuanes L, DeDiego ML (2015) Severe acute respiratory syndrome coronaviruses with mutations in the E protein are attenuated and promising vaccine candidates. J Virol 89:3870–3887. https://doi.org/10.1128/JVI.03566-14

43. Day CW, Baric R, Cai SX, Frieman M, Kumaki Y, Morrey JD, Smee DF, Barnard DL (2009) A new mouse-adapted strain of SARS-CoV as a lethal model for evaluating antiviral agents in vitro and in vivo. Virology 395:210–222

44. Gu H, Chen Q, Yang G, He L, Fan H, Deng Y-Q, Wang Y, Teng Y, Zhao Z, Cui Y, Li Y, Li X-F, Li J, Zhang N-N, Yang X, Chen S, Guo Y, Zhao G, Wang X, Luo D-Y, Wang H, Yang X, Li Y, Han G, He Y, Zhou X, Geng S, Sheng X, Jiang S, Sun S, Qin C-F, Zhou Y (2020) Adaptation of SARS-CoV-2 in BALB/c mice for testing vaccine efficacy. Science 80:369, 1603 LP–1607. https://doi.org/10.1126/science.abc4730

45. Dutta AK (2020) Vaccine against covid-19 disease—present status of development. Indian J Pediatr 87:810–816. https://doi.org/10.1007/s12098-020-03475-w

46. Han S (2015) Clinical vaccine development. Clin Exp Vaccine Res 4:46–53. https://doi.org/10.7774/cevr.2015.4.1.46

47. van Riel D, de Wit E (2020) Next-generation vaccine platforms for COVID-19. Nat Mater 19:810–812. https://doi.org/10.1038/s41563-020-0746-0

48. Draft landscape of COVID-19 candidate vaccines (2020)

49. Minor PD (2015) Live attenuated vaccines: historical successes and current challenges. Virology 479–480:379–392. https://doi.org/10.1016/j.virol.2015.03.032

50. Lamirande EW, DeDiego ML, Roberts A, Jackson JP, Alvarez E, Sheahan T, Shieh W-J, Zaki SR, Baric R, Enjuanes L, Subbarao K (2008) A live attenuated severe acute respiratory syndrome coronavirus is immunogenic and efficacious in golden Syrian hamsters. J Virol 82:7721–7724. https://doi.org/10.1128/JVI.00304-08

51. Netland J, DeDiego ML, Zhao J, Fett C, Álvarez E, Nieto-Torres JL, Enjuanes L, Perlman S (2010) Immunization with an attenuated severe acute respiratory syndrome coronavirus deleted in E protein protects against lethal respiratory disease. Virology 399:120–128. https://doi.org/10.1016/j.virol.2010.01.004

52. Fett C, DeDiego ML, Regla-Nava JA, Enjuanes L, Perlman S (2013) Complete protection against severe acute respiratory syndrome coronavirus-mediated lethal respiratory disease in aged mice by immunization with a mouse-adapted virus lacking E protein. J Virol 87:6551–6559. https://doi.org/10.1128/JVI.00087-13

53. DeDiego ML, Alvarez E, Almazán F, Rejas MT, Lamirande E, Roberts A, Shieh W-J, Zaki SR, Subbarao K, Enjuanes L (2007) A severe acute respiratory syndrome coronavirus that lacks the E gene is attenuated in vitro and in vivo. J Virol 81:1701–1713. https://doi.org/10.1128/JVI.01467-06

54. Menachery VD, Gralinski LE, Mitchell HD, Dinnon KH 3rd, Leist SR, Yount BL Jr, McAnarney ET, Graham RL, Waters KM, Baric RS (2018) Combination attenuation offers strategy for live attenuated coronavirus vaccines. J Virol 92:e00710–18. https://doi.org/10.1128/JVI.00710-18

55. Menachery VD, Gralinski LE, Mitchell HD, Dinnon KH 3rd, Leist SR, Yount BL Jr, Graham RL, McAnarney ET, Stratton KG, Cockrell AS, Debbink K, Sims AC, Waters KM, Baric RS (2017) Middle East respiratory syndrome coronavirus nonstructural protein 16 Is necessary for interferon resistance and viral pathogenesis. MSphere 2:e00346–17. https://doi.org/10.1128/mSphere.00346-17

56. DeDiego ML, Nieto-Torres JL, Jimenez-Guardeño JM, Regla-Nava JA, Castaño-Rodriguez C, Fernandez-Delgado R, Usera F, Enjuanes L (2014) Coronavirus virulence genes with main focus on SARS-CoV envelope gene. Virus Res 194:124–137. https://doi.org/10.1016/j.virusres.2014.07.024

57. Mueller S, Stauft CB, Kalkeri R, Koidei F, Kushnir A, Tasker S, Coleman JR (2020) A codon-pair deoptimized live-attenuated vaccine against respiratory syncytial virus is immunogenic and efficacious in non-human primates. Vaccine 38:2943–2948. https://doi.org/10.1016/j.vaccine.2020.02.056

58. Vignuzzi M, Wendt E, Andino R (2008) Engineering attenuated virus vaccines by controlling replication fidelity. Nat Med 14:154–161. https://doi.org/10.1038/nm1726

59. Jimenez-Guardeño JM, Regla-Nava JA, Nieto-Torres JL, DeDiego ML, Castaño-Rodriguez C, Fernandez-Delgado R, Perlman S, Enjuanes L (2015) Identification of the mechanisms causing reversion to virulence in an attenuated SARS-CoV for the design of a genetically stable vaccine. PLoS Pathog 11:e1005215–e1005215. https://doi.org/10.1371/journal.ppat.1005215

60. Roper RL, Rehm KE (2009) SARS vaccines: where are we? Expert Rev Vaccines 8:887–898. https://doi.org/10.1586/erv.09.43

61. Cryz SJ, Fürer E, Germanier R (1982) Effect of chemical and heat inactivation on the antigenicity and immunogenicity of Vibrio cholerae. Infect Immun 38:21 LP–26

62. Tsunetsugu-Yokota Y (2008) Large-scale preparation of UV-inactivated SARS coronavirus virions for vaccine antigen. Methods Mol Biol 454:119–126. https://doi.org/10.1007/978-1-59745-181-9_11

63. Wang Q, Zhang L, Kuwahara K, Li L, Liu Z, Li T, Zhu H, Liu J, Xu Y, Xie J, Morioka H, Sakaguchi N, Qin C, Liu G (2016) Immunodominant SARS coronavirus epitopes in humans elicited both enhancing and neutralizing effects on infection in non-human primates. ACS Infect Dis 2:361–376. https://doi.org/10.1021/acsinfecdis.6b00006

64. Yang ZY, Werner HC, Kong WP, Leung K, Traggiai E, Lanzavecchia A, Nabel GJ (2005) Evasion of antibody neutralization in emerging severe acute respiratory syndrome coronaviruses. Proc Natl Acad Sci USA 102:797–801. https://doi.org/10.1073/pnas.0409065102

65. Xiong S, Wang Y-F, Zhang M-Y, Liu X-J, Zhang C-H, Liu S-S, Qian C-W, Li J-X, Lu J-H, Wan Z-Y, Zheng H-Y, Yan X-G, Meng M-J, Fan J (2004) Immunogenicity of SARS inactivated vaccine in BALB/c mice. Immunol Lett 95:139–143. https://doi.org/10.1016/j.imlet.2004.06.014

66. Luo F, Liao F-L, Wang H, Tang H-B, Yang Z-Q, Hou W (2018) Evaluation of antibody-dependent enhancement of SARS-CoV infection in rhesus macaques immunized with an inactivated SARS-CoV vaccine. Virol Sin 33:201–204. https://doi.org/10.1007/s12250-018-0009-2

67. Takasuka N, Fujii H, Takahashi Y, Kasai M, Morikawa S, Itamura S, Ishii K, Sakaguchi M, Ohnishi K, Ohshima M, Hashimoto S, Odagiri T, Tashiro M, Yoshikura H, Takemori T, Tsunetsugu-Yokota Y (2004) A subcutaneously injected UV-inactivated SARS coronavirus vaccine elicits systemic humoral immunity in mice. Int Immunol 16:1423–1430. https://doi.org/10.1093/intimm/dxh143

68. Bolles M, Deming D, Long K, Agnihothram S, Whitmore A, Ferris M, Funkhouser W, Gralinski L, Totura A, Heise M, Baric RS (2011) A double-inactivated severe acute respiratory syndrome coronavirus vaccine provides incomplete protection in mice and induces increased eosinophilic proinflammatory pulmonary response upon challenge. J Virol 85:12201–12215. https://doi.org/10.1128/JVI.06048-11

69. Tseng C-T, Sbrana E, Iwata-Yoshikawa N, Newman PC, Garron T, Atmar RL, Peters CJ, Couch RB (2012) Immunization with SARS coronavirus vaccines leads to pulmonary immunopathology on challenge with the SARS virus. PLoS ONE 7:e35421–e35421. https://doi.org/10.1371/journal.pone.0035421

70. Yasui F, Kai C, Kitabatake M, Inoue S, Yoneda M, Yokochi S, Kase R, Sekiguchi S, Morita K, Hishima T, Suzuki H, Karamatsu K, Yasutomi Y, Shida H, Kidokoro M, Mizuno K, Matsushima K, Kohara M (2008) Prior immunization with severe acute respiratory syndrome (SARS)-Associated Coronavirus (SARS-CoV) nucleocapsid protein causes severe pneumonia in mice infected with SARS-CoV. J Immunol 181:6337 LP–6348. https://doi.org/10.4049/jimmunol.181.9.6337

71. Agrawal AS, Tao X, Algaissi A, Garron T, Narayanan K, Peng B-H, Couch RB, Tseng C-TK (2016) Immunization with inactivated Middle East respiratory syndrome coronavirus vaccine leads to lung immunopathology on challenge with live virus. Hum Vaccin Immunother 12:2351–2356. https://doi.org/10.1080/21645515.2016.1177688

72. Iwata-Yoshikawa N, Uda A, Suzuki T, Tsunetsugu-Yokota Y, Sato Y, Morikawa S, Tashiro M, Sata T, Hasegawa H, Nagata N (2014) Effects of Toll-like receptor stimulation on eosinophilic infiltration in lungs of BALB/c mice immunized with UV-inactivated severe acute respiratory syndrome-related coronavirus vaccine. J Virol 88:8597–8614. https://doi.org/10.1128/JVI.00983-14

73. Deng Y, Lan J, Bao L, Huang B, Ye F, Chen Y, Yao Y, Wang W, Qin C, Tan W (2018) Enhanced protection in mice induced by immunization with inactivated whole viruses compare to spike protein of Middle East respiratory syndrome coronavirus. Emerg Microbes Infect 7:60. https://doi.org/10.1038/s41426-018-0056-7

74. Gao Q, Bao L, Mao H, Wang L, Xu K, Yang M, Li Y, Zhu L, Wang N, Lv Z, Gao H, Ge X, Kan B, Hu Y, Liu J, Cai F, Jiang D, Yin Y, Qin C, Li J, Gong X, Lou X, Shi W, Wu D, Zhang H, Zhu L, Deng W, Li Y, Lu J, Li C, Wang X, Yin W, Zhang Y, Qin C (2020) Development of an inactivated vaccine candidate for SARS-CoV-2. Science 369:77–81. https://doi.org/10.1126/science.abc1932

75. Zhang Y, Zeng G, Pan H, Li C, Hu Y, Chu K, Han W, Chen Z, Tang R, Yin W, Chen X, Hu Y, Liu X, Jiang C, Li J, Yang M, Song Y, Wang X, Gao Q, Zhu F (2020) Safety, tolerability, and immunogenicity of an inactivated SARS-CoV-2 vaccine in healthy adults aged 18-59 years: a randomised, double-blind, placebo-controlled, phase 1/2 clinical trial. Lancet Infect Dis. https://doi.org/10.1016/S1473-3099(20)30843-4

76. Xia S, Duan K, Zhang Y, Zhao D, Zhang H, Xie Z, Li X, Peng C, Zhang Y, Zhang W, Yang Y, Chen W, Gao X, You W, Wang X, Wang Z, Shi Z, Wang Y, Yang X, Zhang L, Huang L, Wang Q, Lu J, Yang Y, Guo J, Zhou W, Wan X, Wu C, Wang W, Huang S, Du J, Meng Z, Pan A, Yuan Z, Shen S, Guo W, Yang X (2020) Effect of an inactivated vaccine against SARS-CoV-2 on safety and immunogenicity outcomes: interim analysis of 2 randomized clinical trials. JAMA 324:951–960. https://doi.org/10.1001/jama.2020.15543

77. Wang H, Zhang Y, Huang B, Deng W, Quan Y, Wang W (2020) Development of an inactivated vaccine candidate, BBIBP-CorV, with potent protection against SARS- CoV-2. Cell 182:713–721. https://doi.org/10.1016/j.cell.2020.06.008

78. Xia S, Zhang Y, Wang Y, Wang H, Yang Y, Gao GF, Tan W, Wu G, Xu M, Lou Z, Huang W, Xu W, Huang B, Wang H, Wang W, Zhang W, Li N, Xie Z, Ding L, You W, Zhao Y, Yang X, Liu Y, Wang Q, Huang L, Yang Y, Xu G, Luo B, Wang W, Liu P, Guo W, Yang X (2020) Safety and immunogenicity of an inactivated SARS-CoV-2 vaccine, BBIBP-CorV: a randomised, double-blind, placebo-controlled, phase 1/2 trial. Lancet Infect Dis 21:39–51. https://doi.org/10.1016/S1473-3099(20)30831-8

79. A double-blinded, randomized, placebo-controlled Phase I Clinical trial to evaluate the safety and immunogenicity of COVID-19 inactivated vaccine (Shif-Pharmed) in a healthy population (2020)

80. Bouard D, Alazard-Dany D, Cosset F-L (2009) Viral vectors: from virology to transgene expression. Br J Pharmacol 157:153–165. https://doi.org/10.1038/bjp.2008.349

81. Rauch S, Jasny E, Schmidt KE, Petsch B (2018) New vaccine technologies to combat outbreak situations. Front Immunol 9:1963. https://doi.org/10.3389/fimmu.2018.01963

82. Afrough B, Dowall S, Hewson R (2019) Emerging viruses and current strategies for vaccine intervention. Clin Exp Immunol 196:157–166. https://doi.org/10.1111/cei.13295

83. Gao W, Tamin A, Soloff A, D'Aiuto L, Nwanegbo E, Robbins PD, Bellini WJ, Barratt-Boyes S, Gambotto A (2003) Effects of a SARS-associated coronavirus vaccine in monkeys. Lancet (London, England) 362:1895–1896. https://doi.org/10.1016/S0140-6736(03)14962-8

84. Liu R-Y, Wu L-Z, Huang B-J, Huang J-L, Zhang Y-L, Ke M-L, Wang J-M, Tan W-P, Zhang R-H, Chen H-K, Zeng Y-X, Huang W (2005) Adenoviral expression of a truncated S1 subunit of SARS-CoV spike protein results in specific humoral immune responses against SARS-CoV in rats. Virus Res 112:24–31. https://doi.org/10.1016/j.virusres.2005.02.009

85. See RH, Petric M, Lawrence DJ, Mok CPY, Rowe T, Zitzow LA, Karunakaran KP, Voss TG, Brunham RC, Gauldie J, Finlay BB, Roper RL (2008) Severe acute respiratory syndrome vaccine efficacy in ferrets: whole killed virus and adenovirus-vectored vaccines. J Gen Virol 89:2136–2146. https://doi.org/10.1099/vir.0.2008/001891-0

86. Kim E, Okada K, Kenniston T, Raj VS, AlHajri MM, Farag EABA, AlHajri F, Osterhaus ADME, Haagmans BL, Gambotto A (2014) Immunogenicity of an adenoviral-based Middle East Respiratory Syndrome coronavirus vaccine in BALB/c mice. Vaccine 32:5975–5982. https://doi.org/10.1016/j.vaccine.2014.08.058

87. Guo X, Deng Y, Chen H, Lan J, Wang W, Zou X, Hung T, Lu Z, Tan W (2015) Systemic and mucosal immunity in mice elicited by a single immunization with human adenovirus type 5 or 41 vector-based vaccines carrying the spike protein of Middle East respiratory syndrome coronavirus. Immunology 145:476–484. https://doi.org/10.1111/imm.12462

88. Jung S-Y, Kang KW, Lee E-Y, Seo D-W, Kim H-L, Kim H, Kwon T, Park H-L, Kim H, Lee S-M, Nam J-H (2018) Heterologous prime-boost vaccination with adenoviral vector and protein nanoparticles induces both Th1 and Th2 responses against Middle East respiratory syndrome coronavirus. Vaccine 36:3468–3476. https://doi.org/10.1016/j.vaccine.2018.04.082

89. Folegatti PM, Ewer KJ, Aley PK, Angus B, Becker S, Belij-Rammerstorfer S, Bellamy D, Bibi S, Bittaye M, Clutterbuck EA, Dold C, Faust SN, Finn A, Flaxman AL, Hallis B, Heath P, Jenkin D, Lazarus R, Makinson R, Minassian AM, Pollock KM, Ramasamy M, Robinson H, Snape M, Tarrant R, Voysey M, Green C, Douglas AD, Hill AVS, Lambe T, Gilbert SC, Pollard AJ, Group OCVT (2020) Safety and immunogenicity of the ChAdOx1 nCoV-19 vaccine against SARS-CoV-2: a preliminary report of a phase 1/2, single-blind, randomised controlled trial. Lancet (London, England) 396:467–478. https://doi.org/10.1016/S0140-6736(20)31604-4

90. Dicks MDJ, Spencer AJ, Edwards NJ, Wadell G, Bojang K, Gilbert SC, Hill AVS, Cottingham MG (2012) A novel chimpanzee adenovirus vector with low human seroprevalence: improved systems for vector derivation and comparative immunogenicity. PLoS ONE 7:e40385–e40385. https://doi.org/10.1371/journal.pone.0040385

91. Ramasamy MN, Minassian AM, Ewer KJ, Flaxman AL, Folegatti PM, Owens DR, Voysey M, Aley PK, Angus B, Babbage G, Belij-Rammerstorfer S, Berry L, Bibi S, Bittaye M, Cathie K, Chappell H, Charlton S, Cicconi P, Clutterbuck EA, Colin-Jones R, Dold C, Emary KRW, Fedosyuk S, Fuskova M, Gbesemete D, Green C, Hallis B, Hou MM, Jenkin D, Joe CCD, Kelly EJ, Kerridge S, Lawrie AM, Lelliott A, Lwin MN, Makinson R, Marchevsky NG, Mujadidi Y, Munro APS, Pacurar APS, Plested E, Rand J, Rawlinson T, Rhead S, Robinson H, Ritchie AJ, Ross-Russell AL, Saich S, Singh N, Smith CC, Snape MD, Song R, Tarrant R, Themistocleous Y, Thomas KM, Villafana TL, Warren SC, Watson MEE, Douglas AD, Hill AVS, Lambe T, Gilbert SC, Faust SN, Pollard AJ, Group OCVT (2020) Safety and immunogenicity of ChAdOx1 nCoV-19 vaccine administered in a prime-boost regimen in young and old adults (COV002): a single-blind, randomised, controlled, phase 2/3 trial. Lancet (London, England) S0140-6736(20)32466–1. https://doi.org/10.1016/S0140-6736(20)32466-1

92. Voysey M, Clemens SAC, Madhi SA, Weckx LY, Folegatti PM, Aley PK, Angus B, Baillie VL, Barnabas SL, Bhorat QE, Bibi S, Briner C, Cicconi P, Collins AM, Colin-Jones R, Cutland CL, Darton TC, Dheda K, Duncan CJA, Emary KRW, Ewer KJ, Fairlie L, Faust SN, Feng

S, Ferreira DM, Finn A, Goodman AL, Green CM, Green CA, Heath PT, Hill C, Hill H, Hirsch I, Hodgson SHC, Izu A, Jackson S, Jenkin D, Joe CCD, Kerridge S, Koen A, Kwatra G, Lazarus R, Lawrie AM, Lelliott A, Libri V, Lillie PJ, Mallory R, Mendes AVA, Milan EP, Minassian AM, McGregor A, Morrison H, Mujadidi YF, Nana A, O'Reilly PJ, Padayachee SD, Pittella A, Plested E, Pollock KM, Ramasamy MN, Rhead S, Schwarzbold AV, Singh N, Smith A, Song R, Snape MD, Sprinz E, Sutherland RK, Tarrant R, Thomson EC, Török ME, Toshner M, Turner DPJ, Vekemans J, Villafana TL, Watson MEE, Williams CJ, Douglas AD, Hill AVS, Lambe T, Gilbert SC, Pollard AJ, Aban M, Abayomi F, Abeyskera K, Aboagye J, Adam M, Adams K, Adamson J, Adelaja YA, Adlou S, Ahmed K, Akhalwaya Y, Akhalwaya S, Alcock A, Ali A, Allen ER, Allen L, Almeida TCDSC, Alves MPS, Amorim F, Andritsou F, Anslow R, Appleby M, Arbe-Barnes EH, Ariaans MP, Arns B, Arruda L, Awedetan G, Azi P, Azi L, Babbage G, Bailey C, Baker KF, Baker M, Baker N, Baker P, Baldwin L, Baleanu I, Bandeira D, Bara A, Barbosa MAS, Barker D, Barlow GD, Barnes E, Barr AS, Barrett JR, Barrett J, Bates L, Batten A, Beadon K, Beales E, Beckley R, Belij-Rammerstorfer S, Bell J, Bellamy D, Bellei N, Belton S, Berg A, Bermejo L, Berrie E, Berry L, Berzenyi D, Beveridge A, Bewley KR, Bexhell H, Bhikha S, Bhorat AE, Bhorat ZE, Bijker E, Birch G, Birch S, Bird A, Bird O, Bisnauthsing K, Bittaye M, Blackstone K, Blackwell L, Bletchly H, Blundell CL, Blundell SR, Bodalia P, Boettger BC, Bolam E, Boland E, Bormans D, Borthwick N, Bowring F, Boyd A, Bradley P, Brenner T, Brown P, Brown C, Brown-O'Sullivan C, Bruce S, Brunt E, Buchan R, Budd W, Bulbulia YA, Bull M, Burbage J, Burhan H, Burn A, Buttigieg KR, Byard N, Cabera Puig I, Calderon G, Calvert A, Camara S, Cao M, Cappuccini F, Cardoso JR, Carr M, Carroll MW, Carson-Stevens A, de M Carvalho Y, Carvalho JAM, Casey HR, Cashen P, Castro T, Castro LC, Cathie K, Cavey A, Cerbino-Neto J, Chadwick J, Chapman D, Charlton S, Chelysheva I, Chester O, Chita S, Cho J-S, Cifuentes L, Clark E, Clark M, Clarke A, Clutterbuck EA, Collins SLK, Conlon CP, Connarty S, Coombes N, Cooper C, Cooper R, Cornelissen L, Corrah T, Cosgrove C, Cox T, Crocker WEM, Crosbie S, Cullen L, Cullen D, Cunha DRMF, Cunningham C, Cuthbertson FC, Da Guarda SNF, da Silva LP, Damratoski BE, Danos Z, Dantas MTDC, Darroch P, Datoo MS, Datta C, Davids M, Davies SL, Davies H, Davis E, Davis J, Davis J, De Nobrega MMD, De Oliveira Kalid LM, Dearlove D, Demissie T, Desai A, Di Marco S, Di Maso C, Dinelli MIS, Dinesh T, Docksey C, Dold C, Dong T, Donnellan FR, Dos Santos T, dos Santos TG, Dos Santos EP, Douglas N, Downing C, Drake J, Drake-Brockman R, Driver K, Drury R, Dunachie SJ, Durham BS, Dutra L, Easom NJW, van Eck S, Edwards M, Edwards NJ, El Muhanna OM, Elias SC, Elmore M, English M, Esmail A, Essack YM, Farmer E, Farooq M, Farrar M, Farrugia L, Faulkner B, Fedosyuk S, Felle S, Feng S, Ferreira Da Silva C, Field S, Fisher R, Flaxman A, Fletcher J, Fofie H, Fok H, Ford KJ, Fowler J, Fraiman PHA, Francis E, Franco MM, Frater J, Freire MSM, Fry SH, Fudge S, Furze J, Fuskova M, Galian-Rubio P, Galiza E, Garlant H, Gavrila M, Geddes A, Gibbons KA, Gilbride C, Gill H, Glynn S, Godwin K, Gokani K, Goldoni UC, Goncalves M, Gonzalez IGS, Goodwin J, Goondiwala A, Gordon-Quayle K, Gorini G, Grab J, Gracie L, Greenland M, Greenwood N, Greffrath J, Groenewald MM, Grossi L, Gupta G, Hackett M, Hallis B, Hamaluba M, Hamilton E, Hammersley D, Hanrath AT, Hanumunthadu B, Harris SA, Harris C, Harris T, Harrison TD, Harrison D, Hart TC, Hartnell B, Hassan S, Haughney J, Hawkins S, Hay J, Head I, Henry J, Hermosin Herrera M, Hettle DB, Hill J, Hodges G, Horne E, Hou MM, Houlihan C, Howe E, Howell N, Humphreys J, Humphries HE, Hurley K, Huson C, Hyder-Wright A, Hyamns C, Ikram S, Ishwarbhai A, Ivan M, Iveson P, Iyer V, Jackson F, De Jager J, Jaumdally S, Jeffers H, Jesudason N, Jones B, Jones K, Jones E, Jones C, Jorge MR, Jose A, Joshi A, Júnior EAMS, Kadziola J, Kailath R, Kana F, Karampatsas K, Kasanyinga M, Keen J, Kelly EJ, Kelly DM, Kelly D, Kelly S, Kerr D, de Á Kfouri R, Khan L, Khozoee B, Kidd S, Killen A, Kinch J, Kinch P, King LDW, King TB, Kingham L, Klenerman P, F. Knapper, Knight JC, Knott D, Koleva S, Lang M, Lang G, Larkworthy CW, Larwood JPJ, Law R, Lazarus EM, Leach A, Lees EA, Lemm N-M, Lessa A, Leung S, Li Y, Lias AM, Liatsikos K, Linder A, Lipworth S, Liu S, Liu X, Lloyd A, Lloyd S, Loew L, Lopez Ramon R, Lora L, Lowthorpe V, Luz K, MacDonald JC, MacGregor G, Madhavan M, Mainwaring DO,

Makambwa E, Makinson R, Malahleha M, Malamatsho R, Mallett G, Mansatta K, Maoko T, Mapetla K, Marchevsky NG, Marinou S, Marlow E, Marques GN, Marriott P, Marshall RP, Marshall JL, Martins FJ, Masenya M, Masilela M, Masters SK, Mathew M, Matlebjane H, Matshidiso K, Mazur O, Mazzella A, McCaughan H, McEwan J, McGlashan J, McInroy L, McIntyre Z, McLenaghan D, McRobert N, McSwiggan S, Megson C, Mehdipour S, Meijs W, Mendonça RNÁ, Mentzer AJ, Mirtorabi N, Mitton C, Mnyakeni S, Moghaddas F, Molapo K, Moloi M, Moore M, Moraes-Pinto MI, Moran M, Morey E, Morgans R, Morris S, Morris S, Morris HC, Morselli F, Morshead G, Morter R, Mottal L, Moultrie A, Moya N, Mpelembue M, Msomi S, Mugodi Y, Mukhopadhyay E, Muller J, Munro A, Munro C, Murphy S, Mweu P, Myasaki CH, Naik G, Naker K, Nastouli E, Nazir A, Ndlovu B, Neffa F, Njenga C, Noal H, Noé A, Novaes G, Nugent FL, Nunes G, O'Brien K, O'Connor D, Odam M, Oelofse S, Oguti B, Olchawski V, Oldfield NJ, Oliveira MG, Oliveira C, Oosthuizen A, O'Reilly P, Osborne P, Owen DRJ, Owen L, Owens D, Owino N, Pacurar M, Paiva BVB, Palhares EMF, Palmer S, Parkinson S, Parracho HMRT, Parsons K, Patel D, Patel B, Patel F, Patel K, Patrick-Smith M, Payne RO, Peng Y, Penn EJ, Pennington A, Peralta Alvarez MP, Perring J, Perry N, Perumal R, Petkar S, Philip T, Phillips DJ, Phillips J, Phohu MK, Pickup L, Pieterse S, Piper J, Pipini D, Plank M, Du Plessis J, Pollard S, Pooley J, Pooran A, Poulton I, Powers C, Presa FB, Price DA, Price V, Primeira M, Proud PC, Provstgaard-Morys S, Pueschel S, Pulido D, Quaid S, Rabara R, Radford A, Radia K, Rajapaska D, Rajeswaran T, Ramos ASF, Ramos Lopez F, Rampling T, Rand J, Ratcliffe H, Rawlinson T, Rea D, Rees B, Reiné J, Resuello-Dauti M, Reyes Pabon E, Ribiero CM, Ricamara M, Richter A, Ritchie N, Ritchie AJ, Robbins AJ, Roberts H, Robinson RE, Robinson H, Rocchetti TT, Rocha BP, Roche S, Rollier C, Rose L, Ross Russell AL, Rossouw L, Royal S, Rudiansyah I, Ruiz S, Saich S, Sala C, Sale J, Salman AM, Salvador N, Salvador S, Sampaio M, Samson AD, Sanchez-Gonzalez A, Sanders H, Sanders K, Santos E, Santos Guerra MFS, Satti I, Saunders JE, Saunders C, Sayed A, Schim van der Loeff I, Schmid AB, Schofield E, Screaton G, Seddiqi S, Segireddy RR, Senger R, Serrano S, Shah R, Shaik I, Sharpe HE, Sharrocks K, Shaw R, Shea A, Shepherd A, Shepherd JG, Shiham F, Sidhom E, Silk SE, da Silva Moraes AC, Silva-Junior G, Silva-Reyes L, Silveira AD, Silveira MBV, Sinha J, Skelly DT, Smith DC, Smith N, Smith HE, Smith DJ, Smith CC, Soares A, Soares T, Solórzano C, Sorio GL, Sorley K, Sosa-Rodriguez T, Souza CMCDL, Souza BSDF, Souza AR, Spencer AJ, Spina F, Spoors L, Stafford L, Stamford I, Starinskij I, Stein R, Steven J, Stockdale L, Stockwell LV, Strickland LH, Stuart AC, Sturdy A, Sutton N, Szigeti A, Tahiri-Alaoui A, Tanner R, Taoushanis C, Tarr AW, Taylor K, Taylor U, Taylor IJ, Taylor J, te Water Naude R, Themistocleous Y, Themistocleous A, Thomas M, Thomas K, Thomas TM, Thombrayil A, Thompson F, Thompson A, Thompson K, Thompson A, Thomson J, Thornton-Jones V, Tighe PJ, Tinoco LA, Tiongson G, Tladinyane B, Tomasicchio M, Tomic A, Tonks S, Tran N, Tree J, Trillana G, Trinham C, Trivett R, Truby A, Tsheko BL, Turabi A, Turner R, Turner C, Ulaszewska M, Underwood BR, Varughese R, Verbart D, Verheul M, Vichos I, Vieira T, Waddington CS, Walker L, Wallis E, Wand M, Warbick D, Wardell T, Warimwe G, Warren SC, Watkins B, Watson E, Webb S, Webb-Bridges A, Webster A, Welch J, Wells J, West A, White C, White R, Williams P, Williams RL, Winslow R, Woodyer M, Worth AT, Wright D, Wroblewska M, Yao A, Zimmer R, Zizi D, Zuidewind P (2020) Safety and efficacy of the ChAdOx1 nCoV-19 vaccine (AZD1222) against SARS-CoV-2: an interim analysis of four randomised controlled trials in Brazil, South Africa, and the UK. Lancet. https://doi.org/10.1016/s0140-6736(20)32661-1

93. Zhu F-C, Guan X-H, Li Y-H, Huang J-Y, Jiang T, Hou L-H, Li J-X, Yang B-F, Wang L, Wang W-J, Wu S-P, Wang Z, Wu X-H, Xu J-J, Zhang Z, Jia S-Y, Wang B-S, Hu Y, Liu J-J, Zhang J, Qian X-A, Li Q, Pan H-X, Jiang H-D, Deng P, Gou J-B, Wang X-W, Wang X-H, Chen W (2020) Immunogenicity and safety of a recombinant adenovirus type-5-vectored COVID-19 vaccine in healthy adults aged 18 years or older: a randomised, double-blind, placebo-controlled, phase 2 trial. Lancet 396:479–488. https://doi.org/10.1016/S0140-6736(20)31605-6

94. Zhang C, Maruggi G, Shan H, Li J (2019) Advances in mRNA Vaccines for Infectious Diseases. Front Immunol 10:594. https://doi.org/10.3389/fimmu.2019.00594

95. Pardi N, Hogan MJ, Porter FW, Weissman D (2018) mRNA vaccines-a new era in vaccinology. Nat Rev Drug Discov 17:261–279. https://doi.org/10.1038/nrd.2017.243

96. Kauffman KJ, Webber MJ, Anderson DG (2016) Materials for non-viral intracellular delivery of messenger RNA therapeutics. J Control Release 240:227–234. https://doi.org/10.1016/j.jconrel.2015.12.032

97. Mulligan MJ, Lyke KE, Kitchin N, Absalon J, Gurtman A, Lockhart S, Neuzil K, Raabe V, Bailey R, Swanson KA, Li P, Koury K, Kalina W, Cooper D, Fontes-Garfias C, Shi P-Y, Türeci Ö, Tompkins KR, Walsh EE, Frenck R, Falsey AR, Dormitzer PR, Gruber WC, Şahin U, Jansen KU (2020) Phase I/II study of COVID-19 RNA vaccine BNT162b1 in adults. Nature 586:589–593. https://doi.org/10.1038/s41586-020-2639-4

98. Karikó K, Muramatsu H, Welsh FA, Ludwig J, Kato H, Akira S, Weissman D (2008) Incorporation of pseudouridine into mRNA yields superior nonimmunogenic vector with increased translational capacity and biological stability. Mol Ther 16:1833–1840. https://doi.org/10.1038/mt.2008.200

99. Sahin U, Muik A, Derhovanessian E, Vogler I, Kranz LM, Vormehr M, Baum A, Pascal K, Quandt J, Maurus D, Brachtendorf S, Lörks V, Sikorski J, Hilker R, Becker D, Eller A-K, Grützner J, Boesler C, Rosenbaum C, Kühnle M-C, Luxemburger U, Kemmer-Brück A, Langer D, Bexon M, Bolte S, Karikó K, Palanche T, Fischer B, Schultz A, Shi P-Y, Fontes-Garfias C, Perez JL, Swanson KA, Loschko J, Scully IL, Cutler M, Kalina W, Kyratsous CA, Cooper D, Dormitzer PR, Jansen KU, Türeci Ö (2020) COVID-19 vaccine BNT162b1 elicits human antibody and TH1 T cell responses. Nature 586:594–599. https://doi.org/10.1038/s41586-020-2814-7

100. Polack FP, Thomas SJ, Kitchin N, Absalon J, Gurtman A, Lockhart S, Perez JL, Pérez Marc G, Moreira ED, Zerbini C, Bailey R, Swanson KA, Roychoudhury S, Koury K, Li P, Kalina WV, Cooper D, Frenck Jr RW, Hammitt LL, Türeci Ö, Nell H, Schaefer A, Ünal S, Tresnan DB, Mather S, Dormitzer PR, Şahin U, Jansen KU, Gruber WC, Group CCT (2020) Safety and efficacy of the BNT162b2 mRNA covid-19 vaccine. N Engl J Med. https://doi.org/10.1056/nejmoa2034577

101. Wrapp D, Wang N, Corbett KS, Goldsmith JA, Hsieh C, Abiona O, Graham BS, Mclellan JS (2020) Cryo-EM structure of the 2019-nCoV spike in the prefusion conformation. 1263:1260–1263

102. Jackson LA, Anderson EJ, Rouphael NG, Roberts PC, Makhene M, Coler RN, McCullough MP, Chappell JD, Denison MR, Stevens LJ, Pruijssers AJ, McDermott A, Flach B, Doria-Rose NA, Corbett KS, Morabito KM, O'Dell S, Schmidt SD, Swanson PA, Padilla M, Mascola JR, Neuzil KM, Bennett H, Sun W, Peters E, Makowski M, Albert J, Cross K, Buchanan W, Pikaart-Tautges R, Ledgerwood JE, Graham BS, Beigel JH (2020) An mRNA Vaccine against SARS-CoV-2—preliminary report. N Engl J Med. https://doi.org/10.1056/nejmoa2022483

103. Corbett KS, Edwards DK, Leist SR, Abiona OM, Boyoglu-Barnum S, Gillespie RA, Himansu S, Schäfer A, Ziwawo CT, DiPiazza AT, Dinnon KH, Elbashir SM, Shaw CA, Woods A, Fritch EJ, Martinez DR, Bock KW, Minai M, Nagata BM, Hutchinson GB, Wu K, Henry C, Bahl K, Garcia-Dominguez D, Ma L, Renzi I, Kong W-P, Schmidt SD, Wang L, Zhang Y, Phung E, Chang LA, Loomis RJ, Altaras NE, Narayanan E, Metkar M, Presnyak V, Liu C, Louder MK, Shi W, Leung K, Yang ES, West A, Gully KL, Stevens LJ, Wang N, Wrapp D, Doria-Rose NA, Stewart-Jones G, Bennett H, Alvarado GS, Nason MC, Ruckwardt TJ, McLellan JS, Denison MR, Chappell JD, Moore IN, Morabito KM, Mascola JR, Baric RS, Carfi A, Graham BS (2020) SARS-CoV-2 mRNA vaccine design enabled by prototype pathogen preparedness. Nature 586:567–571. https://doi.org/10.1038/s41586-020-2622-0

104. Moderna (2020) Moderna's COVID-19 vaccine candidate meets its primary efficacy endpoint in the first interim analysis of the phase 3 COVE study

105. Hobernik D, Bros M (2018) DNA vaccines-how far from clinical use? Int J Mol Sci 19:3605. https://doi.org/10.3390/ijms19113605

106. Silveira MM, Oliveira TL, Schuch RA, McBride AJA, Dellagostin OA, Hartwig DD (2017) DNA vaccines against leptospirosis: a literature review. Vaccine 35:5559–5567. https://doi.org/10.1016/j.vaccine.2017.08.067

107. Li L, Petrovsky N (2016) Molecular mechanisms for enhanced DNA vaccine immunogenicity. Expert Rev Vaccines 15:313–329. https://doi.org/10.1586/14760584.2016.1124762
108. Yang Z-Y, Kong W-P, Huang Y, Roberts A, Murphy BR, Subbarao K, Nabel GJ (2004) A DNA vaccine induces SARS coronavirus neutralization and protective immunity in mice. Nature 428:561–564. https://doi.org/10.1038/nature02463
109. Kim TW, Lee JH, Hung C-F, Peng S, Roden R, Wang M-C, Viscidi R, Tsai Y-C, He L, Chen P-J, Boyd DAK, Wu T-C (2004) Generation and characterization of DNA vaccines targeting the nucleocapsid protein of severe acute respiratory syndrome coronavirus. J Virol 78:4638–4645. https://doi.org/10.1128/jvi.78.9.4638-4645.2004
110. Zhao P, Cao J, Zhao L-J, Qin Z-L, Ke J-S, Pan W, Ren H, Yu J-G, Qi Z-T (2005) Immune responses against SARS-coronavirus nucleocapsid protein induced by DNA vaccine. Virology 331:128–135. https://doi.org/10.1016/j.virol.2004.10.016
111. Okada M, Okuno Y, Hashimoto S, Kita Y, Kanamaru N, Nishida Y, Tsunai Y, Inoue R, Nakatani H, Fukamizu R, Namie Y, Yamada J, Takao K, Asai R, Asaki R, Kase T, Takemoto Y, Yoshida S, Peiris JSM, Chen P-J, Yamamoto N, Nomura T, Ishida I, Morikawa S, Tashiro M, Sakatani M (2007) Development of vaccines and passive immunotherapy against SARS corona virus using SCID-PBL/hu mouse models. Vaccine 25:3038–3040. https://doi.org/10.1016/j.vaccine.2007.01.032
112. Woo PCY, Lau SKP, Tsoi H-W, Chen Z-W, Wong BHL, Zhang L, Chan JKH, Wong L-P, He W, Ma C, Chan K-H, Ho DD, Yuen K-Y (2005) SARS coronavirus spike polypeptide DNA vaccine priming with recombinant spike polypeptide from Escherichia coli as booster induces high titer of neutralizing antibody against SARS coronavirus. Vaccine 23:4959–4968. https://doi.org/10.1016/j.vaccine.2005.05.023
113. Zakhartchouk AN, Liu Q, Petric M, Babiuk LA (2005) Augmentation of immune responses to SARS coronavirus by a combination of DNA and whole killed virus vaccines. Vaccine 23:4385–4391. https://doi.org/10.1016/j.vaccine.2005.04.011
114. Muthumani K, Falzarano D, Reuschel EL, Tingey C, Flingai S, Villarreal DO, Wise M, Patel A, Izmirly A, Aljuaid A, Seliga AM, Soule G, Morrow M, Kraynyak KA, Khan AS, Scott DP, Feldmann F, LaCasse R, Meade-White K, Okumura A, Ugen KE, Sardesai NY, Kim JJ, Kobinger G, Feldmann H, Weiner DB (2015) A synthetic consensus anti-spike protein DNA vaccine induces protective immunity against Middle East respiratory syndrome coronavirus in nonhuman primates. Sci Transl Med 7:301ra132-301ra132. https://doi.org/10.1126/scitranslmed.aac7462
115. Modjarrad K, Roberts CC, Mills KT, Castellano AR, Paolino K, Muthumani K, Reuschel EL, Robb ML, Racine T, Oh M-D, Lamarre C, Zaidi FI, Boyer J, Kudchodkar SB, Jeong M, Darden JM, Park YK, Scott PT, Remigio C, Parikh AP, Wise MC, Patel A, Duperret EK, Kim KY, Choi H, White S, Bagarazzi M, May JM, Kane D, Lee H, Kobinger G, Michael NL, Weiner DB, Thomas SJ, Maslow JN (2019) Safety and immunogenicity of an anti-Middle East respiratory syndrome coronavirus DNA vaccine: a phase 1, open-label, single-arm, dose-escalation trial. Lancet Infect Dis 19:1013–1022. https://doi.org/10.1016/S1473-3099(19)30266-X
116. Al-Amri SS, Abbas AT, Siddiq LA, Alghamdi A, Sanki MA, Al-Muhanna MK, Alhabbab RY, Azhar EI, Li X, Hashem AM (2017) Immunogenicity of candidate MERS-CoV DNA vaccines based on the spike protein. Sci Rep 7:44875. https://doi.org/10.1038/srep44875
117. Smith TRF, Patel A, Ramos S, Elwood D, Zhu X, Yan J, Gary EN, Walker SN, Schultheis K, Purwar M, Xu Z, Walters J, Bhojnagarwala P, Yang M, Chokkalingam N, Pezzoli P, Parzych E, Reuschel EL, Doan A, Tursi N, Vasquez M, Choi J, Tello-Ruiz E, Maricic I, Bah MA, Wu Y, Amante D, Park DH, Dia Y, Ali AR, Zaidi FI, Generotti A, Kim KY, Herring TA, Reeder S, Andrade VM, Buttigieg K, Zhao G, Wu J-M, Li D, Bao L, Liu J, Deng W, Qin C, Brown AS, Khoshnejad M, Wang N, Chu J, Wrapp D, McLellan JS, Muthumani K, Wang B, Carroll MW, Kim JJ, Boyer J, Kulp DW, Humeau LMPF, Weiner DB, Broderick KE (2020) Immunogenicity of a DNA vaccine candidate for COVID-19. Nat Commun 11:2601. https://doi.org/10.1038/s41467-020-16505-0

118. Diehl MC, Lee JC, Daniels SE, Tebas P, Khan AS, Giffear M, Sardesai NY, Bagarazzi ML (2013) Tolerability of intramuscular and intradermal delivery by CELLECTRA(®) adaptive constant current electroporation device in healthy volunteers. Hum Vaccin Immunother 9:2246–2252. https://doi.org/10.4161/hv.24702
119. Kared H, Redd AD, Bloch EM, Bonny TS, Sumatoh H, Kairi F, Carbajo D, Abel B, Newell EW, Bettinotti MP, Benner SE, Patel EU, Littlefield K, Laeyendecker O, Shoham S, Sullivan D, Casadevall A, Pekosz A, Nardin A, Fehlings M, Tobian AA, Quinn TC (2020) CD8+ T cell responses in convalescent COVID-19 individuals target epitopes from the entire SARS-CoV-2 proteome and show kinetics of early differentiation. BioRxiv Prepr Serv Biol. https://doi.org/10.1101/2020.10.08.330688
120. Wang N, Shang J, Jiang S, Du L (2020) Subunit vaccines against emerging pathogenic human coronaviruses. Front Microbiol 11:298. https://doi.org/10.3389/fmicb.2020.00298
121. Chen W-H, Tao X, Agrawal AS, Algaissi A, Peng B-H, Pollet J, Strych U, Bottazzi ME, Hotez PJ, Lustigman S, Du L, Jiang S, Tseng C-TK (2020) Yeast-expressed SARS-CoV recombinant receptor-binding domain (RBD219-N1) formulated with aluminum hydroxide induces protective immunity and reduces immune enhancement. Vaccine 38:7533–7541. https://doi.org/10.1016/j.vaccine.2020.09.061
122. Capell T, Twyman RM, Armario-Najera V, Ma JK-C, Schillberg S, Christou P (2020) Potential applications of plant biotechnology against SARS-CoV-2. Trends Plant Sci 25:635–643. https://doi.org/10.1016/j.tplants.2020.04.009
123. Zheng N, Xia R, Yang C, Yin B, Li Y, Duan C, Liang L, Guo H, Xie Q (2009) Boosted expression of the SARS-CoV nucleocapsid protein in tobacco and its immunogenicity in mice. Vaccine 27:5001–5007. https://doi.org/10.1016/j.vaccine.2009.05.073
124. Deng M, Hu Z, Wang H, Deng F (2012) Developments of subunit and VLP vaccines against influenza a virus. Virol Sin 27:145–153. https://doi.org/10.1007/s12250-012-3241-1
125. Cao Y, Zhu X, Hossen MN, Kakar P, Zhao Y, Chen X (2018) Augmentation of vaccine-induced humoral and cellular immunity by a physical radiofrequency adjuvant. Nat Commun 9:3695. https://doi.org/10.1038/s41467-018-06151-y
126. Suthar MS, Zimmerman MG, Kauffman RC, Mantus G, Linderman SL, Hudson WH, Vanderheiden A, Nyhoff L, Davis CW, Adekunle O, Affer M, Sherman M, Reynolds S, Verkerke HP, Alter DN, Guarner J, Bryksin J, Horwath MC, Arthur CM, Saakadze N, Smith GH, Edupuganti S, Scherer EM, Hellmeister K, Cheng A, Morales JA, Neish AS, Stowell SR, Frank F, Ortlund E, Anderson EJ, Menachery VD, Rouphael N, Mehta AK, Stephens DS, Ahmed R, Roback JD, Wrammert J (2020) Rapid generation of neutralizing antibody responses in COVID-19 patients. Cell Rep Med 1: https://doi.org/10.1016/j.xcrm.2020.100040
127. Lan J, Yao Y, Deng Y, Chen H, Lu G, Wang W, Bao L, Deng W, Wei Q, Gao GF, Qin C, Tan W (2015) Recombinant receptor binding domain protein induces partial protective immunity in rhesus macaques against Middle East respiratory syndrome coronavirus challenge. EBioMedicine 2:1438–1446. https://doi.org/10.1016/j.ebiom.2015.08.031
128. He Y, Zhou Y, Siddiqui P, Jiang S (2004) Inactivated SARS-CoV vaccine elicits high titers of spike protein-specific antibodies that block receptor binding and virus entry. Biochem Biophys Res Commun 325:445–452. https://doi.org/10.1016/j.bbrc.2004.10.052
129. Graham BS (2020) Rapid COVID-19 vaccine development. Science 80:368, 945 LP–946. https://doi.org/10.1126/science.abb8923
130. Du L, Zhao G, Chan CCS, Sun S, Chen M, Liu Z, Guo H, He Y, Zhou Y, Zheng B-J, Jiang S (2009) Recombinant receptor-binding domain of SARS-CoV spike protein expressed in mammalian, insect and E. coli cells elicits potent neutralizing antibody and protective immunity. Virology 393:144–150. https://doi.org/10.1016/j.virol.2009.07.018
131. He Y, Li J, Heck S, Lustigman S, Jiang S (2006) Antigenic and immunogenic characterization of recombinant baculovirus-expressed severe acute respiratory syndrome coronavirus spike protein: implication for vaccine design. J Virol 80:5757–5767. https://doi.org/10.1128/JVI.00083-06
132. Kam YW, Kien F, Roberts A, Cheung YC, Lamirande EW, Vogel L, Chu SL, Tse J, Guarner J, Zaki SR, Subbarao K, Peiris M, Nal B, Altmeyer R (2007) Antibodies against trimeric S

glycoprotein protect hamsters against SARS-CoV challenge despite their capacity to mediate FcgammaRII-dependent entry into B cells in vitro. Vaccine 25:729–740. https://doi.org/10.1016/j.vaccine.2006.08.011

133. Jaume M, Yip M, Kam Y, Cheung C, Kien F, Roberts A (2012) SARS-CoV subunit vaccine: antibody-mediated neutralisation and enhancement. Hong Kong Med J 18:31–36

134. He Y, Zhou Y, Liu S, Kou Z, Li W, Farzan M, Jiang S (2004) Receptor-binding domain of SARS-CoV spike protein induces highly potent neutralizing antibodies: implication for developing subunit vaccine. Biochem Biophys Res Commun 324:773–781. https://doi.org/10.1016/j.bbrc.2004.09.106

135. Du L, Zhao G, Chan CCS, Li L, He Y, Zhou Y, Zheng B-J, Jiang S (2010) A 219-mer CHO-expressing receptor-binding domain of SARS-CoV S protein induces potent immune responses and protective immunity. Viral Immunol 23:211–219. https://doi.org/10.1089/vim.2009.0090

136. Du L, Zhao G, He Y, Guo Y, Zheng B-J, Jiang S, Zhou Y (2007) Receptor-binding domain of SARS-CoV spike protein induces long-term protective immunity in an animal model. Vaccine 25:2832–2838. https://doi.org/10.1016/j.vaccine.2006.10.031

137. Guo Y, Sun S, Wang K, Zhang S, Zhu W, Chen Z (2005) Elicitation of immunity in mice after immunization with the S2 subunit of the severe acute respiratory syndrome coronavirus. DNA Cell Biol 24:510–515. https://doi.org/10.1089/dna.2005.24.510

138. Liu S-J, Leng C-H, Lien S-P, Chi H-Y, Huang C-Y, Lin C-L, Lian W-C, Chen C-J, Hsieh S-L, Chong P (2006) Immunological characterizations of the nucleocapsid protein based SARS vaccine candidates. Vaccine 24:3100–3108. https://doi.org/10.1016/j.vaccine.2006.01.058

139. Qiu M, Shi Y, Guo Z, Chen Z, He R, Chen R, Zhou D, Dai E, Wang X, Si B, Song Y, Li J, Yang L, Wang J, Wang H, Pang X, Zhai J, Du Z, Liu Y, Zhang Y, Li L, Wang J, Sun B, Yang R (2005) Antibody responses to individual proteins of SARS coronavirus and their neutralization activities. Microbes Infect 7:882–889. https://doi.org/10.1016/j.micinf.2005.02.006

140. He Y, Zhou Y, Siddiqui P, Niu J, Jiang S (2005) Identification of immunodominant epitopes on the membrane protein of the severe acute respiratory syndrome-associated coronavirus. J Clin Microbiol 43:3718–3726. https://doi.org/10.1128/JCM.43.8.3718-3726.2005

141. Tai W, Zhao G, Sun S, Guo Y, Wang Y, Tao X, Tseng C-TK, Li F, Jiang S, Du L, Zhou Y (2016) A recombinant receptor-binding domain of MERS-CoV in trimeric form protects human dipeptidyl peptidase 4 (hDPP4) transgenic mice from MERS-CoV infection. Virology 499:375–382. https://doi.org/10.1016/j.virol.2016.10.005

142. Tai W, Wang Y, Fett CA, Zhao G, Li F, Perlman S, Jiang S, Zhou Y, Du L (2016) Recombinant receptor-binding domains of Multiple Middle East respiratory syndrome coronaviruses (MERS-CoVs) induce cross-neutralizing antibodies against divergent human and camel MERS-CoVs and antibody escape mutants. J Virol 91:e01651–16. https://doi.org/10.1128/JVI.01651-16

143. Coleman CM, Liu YV, Mu H, Taylor JK, Massare M, Flyer DC, Smith GE, Frieman MB (2014) Purified coronavirus spike protein nanoparticles induce coronavirus neutralizing antibodies in mice. Vaccine 32:3169–3174. https://doi.org/10.1016/j.vaccine.2014.04.016

144. Tian J-H, Patel N, Haupt R, Zhou H, Weston S, Hammond H, Lague J, Portnoff AD, Norton J, Guebre-Xabier M, Zhou B, Jacobson K, Maciejewski S, Khatoon R, Wisniewska M, Moffitt W, Kluepfel-Stahl S, Ekechukwu B, Papin J, Boddapati S, Wong CJ, Piedra PA, Frieman MB, Massare MJ, Fries L, Lövgren Bengtsson K, Stertman L, Ellingsworth L, Glenn G, Smith G (2020) SARS-CoV-2 spike glycoprotein vaccine candidate NVX-CoV2373 elicits immunogenicity in baboons and protection in mice. BioRxiv. https://doi.org/10.1101/2020.06.29.178509

145. Guebre-Xabier M, Patel N, Tian J-H, Zhou B, Maciejewski S, Lam K, Portnoff AD, Massare MJ, Frieman MB, Piedra PA, Ellingsworth L, Glenn G, Smith G (2020) NVX-CoV2373 vaccine protects cynomolgus macaque upper and lower airways against SARS-CoV-2 challenge. Vaccine S0264-410X(20)31373–6. https://doi.org/10.1016/j.vaccine.2020.10.064

146. Keech C, Albert G, Cho I, Robertson A, Reed P, Neal S, Plested JS, Zhu M, Cloney-Clark S, Zhou H, Smith G, Patel N, Frieman MB, Haupt RE, Logue J, McGrath M, Weston S, Piedra PA, Desai C, Callahan K, Lewis M, Price-Abbott P, Formica N, Shinde V, Fries L, Lickliter JD, Griffin P, Wilkinson B, Glenn GM (2020) Phase 1-2 trial of a SARS-CoV-2 recombinant spike protein nanoparticle vaccine. N Engl J Med 383:2320–2332. https://doi.org/10.1056/NEJMoa2026920

147. Al-Barwani F, Donaldson B, Pelham SJ, Young SL, Ward VK (2014) Antigen delivery by virus-like particles for immunotherapeutic vaccination. Ther Deliv 5:1223–1240. https://doi.org/10.4155/tde.14.74

148. Donaldson B, Lateef Z, Walker GF, Young SL, Ward VK (2018) Virus-like particle vaccines: immunology and formulation for clinical translation. Expert Rev Vaccines 17:833–849. https://doi.org/10.1080/14760584.2018.1516552

149. Lu X, Chen Y, Bai B, Hu H, Tao L, Yang J, Chen J, Chen Z, Hu Z, Wang H (2007) Immune responses against severe acute respiratory syndrome coronavirus induced by virus-like particles in mice. Immunology 122:496–502. https://doi.org/10.1111/j.1365-2567.2007.02676.x

150. Naskalska A, Dabrowska A, Nowak P, Szczepanski A, Jasik K, Milewska A, Ochman M, Zeglen S, Rajfur Z, Pyrc K (2018) Novel coronavirus-like particles targeting cells lining the respiratory tract. PLoS ONE 13:e0203489–e0203489. https://doi.org/10.1371/journal.pone.0203489

151. Lokugamage KG, Yoshikawa-Iwata N, Ito N, Watts DM, Wyde PR, Wang N, Newman P, Kent Tseng C-T, Peters CJ, Makino S (2007) Chimeric coronavirus-like particles carrying severe acute respiratory syndrome coronavirus (SCoV) S protein protect mice against challenge with SCoV. Vaccine 26:797–808

152. Liu YV, Massare MJ, Barnard DL, Kort T, Nathan M, Wang L, Smith G (2011) Chimeric severe acute respiratory syndrome coronavirus (SARS-CoV) S glycoprotein and influenza matrix 1 efficiently form virus-like particles (VLPs) that protect mice against challenge with SARS-CoV. Vaccine 29:6606–6613

153. Wang C, Zheng X, Gai W, Zhao Y, Wang H, Wang H, Feng N, Chi H, Qiu B, Li N, Wang T, Gao Y, Yang S, Xia X (2017) MERS-CoV virus-like particles produced in insect cells induce specific humoural and cellular imminity in rhesus macaques. Oncotarget 8:12686–12694. https://doi.org/10.18632/oncotarget.8475

154. Wang C, Zheng X, Gai W, Wong G, Wang H, Jin H, Feng N, Zhao Y, Zhang W, Li N, Zhao G, Li J, Yan J, Gao Y, Hu G, Yang S, Xia X (2017) Novel chimeric virus-like particles vaccine displaying MERS-CoV receptor-binding domain induce specific humoral and cellular immune response in mice. Antiviral Res 140:55–61. https://doi.org/10.1016/j.antiviral.2016.12.019

155. Ward BJ, Gobeil P, Séguin A, Atkins J, Boulay I, Charbonneau P-Y, Couture M, D'Aoust M-A, Dhaliwall J, Finkle C, Hager K, Mahmood A, Makarkov A, Cheng M, Pillet S, Schimke P, St-Martin S, Trépanier S, Landry N (2020) Phase 1 trial of a candidate recombinant virus-like particle vaccine for covid-19 disease produced in plants. MedRxiv. https://doi.org/10.1101/2020.11.04.20226282

156. COVID-19 Medicago's Development Programs (2020)

157. Moorlag SJCFM, Arts RJW, van Crevel R, Netea MG (2019) Non-specific effects of BCG vaccine on viral infections. Clin Microbiol Infect 25:1473–1478. https://doi.org/10.1016/j.cmi.2019.04.020

158. Angelidou A, Diray-Arce J, Conti MG, Smolen KK, van Haren SD, Dowling DJ, Husson RN, Levy O (2020) BCG as a case study for precision vaccine development: lessons from vaccine heterogeneity, trained immunity, and immune ontogeny. Front Microbiol 11:332. https://doi.org/10.3389/fmicb.2020.00332

159. Netea MG, Domínguez-Andrés J, Barreiro LB, Chavakis T, Divangahi M, Fuchs E, Joosten LAB, van der Meer JWM, Mhlanga MM, Mulder WJM, Riksen NP, Schlitzer A, Schultze JL, Stabell Benn C, Sun JC, Xavier RJ, Latz E (2020) Defining trained immunity and its role

in health and disease. Nat Rev Immunol 20:375–388. https://doi.org/10.1038/s41577-020-0285-6

160. Urbán S, Paragi G, Burián K, McLean GR, Virok DP (2020) Identification of similar epitopes between severe acute respiratory syndrome coronavirus-2 and Bacillus Calmette-Guérin: potential for cross-reactive adaptive immunity. Clin Transl Immunol 9:e1227–e1227. https://doi.org/10.1002/cti2.1227

161. Sharma A, Kumar Sharma S, Shi Y, Bucci E, Carafoli E, Melino G, Bhattacherjee A, Das G (2020) BCG vaccination policy and preventive chloroquine usage: do they have an impact on COVID-19 pandemic? Cell Death Dis 11:516. https://doi.org/10.1038/s41419-020-2720-9

162. Hegarty PK, Sfakianos JP, Giannarini G, DiNardo AR, Kamat AM (2020) COVID-19 and Bacillus Calmette-Guérin: what is the link? Eur Urol Oncol 3:259–261. https://doi.org/10.1016/j.euo.2020.04.001

163. Covián C, Retamal-Díaz A, Bueno SM, Kalergis AM (2020) Could BCG vaccination induce protective trained immunity for SARS-CoV-2? Front Immunol 11:970. https://doi.org/10.3389/fimmu.2020.00970

164. Kumar J, Meena J (2020) Demystifying BCG vaccine and COVID-19 relationship. Indian Pediatr 57:588–589. https://doi.org/10.1007/s13312-020-1872-0

165. Riccò M, Gualerzi G, Ranzieri S, Bragazzi NL (2020) Stop playing with data: there is no sound evidence that Bacille Calmette-Guérin may avoid SARS-CoV-2 infection (for now). Acta Biomed 91:207–213. https://doi.org/10.23750/abm.v91i2.9700

Chapter 6
COVID-19 Diagnosis: A Comprehensive Review of Current Testing Platforms; Part A

Sareh Arjmand, Behrad Ghiasi, Samin Haghighi Poodeh, Fataneh Fatemi, Zahra Hassani Nejad, and Seyed Ehsan Ranaei Siadat

6.1 Introduction

The Coronavirus Disease 2019 (COVID-19) pandemic has greatly impacted the scientific community since its emergence in late 2019. The highly contagious nature of the Severe Acute Respiratory Syndrome Coronavirus 2 (SARS-CoV-2) and the fast-growing number of COVID-19 cases have confronted the world medical community with the urgent need for rapid, reliable, accessible, and inexpensive diagnostic tests. Rapid appropriate testing gives the medical systems the opportunity to recognize active cases as early as possible so that they can be quarantined and contact traced. Besides other protective measures like social distancing and the use of masks, this process is fundamental to breaking the virus' transmission chain and slowing the slop of virus contamination and death curve. Hence, health systems and researchers will have more time to find effective treatments and develop vaccines.

SARS-CoV-2 is an enveloped RNA virus, together with the other two human coronaviruses, Severe Acute Respiratory Syndrome (SARS) and Middle East Respiratory Syndrome (MERS) coronaviruses, belongs to the beta group of the *Orthocoronavirinae* family. SARS-CoV-2 carries a positive single-stranded RNA and consists of four main structural proteins, including a positive-sense single-stranded RNA. Accordingly, the detection of nucleic acids (molecular tests) and secreted antibodies

S. Arjmand (✉) · B. Ghiasi · S. Haghighi Poodeh · F. Fatemi (✉)
Protein Research Center, Shahid Beheshti University, Tehran, Iran
e-mail: s_arjmand@sbu.ac.ir

F. Fatemi
e-mail: f_fatemi@sbu.ac.ir

Z. Hassani Nejad
Institute of Biochemistry and Biophysics, University of Tehran, Tehran, Iran

S. E. R. Siadat
Sobhan Recombinant Protein, No. 22, 2nd Noavari St, Pardis Technology Park, 20th Km of Damavand Road, Tehran, Iran

© The Author(s), under exclusive license to Springer Nature Singapore Pte Ltd. 2021
M. Rahmandoust and S.-O. Ranaei-Siadat (eds.), *COVID-19*,
https://doi.org/10.1007/978-981-16-3108-5_6

against the structural proteins (serological tests) are the two main approaches in detecting the virus in patients' samples [1].

Currently, viral ribonucleic acid detection using real-time polymerase chain reaction (PCR) is the "gold standard" method to confirm COVID-19 cases. This method has sufficient specificity and sensitivity to help the physician with the early diagnosis of the infection [2]. However, due to practical issues in the preparation and transportation of specimens, the performance of different testing kits, as well as operator's skills and expertise, such a criterion-referenced method has been shown to elicit considerable false-negative outcomes [3].

Furthermore, the commercial PCR-based kits are less sensitive to identify the virus at its initial stage of infection [4]. Since COVID-19 is highly associated with lung damage, chest computed tomography (CT) is a beneficial imaging modality that has been used since the beginning of the current pandemic. Nevertheless, due to its low specificity, this imaging method shows incomplete clinical performance for the proper diagnosis of COVID-19 [5]. Thus, a combination of real-time PCR and CT imaging has been applied as a more reliable diagnostic modality in confirming positive cases [6]. In addition, serological tests based on the detection of the IgM and IgG antibodies from patients' samples have been used for understanding the infection history [7].

Researchers around the world are on the urge to develop novel methods to accelerate the development of innovative tools for point-of-care diagnosis of the low doses of SARS-CoV-2 at its early stages of infection. This will help to reduce the transmission rate of the virus and thereby help the health care systems to cope with the disease. In this commentary, we review the current molecular diagnostic methods for SARS-CoV-2. We also highlight the recent advances and future developments in rapid and sensitive testing of the patients.

6.2 Nucleic Acid Testing

Coronaviruses possess large RNA genomes, ranging from 26 to 32 kb in length [8]. Coronaviruses' abundant presence in the pool of viral quasispecies increases the possibility of adaptive mutations and interspecies recombination [9]. The first metagenomic RNA sequencing of the SARS-CoV-2 genome deposited in the National Center for Biotechnology Information (NCBI) (MN908947·3) revealed that 29,903 nucleotides were present in its viral genome. The sequence shared high levels (82%) of similarity with the previously encountered SARS-CoV and MERS-CoV, which implied a common pathogenesis mechanism [10]. At the time of writing this paper, 92 complete genomes have been deposited in NCBI Assembly for SARS-CoV-2.

The advents in molecular biology revolutionized the nucleic acid-based technologies for pathogen detection. So far, various methods have been described for specific detection of the SARS-CoV-2 genome. The overall SARS-CoV-2 genome architecture is shown in Fig. 6.1. It is noteworthy that hitherto, different parts of the

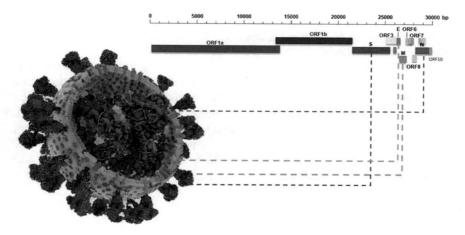

Fig. 6.1 Schematic representation of SARS-CoV-2 genome structure, showing the protein-coding regions. It consists of two overlapping ORFs (ORF1a and ORF1b), Spike (S), ORF3a, Envelope (E), Membrane (M), ORF6, ORF7a, ORF7b, ORF8, Nucleocapsid (N), and ORF10

SARS-CoV-2 genome, which code for essential viral components, have been used for molecular viral detection.

6.2.1   Sequencing

Sequencing is the first and primary step for identifying and classifying unknown organisms (such as SARS-CoV-2). Furthermore, sequencing provides the key information necessary for setting up other nucleic acid detection methods such as real-time PCR (designing of probes and primers requires prior information of the sequence). Additionally, sequencing gives insights into the virus's evolution, potentially paving the way for therapies and vaccine development [11]. High-throughput sequencing is widely used for new virus discovery and detection of already known viruses [12]. This method includes next-generation short-read and third-generation long-read strategies. The short-read sequencer, such as Illumina's NovaSeq, HiSeq, NextSeq, and MiSeq instruments, produce reads shorter than 1000 bp. The workflow includes three consequent steps: genome fragmentation and library preparation, amplifying the fragments, and sequencing of the amplified products based on sequencing-by-synthesis approach [13, 14].

Next-generation sequencing (NGS) does not need any prior knowledge of the pathogen and was the method of choice used for the first genome sequencing of SARS-CoV-2 [15]. In the recently generated long-read sequencing technology, the long reads (>10 kb) are generated from the genome that improves the de novo assembly. Furthermore, the sequencing and library preparation is conducted without the need for amplification, therefore eliminating PCR-related bias in sequencing

[16]. The long-read technology platforms such as PacBio and Oxford nanopore enable scientists to accurately sequence the SARS-CoV-2 genome for strain and quasispecies resolution [17].

Despite being highly accurate and sensitive, sequencing is not currently used as the optimal diagnostic tool for SARS-CoV-2 detection, and its applications have been limited to mutational analysis, phylogenetic studies, and classification of new sequences [18]. In other words, the need for cutting-edge equipment, expert operators, the high cost, and the experimentation time (turnaround time) makes sequencing less favorable for large-scale testing needed in pandemic conditions [19]. Further investigation for automation and simplification is necessary to reach sequencing into widespread clinical applications.

6.2.2   Real-Time PCR

Real-time PCR simply refers to simultaneous DNA amplification and monitoring in a closed system. This molecular method's main benefits include the minimized false-positive results, quantitative determination of starting DNA in the sample, rapidity, sensitivity, and easy standardization [20, 21]. Since it can detect very small amounts of nucleic acid sequences, real-time PCR has emerged as a robust and widely used technique for various biological investigations and clinical applications. In recent years, this method has been used extensively to evaluate cancer status and quantify pathogen agent load, including different threatening emerging viruses such as Zika, Ebola, HIV, influenza, hepatitis, and SARS [22]. Currently, real-time PCR is recommended by the World Health Organization (WHO) as the "gold standard" method for SARS-CoV-2 detection. By early sequencing of the viral genome, several primers and probes have been designed that target different gene loci (specific for SARS-CoV-2). The target sequences include structural encoding protein genes (spike (S), envelope (E), transmembrane (M), helicase (Hel), and nucleocapsid (N)), and the species-specific genes required for viral replication (RNA-dependent RNA polymerase (RdRp), hemagglutinin-esterase (HE), and open reading frames 1a (ORF1a) and ORF1b) [23, 24]. In brief, samples are collected from patients or suspected cases and stored in a transport medium. After lysis and RNA extraction, the isolated genome is reverse transcribed to complementary DNA (cDNA) and amplified via several rounds of PCR. The amplified viral target sequence is detected and quantified using fluorescent or colorimetric data. Each step of the process can be optimized or modified independently. The workflow of real-time PCR is summarized in Fig. 6.2.

The master mix is intended for different real-time PCR methods to reduce the number of pipetting steps and, consequently, cross-well contamination. A master mix typically contains a mixture of a DNA polymerase, a mixture of reverse transcriptase enzymes, and one or more primers that target specific locations within the viral genome. During the amplification process, detection can be performed using several approaches. SYBRGreen dyes, for instance, can be used as a nonspecific DNA intercalating dye. However, due to its non-specificity, any DNA amplification

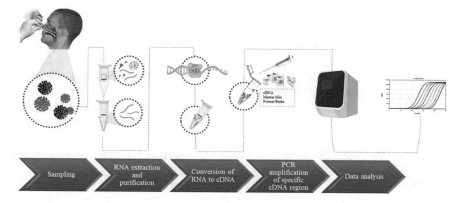

Fig. 6.2 Real-time PCR workflow for RNA virus detection

can intensify fluorescent readout [25]. Using TaqMan probes provides higher specificity due to containing a 3′ quencher and 5′ fluorophore that anneals to specific sequences within the DNA strands. Subsequently, annealed probes get degraded by the 5′ to 3′ exonuclease activity of Taq polymerase, and the fluorophores liberate from its quencher [26]. Normally, commercial factories are capable of providing master mix on a large scale. However, in a pandemic situation, disruption or delay in the supply chain can occur everywhere at any time. Overcoming this obstacle needs the development of a totally open-source homemade master mix. For example, Bhadra et al. devised a real-time PCR assay that relies on a thermostable reverse transcriptase/DNA polymerase (RTX) instead of a commercial master mix. Notably, RTX can be expressed in *Escherichia coli* strains such as BL21 and purified by histidine-tag. The required buffers can be easily prepared in every laboratory [27]. At least two target sequences are selected for SARS-CoV-2 detection; one for a conserved region among beta coronaviruses and the other specific for SARS-COV-2. Currently, several protocols have been provided by WHO that target different gene loci in the SARS-CoV-2 genome. For instance, in Germany RdRP, E, and N genes, in France, two target regions in RdRp sequence, in Japan S gene, in US three targets in N gene sequence, and in China ORF1ab and N gene are used for screening [28].

Some studies have compared the sensitivity of different primers and probes; the reports indicate the lowest sensitivity for the RdRp-SARS and the highest sensitivity for the E gene primer-probe; while other primers and probes show a high level of similarity in sensitivity [29, 30]. However, Vogels et al. declared some limitations in their findings, such as the thermocycler's conditions, the concentration of PCR kit components, and standardization of the primer and probes to be compared directly, which might affect their function.

Although real-time PCR is a well-established method, significant pitfalls have been reported. False-positive results produce an extra burden on the healthcare system, and false-negative results can endanger the quarantine controlling pandemic protocols. False results generally rise from vulnerability in the preanalytical steps (such as sampling and handling) and analytical steps (such as wrong collection time and low sample viral load) [31]. Combining other diagnostic methods with real-time PCR is a potential solution to increase the results' validity [32].

Sampling is the mutual step in all diagnostic methods. The US Centers for Disease Control and Prevention (CDC) recommends using swabs with a plastic or aluminum shaft and synthetic tips such as Dacron or nylon to prevent contaminating specimens with real-time PCR inhibitors that might come from sampling swabs. Moreover, using flocked swabs is suggested for sampling because it can considerably increase the total number of collected respiratory cells, which will lead to higher diagnostic sensitivity. The swab collections are kept in a viral transport medium prior to initiating the test [33].

Different specimens have been analyzed for the sensitivity of SARS-CoV-2 detection. The highest detection rate was observed for bronchoalveolar lavage fluid, saliva, and respiratory swabs with 98.3%, 91.7%, and 77.9%, respectively. The lowest sensitivity was reported for blood and urine with 1% and 0.73%. Among the respiratory swabs (usually used in clinical applications), nasal, throat, nasopharyngeal, oropharyngeal, and pharyngeal swabs showed better detection rates, respectively [34, 35]. The world health organization (WHO) recommends using upper respiratory tract samples for real-time PCR since it is less risky for the person who collects samples. The swabs taken from the nasopharynx and oropharynx are the most common sample types tested for SARS-CoV-2 diagnostic real-time PCR. If both samples are collected, they are mixed and used in a single real-time reaction to save reagents [36]. For the symptomatic individuals with negative real-time PCR, sampling from the lower respiratory tract is recommended for the second round of tests [37]. The highest SARS-CoV-2 load in the lower respiratory tract samples is observed five days post symptoms onset [38]. It is probably due to the abundant expression of Angiotensin-converting enzyme 2 (ACE2)—the main SARS-CoV-2 entry receptor—in alveolar type II epithelial cells [39]. After sample collection, the viral transport mediums are shipped to the laboratory in appropriate buffers and kept at 2–8 °C for no longer than three days. For a longer storage period, lower temperatures are used. The handling and storage conditions are critical for precise diagnosis, and potential RNA distortion in these steps may lead to false-negative results.

Although RNA extraction and purification steps are optional for real-time PCR, most protocols suggest performing these steps to enhance the sensitivity of the test. RNA extraction and purification are time-consuming and require trained personnel and expensive equipment. In addition, the reagents required for these procedures may work as an obstacle in real-time PCR testing, especially in low-income countries. A potential solution that can be considered is the use of handmade reagents. However, the residual organic solvents and salts in these reagents could function as an inhibitor in the subsequent real-time PCR steps. In the case of COVID-19 and high demands for large-scale testing, this obstacle can lead to more problems. Scallan and Aitken et al. developed a strategy to reduce the dependency on commercial kits and reagents. They found a high potential for using homemade solutions that contain 4–5 M guanidinium thiocyanate for sample lysis and RNA recovery [40, 41]. In some other studies, virus-inactivating reagents were suggested for this purpose [42, 43].

Another critical point essential for an excellent real-time PCR performance is the primer and probe designing. Since RNA is cleaved into shorter fragments after heat inactivation, it is recommended that primers/probes be set for the short amplicons [44].

6.2.3   Nested Real-Time PCR

The sensitivity of real-time PCR can be further increased by performing "nested" real-time PCR. Several research groups have evaluated the performance of nested real-time PCR for precise SARS-CoV-2 detection, and their results showed the method's ability to detect low viral load in the early stage of infection. However, the process of nested real-time PCR requires additional steps (addition of nested primers) during the PCR amplification, which may increase the risk of cross-contamination. Wang et al. designed a one-step single-tube nested real-time PCR for targeting N genes and ORF1ab with the ability to recognize one copy per reaction. Their design's uniqueness is that the two annealing steps occur independently in a single closed tube. The method has enough sensitivity to detect the virus in the samples with very low viral load, such as urine and blood. It is noteworthy that using such samples has a lower risk of exposure for healthcare individuals [45]. In another study, four sample pools, containing 49 negative and one positive specimen (with low viral concentration) were prepared. The single-tube nested real-time PCR detects two of them as positive and real-time one sample [46]. Compared to conventional real-time PCR, single-tube nested real-time PCR has a higher risk of producing false-positive results and needs more optimizations and modifications.

6.2.4   Droplet Digital PCR (ddPCR)

Despite being widely implemented for clinical viral load testing, real-time PCR is an indirect method that quantifies amplicon depending on the relationship between the cycle of threshold (C_t) of the sample to a standard curve. Furthermore, multiple steps required for its handling and analysis (e.g., sampling, RNA extraction, and quality control, enzymatic efficiency, internal control, and standard curve preparation) prone the real-time PCR to errors, misinterpretation, and lab result variations [47, 48]. This warrants the need for more direct quantification methods.

Droplet digital PCR (ddPCR) is a recently introduced method of direction quantification with no need for a calibration curve. In ddPCR, a sample containing the target nucleic acid is partitioned randomly into thousands of nano-liter-sized droplets, such that each droplet contains one or no copies of nucleic acid. In each droplet, an individual PCR is occurred using fluorescent Taqman probes. All droplets are then recorded as positive or negative, depending on the fluorescent signal. The fraction of positive droplets is then converted into a concentration measurement by applying

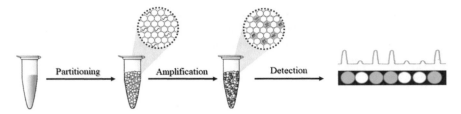

Fig. 6.3 Schematic representation of ddPCR workflow

the Poisson algorithm (Fig. 6.3) [49, 50]. Compared to real-time PCR, ddPCR offers superior sensitivity and absolute quantification of target nucleic acid, making it a valuable tool for clinical detection of viruses in samples with low viral load and to reduce false-negative results [51, 52]. For SARS-CoV-2, ddPCR showed a lower limit of detection than the standard real-time PCR. Suo et al. were able to detect 26 positive samples from COVID-19 outpatients with real-time PCR using the optimized ddPCR [53]. In another study, the sensitivity of ddPCR was compared with two methods of real-time PCR (SYBR-Green and TaqMan probe real-time PCR) approved by WHO for SARS-CoV-2 detection. The SYBRGreen real-time PCR failed to diagnose the positive samples with low viral load, and the TaqMan probe real-time PCR showed positive signals at very late C_t values. On the contrary, the ddPCR (using fluorescent chemistry or probe) is able to detect the samples' positive results with the very low viral load even at a 10-fold diluted concentration [54].

Furthermore, ddPCR was shown to be feasible for detecting viral surface contamination, which is an advantage over real-time PCR. In Lv and colleagues' study, the SARS-CoV-2 RNA residue at multiple sites using real-time PCR and ddPCR were detected. No positive results were obtained by real-time PCR, while using ddPCR, 13 of 61 samples were positive for SARS-CoV-2. The highest density of SARS-CoV-2 RNA was found on the outer gloves of operators and refrigerator's door handle [55]. Even though ddPCR is highly sensitive for accurate diagnosis of COVID-19 patients or infected surfaces, it should be kept in mind that this method is almost time-consuming and needs expert operators and specific facilities that may not be affordable in low-income settings. Thus, further simplification of these methods is necessary to reach the clinical application [56].

6.2.5   Isothermal Amplification Techniques

A major bottleneck to widespread real-time testing in the emerging pandemic is the need for a costly thermocycler, which hinders its use in low-resource setting laboratories. Isothermal amplification is a promising alternative that dramatically simplifies the process of amplification and does not require the thermal cycler instrument. Several isothermal amplification techniques, first reported by Notomi et al., have been developed over time [57]. Among them, recombinase polymerase amplification (RPA) and loop-mediated isothermal amplification (LAMP) have been used more frequently for rapid and sensitive detection of nucleic acids. The isothermal amplification techniques can be coupled with reverse-transcription to detect RNA targets (RT-LAMP), such as influenza, Zika, Ebola, and SARS viruses [58, 59]. The LAMP cycles proceed at a constant temperature (usually 60–65 °C for 60 min). Using a strand displacement reaction LAMP uses four different primers, specifically recognizing six distinct regions of the target sequence, therefore being highly sequence-specific. The four primers are as follows: Forward Inner Primer (FIP), Forward Outer Primer (FOP): The FOP (also called F3 Primer), Backward Inner Primer (BIP), and Backward Outer Primer (BOP). The inner primers synthesize an initial DNA strand, subsequently displaced by synthesis primed by outer primers using a strand-displacing DNA polymerase, and released as a single-stranded DNA. The reverse complementary sequence in the 5′ and 3′ ends anneal with a sequence in the displaced ssDNA strand, forming a loop. Repeated priming and strand displacement cycles generate stem-loop DNA structures with several inverted repeats of the target and cauliflower-like structures containing multiple loops. The final amplified products are detected using fluorescent or colorimetric dyes [60]. The primer optimization is a key step in the LAMP method. Moreover, due to the probability of independent primer set cross interaction, LAMP is a difficult method for multiplexing [61].

RPA is another isothermal method, which by adding a reverse transcriptase, can be used as an alternative for real-time PCR in RNA virus detection (RT-RPA). The process is operated optimally at 36–42 °C, and more slowly at room temperature, making it an excellent candidate for the low-cost point-of-care test in limited-resource settings. The RT-RPA has been used for the detection of many viruses [62]. In the RT-PPA, the viral RNA is converted to the dsDNA. The RPA process relies on three enzymes: recombinase, single-stranded DNA-binding protein (SSBP), and strand-displacing DNA polymerase. The primers are inserted at homologous target sequences using recombinase. Then, the SSBP stabilizes the displaced single-stranded DNA and prevents the primers' dissociation. The 3′ ends of the primers are now accessible to the strand-displacing DNA polymerase for elongation [63, 64]. Multiplexing is feasible with RT-RPA, which is useful to interrogate multiple loci when a limited amount of sample is available [65]. The Schematic of the typical LAMP and RPA methods are shown in Fig. 6.4.

Thi et al. tested several hundreds of RNA samples isolated from pharyngeal swabs of individuals tested for COVID-19. They found that RT-LAMP can detect the viral RNA more simply but with less sensitivity. They developed a method with no prior requirement for RNA isolation (using a direct swab to RT-LAMP assay) [66]. Huang et al. have designed four sets of LAMP primers for targeting N, S, and ORF1ab genes and were able to detect SARS-CoV-2 in 30 min at 65 °C. The amplification results were simply visualized with the naked eye due to visible color change based on a decrease in pH. This method was reported to have the sensitivity of detecting 80 copies of the targeted viral genome per ml. However, it can increase false-positive results based on carry-over contamination, common in the LAMP method [67]. In

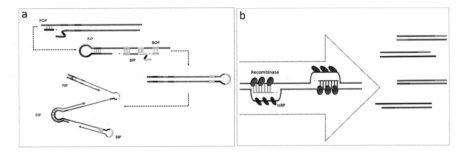

Fig. 6.4 The schematic diagrams of two isothermal amplification methods: **a** LAMP and **b** RPA. In the LAMP method, two pairs of primers (FOP/BOP and FIP/BIP) detect six regions on the target sequence, which results in a double-loop stem structure. The primer design in RPA is similar to that of standard PCR, and the amplification is conducted using three enzymes: recombinase, SSBP, and strand-displacing DNA polymerase

another study, Park et al. selected five sets of primers from the S gene, non-structural protein 3 (Nsp3), and ORF8 for RT-LAMP detection of SARS-CoV-2. The results showed the relatively low sensitivity of detection for S gene and ORF8 primers; however, the primers designed for Nsp3 were able to detect the viral RNA in samples with 100 viral RNA copies per reaction with no cross-reactivity with other human SARS-CoVs [68]. Behrman et al. designed an RT-PRA experiment that targets the N gene of SARS-CoV-2. This assay enabled the detection of 7.74 RNA copies per reaction in as little as 15-20 min, one of the fastest detection methods for SARS-CoV-2. The experiment showed no cross-reactivity with other tested coronaviruses [69]. To improve the SARS-CoV-2 detection, El-Tholoth and colleagues developed a two-stage isothermal method called Penn-RAMP that combined LAMP (for more specificity) and RPA (for more sensitivity). Both tests were performed in a closed-tube. The results showed that the COVID-19 RAMP has ten folds better sensitivity than the COVID-19 LAMP and COVID-19 RT-PCR using purified targets and 100 folds better sensitivity with rapidly prepared samples. The method has the potential to be used with minimal instrumentation and training [70].

6.2.6   CRISPR-Cas Systems

CRISPR (clustered regularly interspaced short palindromic repeats) are genomic loci in bacteria and archaea, accompanied by a set of homologous genes (*cas* genes) making up the CRISPR-associated system. This system functions as an adaptive immune system, protecting the prokaryotic microorganisms by inducing blunt double-stranded breaks in the invading DNA. Now, CRISPR-Cas9 mediated genome editing has revolutionized the biomedical sciences [71]. One of the CRISPR-Cas system's current applications is the diagnosis of viral infection, microbes, and diseases [72]. The CRISPR-Cas system is categorized based on the effector molecules to classes 1 and 2, which are further subdivided into six types and multiple subtypes. Class 2 of the CRISPR system is described by the presence of a single effector molecule and contains three types. Cas 9 and 12 are the endonucleases of type II and V, respectively, that target ssDNA molecules, and Cas 13 is the endonuclease of type VI and target ssRNA molecule. Diagnostic tests using Cas12 and Cas13 enzymes, dubbed DETECTR (*DNA* endonuclease-targeted *CRISPR* trans reporter) and SHER-LOCK (specific high-sensitivity enzymatic reporter unlocking), have already been developed to detect the SARS-CoV-2 virus nucleotide [72].

Targeting both Cas 12 and 13 is directed by a CRISPR RNA (crRNA) that matches a specific region in the target sequence. The presence of the target RNA activates the nuclease activity of Cas13. The collateral cleavage of Cas12 and 13 is the key issue for optical readout by which RNA reporter labels get cut and release the fluorescent signals [73]. Cas12 requires a protospacer adjacent motif (PAM) to facilitate the binding of crRNA to ssDNA. CRISPR-based diagnosis has many potencies for different readout approaches; for example, instead of using fluorophore and quencher that provides real-time monitoring, fluorescein amidite (FAM) can be used for a simple readout on a test strip. The production of signals occurs only in the existence of correct sequences produced by isothermal amplification; thus, CRISPR-based detection has significantly higher specificity compared to methods that use nonspecific detectors such as pH indicators or fluorescent dyes. However, the sensitivity is not suited for detecting viral RNA in the collected specimens [74]. Hence, researchers propose combining conventional amplification methods with the CRISPR system to achieve both specificity and sensitivity. For instance, Wang et al. introduced a CRISPR/Cas12a detection system for the diagnosis of COVID-19 that releases 485 nm green fluorescence light that the naked eye can easily observe. They have designed and compared 15 crRNAs (designed on four domains of the E and N genes, orf1b and orf1a), and all of them were validated except E-crRNA1. In order to have enough DNA for the detection by Cas12a, Wang et al. used the combination of reverse transcript recombinase-aided amplification (RT-RAA) and the CRISPR system. CRISPR/Cas12a reaction occurred at 37 °C after RT-RAA had amplified the target gene at 39 °C in 30 min. The results indicate that crRNA that targets the E gene has the highest sensitivity (detecting ten copies of synthetic viral genome in a reaction) [75]. The combination of LAMP and DETECTR system also was suggested to increase the sensitivity of detection for SARS-CoV-2 [76].

Zhang et al. modified the SHERLOCK protocol for the diagnosis of SARS-CoV-2 by designing primers for targeting the Orf1ab and S gene. This method includes three steps: (1) using an RPA kit for amplifying the extracted nucleic acids, (2) detection of viral RNA by Cas13, and (3) using a paper dipstick for visual readout. The whole protocol can be completed in one hour. In another study, the SHER-LOCK method has been developed into a one-step diagnosis called STOP (SHER-LOCK testing in one pot) that reduces the chance of contamination and multiple handling steps that the SHERLOCK method requires. In this method, LAMP is used instead of RPA to overcome the challenges such as limitations in the supply chain for providing conventional RPA reagents. Furthermore, LAMP facilitates the design of one-step and sensitive detection. Compared to RPA, LAMP's operation occurs at higher temperatures; therefore, a thermostable Cas enzyme is needed. The results can be observed after 40 min through the fluorescence readout, which expands to 70 min for a visual readout by lateral flow assay [77].

6.3 Conclusion

The occurrence of COVID-19 pandemic accelerates many types of research in the field of therapy, vaccine, and rapid diagnosis. Despite the lack of promising results in the therapeutic field, rapid vaccine development and the emergence of innovative detection methods would greatly help crisis management. So far, limited methods have been approved by regulatory agencies for the commercial detection of SARS-CoV-2. However, there is no doubt that amongst the very diverse technologies that are testing in laboratories worldwide, some of their best will soon enter the market and dramatically affect the virus detection cost and time. We categorized and reviewed some innovative methods based on nucleic acid detection in this commentary that can be the next technologies for large-scale diagnosis. In the next chapter, the other methods will be discussed.

References

1. Zhou L, Li Z, Zhou J, Li H, Chen Y, Huang Y, Xie D, Zhao L, Fan M, Hashmi S (2020) A rapid, accurate and machine-agnostic segmentation and quantification method for CT-based covid-19 diagnosis. IEEE Trans Med Imaging 39(8):2638–2652
2. Tahamtan A, Ardebili A (2020) Real-time RT-PCR in COVID-19 detection: issues affecting the results. Taylor & Francis
3. Mo X, Qin W, Fu Q, Guan M (2020) Understanding the influence factors in viral nucleic acid test of 2019 novel coronavirus. Chin J Lab Med 43(3)
4. Punn NS, Agarwal S (2020) Automated diagnosis of COVID-19 with limited posteroanterior chest X-ray images using fine-tuned deep neural networks. arXiv preprint. arXiv:2004.11676
5. Zali A, Sohrabi M-R, Mahdavi A, Khalili N, Taheri MS, Maher A, Sadoughi M, Zarghi A, Ziai SA, Shabestari AA (2020) Correlation between low-dose chest computed tomography and RT-PCR results for the diagnosis of COVID-19: a report of 27824 cases in Tehran, Iran. Acad Radiol
6. Huang P, Liu T, Huang L, Liu H, Lei M, Xu W, Hu X, Chen J, Liu B (2020) Use of chest CT in combination with negative RT-PCR assay for the 2019 novel coronavirus but high clinical suspicion. Radiology 295(1):22–23
7. Peeling RW, Wedderburn CJ, Garcia PJ, Boeras D, Fongwen N, Nkengasong J, Sall A, Tanuri A, Heymann DL (2020) Serology testing in the COVID-19 pandemic response. Lancet Infect Dis
8. Lu R, Zhao X, Li J, Niu P, Yang B, Wu H, Wang W, Song H, Huang B, Zhu N (2020) Genomic characterisation and epidemiology of 2019 novel coronavirus: implications for virus origins and receptor binding. The Lancet 395(10224):565–574
9. Alluwaimi AM, Alshubaith IH, Al-Ali AM, Abohelaika S (2020) The coronaviruses of animals and birds: their zoonosis, vaccines, and models for SARS-CoV and SARS-CoV2. Front Veterin Sci 7:655

10. Naqvi AAT, Fatima K, Mohammad T, Fatima U, Singh IK, Singh A, Atif SM, Hariprasad G, Hasan GM, Hassan MI (2020) Insights into SARS-CoV-2 genome, structure, evolution, pathogenesis and therapies: structural genomics approach. Biochim Biophys Acta (BBA) Mol Basis Dis 165878

11. Yao H-P, Lu X, Chen Q, Xu K, Chen Y, Cheng L, Liu F, Wu Z, Wu H, Jin C (2020) Patient-derived mutations impact pathogenicity of SARS-CoV-2. CELL-D-20-01124

12. Petty TJ, Cordey S, Padioleau I, Docquier M, Turin L, Preynat-Seauve O, Zdobnov EM, Kaiser L (2014) Comprehensive human virus screening using high-throughput sequencing with a user-friendly representation of bioinformatics analysis: a pilot study. J Clin Microbiol 52(9):3351–3361

13. Levy SE, Myers RM (2016) Advancements in next-generation sequencing. Annu Rev Genomics Hum Genet 17(1):95–115

14. Burgess DJ (2018) Next regeneration sequencing for reference genomes. Nat Rev Genet 19(3):125

15. Zhu N, Zhang D, Wang W, Li X, Yang B, Song J, Zhao X, Huang B, Shi W, Lu R (2020) A novel coronavirus from patients with pneumonia in China. New England J Med

16. Pollard MO, Gurdasani D, Mentzer AJ, Porter T, Sandhu MS (2018) Long reads: their purpose and place. Hum Mol Genet 27(R2):R234–R241

17. Crits-Christoph A, Kantor RS, Olm MR, Whitney ON, Al-Shayeb B, Lou YC, Flamholz A, Kennedy LC, Greenwald H, Hinkle A (2020) Genome sequencing of sewage detects regionally prevalent SARS-CoV-2 variants. medRxiv

18. Kim J-S, Jang J-H, Kim J-M, Chung Y-S, Yoo C-K, Han M-G (2020) Genome-wide identification and characterization of point mutations in the SARS-CoV-2 genome. Osong Public Health Res Perspect 11(3):101

19. Chiu CY, Miller SA (2019) Clinical metagenomics. Nat Rev Genet 20(6):341–355

20. Watzinger F, Ebner K, Lion T (2006) Detection and monitoring of virus infections by real-time PCR. Mol Aspects Med 27(2–3):254–298

21. Valasek MA, Repa JJ (2005) The power of real-time PCR. Adv Physiol Educ 29(3):151–159

22. Bukasov R, Dossym D, Filchakova O (2020) Detection of RNA viruses from influenza and HIV to Ebola and SARS-CoV-2: a review. Anal Methods

23. Corman VM, Landt O, Kaiser M, Molenkamp R, Meijer A, Chu DK, Bleicker T, Brünink S, Schneider J, Schmidt ML (2020) Detection of 2019 novel coronavirus (2019-nCoV) by real-time RT-PCR. Eurosurveillance 25(3):2000045

24. Toptan T, Hoehl S, Westhaus S, Bojkova D, Berger A, Rotter B, Hoffmeier K, Cinatl J, Ciesek S, Widera M (2020) Optimized qRT-PCR approach for the detection of intra-and extra-cellular SARS-CoV-2 RNAs. Int J Mol Sci 21(12):4396

25. Ponchel F, Toomes C, Bransfield K, Leong FT, Douglas SH, Field SL, Bell SM, Combaret V, Puisieux A, Mighell AJ, Robinson PA, Inglehearn CF, Isaacs JD, Markham AF (2003) Real-time PCR based on SYBR-Green I fluorescence: an alternative to the TaqMan assay for a relative quantification of gene rearrangements, gene amplifications and micro gene deletions. BMC Biotechnol 3(1):18

26. Tajadini M, Panjehpour M, Javanmard SH (2014) Comparison of SYBR Green and TaqMan methods in quantitative real-time polymerase chain reaction analysis of four adenosine receptor subtypes. Adv Biomed Res 3

27. Bhadra S, Maranhao AC, Ellington AD (2020) A one-enzyme RT-qPCR assay for SARS-CoV-2, and procedures for reagent production. bioRxiv

28. Udugama B, Kadhiresan P, Kozlowski HN, Malekjahani A, Osborne M, Li VYC, Chen H, Mubareka S, Gubbay JB, Chan WCW (2020) Diagnosing COVID-19: the disease and tools for detection. ACS Nano 14(4):3822–3835

29. Nalla AK, Casto AM, Huang M-LW, Perchetti GA, Sampoleo R, Shrestha L, Wei Y, Zhu H, Jerome KR, Greninger AL (2020) Comparative performance of SARS-CoV-2 detection assays using seven different primer-probe sets and one assay kit. J Clin Microbiol 58(6):e00557-20

30. Vogels CBF, Brito AF, Wyllie AL, Fauver JR, Ott IM, Kalinich CC, Petrone ME, Casanovas –Massana A, Catherine Muenker M, Moore AJ, Klein J, Lu P, Lu-Culligan A, Jiang X, Kim

DJ, Kudo E, Mao T, Moriyama M, Oh JE, Park A, Silva J, Song E, Takahashi T, Taura M, Tokuyama M, Venkataraman A, Weizman O-E, Wong P, Yang Y, Cheemarla NR, White EB, Lapidus S, Earnest R, Geng B, Vijayakumar P, Odio C, Fournier J, Bermejo S, Farhadian S, Dela Cruz CS, Iwasaki A, Ko AI, Landry ML, Foxman EF, Grubaugh ND (2020) Analytical sensitivity and efficiency comparisons of SARS-CoV-2 RT–qPCR primer–probe sets. Nat Microbiol 5(10):1299–1305

31. Lippi G, Simundic A-M, Plebani M (2020) Potential preanalytical and analytical vulnerabilities in the laboratory diagnosis of coronavirus disease 2019 (COVID-19). Clin Chem Lab Med (CCLM) 58(7):1070–1076

32. Wang Y, Kang H, Liu X, Tong Z (2020) Combination of RT-qPCR testing and clinical features for diagnosis of COVID-19 facilitates management of SARS-CoV-2 outbreak. J Med Virol 92(6):538–539

33. Daley P, Castriciano S, Chernesky M, Smieja M (2006) Comparison of flocked and Rayon Swabs for collection of respiratory epithelial cells from uninfected volunteers and symptomatic patients. J Clin Microbiol 44(6):2265–2267

34. Wang W, Xu Y, Gao R, Lu R, Han K, Wu G, Tan W (2020) Detection of SARS-CoV-2 in different types of clinical specimens. JAMA 323(18):1843–1844

35. Muhammad A, Ameer H, Haider SA, Ali I (2021) Detection of SARS-CoV-2 using real-time polymerase chain reaction in different clinical specimens: a critical review. Allergologia et Immunopathologia 49(1):159–164

36. Patel R, Babady E, Theel ES, Storch GA, Pinsky BA, St. George K, Smith TC, Bertuzzi S (2020) Report from the American Society for Microbiology COVID-19 international summit, 23 March 2020: value of diagnostic testing for SARS–CoV-2/COVID-19. mBio 11(2):e00722-20

37. Hong KH, Lee SW, Kim TS, Huh HJ, Lee J, Kim SY, Park J-S, Kim GJ, Sung H, Roh KH, Kim J-S, Kim HS, Lee S-T, Seong M-W, Ryoo N, Lee H, Kwon KC, Yoo CK, K.S.f.L.M.C.-T. Force, t.K.C.f.D.C. the Center for Laboratory Control of Infectious Diseases, Prevention (2019) Guidelines for laboratory diagnosis of coronavirus disease 2019 (COVID-19) in Korea. Ann Lab Med 40(5):351–360

38. Rothe C, Schunk M, Sothmann P, Bretzel G, Froeschl G, Wallrauch C, Zimmer T, Thiel V, Janke C, Guggemos W, Seilmaier M, Drosten C, Vollmar P, Zwirglmaier K, Zange S, Wölfel R, Hoelscher M (2020) Transmission of 2019-nCoV infection from an asymptomatic contact in Germany. N Engl J Med 382(10):970–971

39. Liu Z, Xiao X, Wei X, Li J, Yang J, Tan H, Zhu J, Zhang Q, Wu J, Liu L (2020) Composition and divergence of coronavirus spike proteins and host ACE2 receptors predict potential intermediate hosts of SARS-CoV-2. J Med Virol 92(6):595–601

40. Scallan MF, Dempsey C, MacSharry J, O'Callaghan I, O'Connor PM, Horgan CP, Durack E, Cotter PD, Hudson S, Moynihan HA, Lucey B (2020) Validation of a Lysis buffer containing 4 M guanidinium thiocyanate (GITC)/Triton X-100 for extraction of SARS-CoV-2 RNA for COVID-19 testing: comparison of formulated Lysis buffers containing 4 to 6 M GITC, Roche external lysis buffer and Qiagen RTL lysis buffer. bioRxiv

41. Aitken J, Ambrose K, Barrell S, Beale R, Bineva-Todd G, Biswas D, Byrne R, Caidan S, Cherepanov P, Churchward L, Clark G, Crawford M, Cubitt L, Dearing V, Earl C, Edwards A, Ekin C, Fidanis E, Gaiba A, Gamblin S, Gandhi S, Goldman J, Goldstone R, Grant PR, Greco M, Heaney J, Hindmarsh S, Houlihan CF, Howell M, Hubank M, Hughes D, Instrell R, Jackson D, Jamal-Hanjani M, Jiang M, Johnson M, Jones L, Kanu N, Kassiotis G, Kirk S, Kjaer S, Levett A, Levett L, Levi M, Lu W-T, MacRae JI, Matthews J, McCoy L, Moore C, Moore D, Nastouli E, Nicod J, Nightingale L, Olsen J, OReilly N, Pabari A, Papayannopoulos V, Patel N, Peat N, Pollitt M, Ratcliffe P, Reis e Sousa C, Rosa A, Rosenthal R, Roustan C, Rowan A, Shin GY, Snell DM, Song O-R, Spyer M, Strange A, Swanton C, Turner JMA, Turner M, Wack A, Walker PA, Ward S, Wong WK, Wright J, Wu M (2020) Scalable and resilient SARS-CoV-2 testing in an Academic Centre. medRxiv

42. Kalnina L, Mateu-Regué À, Oerum S, Hald A, Gerstoft J, Oerum H, Nielsen FC, Iversen AKN (2020) A simple, safe and sensitive method for SARS-CoV-2 inactivation and RNA extraction for RT-qPCR. bioRxiv

43. Pastorino B, Touret F, Gilles M, de Lamballerie X, Charrel RN (2020) Evaluation of heating and chemical protocols for inactivating SARS-CoV-2. bioRxiv
44. Smyrlaki I, Ekman M, Lentini A, Rufino de Sousa N, Papanicolaou N, Vondracek M, Aarum J, Safari H, Muradrasoli S, Rothfuchs AG, Albert J, Högberg B, Reinius B (2020) Massive and rapid COVID-19 testing is feasible by extraction-free SARS-CoV-2 RT-PCR. Nat Commun 11(1):4812
45. Wang J, Cai K, Zhang R, He X, Shen X, Liu J, Xu J, Qiu F, Lei W, Wang J, Li X, Gao Y, Jiang Y, Xu W, Ma X (2020) Novel one-step single-tube nested quantitative real-time PCR assay for highly sensitive detection of SARS-CoV-2. Anal Chem 92(13):9399–9404
46. Yip CC-Y, Sridhar S, Leung K-H, Ng AC-K, Chan K-H, Chan JF-W, Tsang OT-Y, Hung IF-N, Cheng VC-C, Yuen K-Y (2020) Development and evaluation of novel and highly sensitive single-tube nested real-time RT-PCR assays for SARS-CoV-2 detection. Int J Mol Sci 21(16):5674
47. Kosinová L, Cahová M, Fábryová E, Týcová I, Koblas T, Leontovyč I, Saudek F, Kříž J (2016) Unstable expression of commonly used reference genes in rat pancreatic islets early after isolation affects results of gene expression studies. PLoS ONE 11(4):
48. Caliendo AM, Shahbazian MD, Schaper C, Ingersoll J, Abdul-Ali D, Boonyaratanakornkit J, Pang X-L, Fox J, Preiksaitis J, Schönbrunner ER (2009) A commutable cytomegalovirus calibrator is required to improve the agreement of viral load values between laboratories. Clin Chem 55(9):1701–1710
49. Hayden R, Gu Z, Ingersoll J, Abdul-Ali D, Shi L, Pounds S, Caliendo A (2013) Comparison of droplet digital PCR to real-time PCR for quantitative detection of cytomegalovirus. J Clin Microbiol 51(2):540–546
50. Brunetto GS, Massoud R, Leibovitch EC, Caruso B, Johnson K, Ohayon J, Fenton K, Cortese I, Jacobson S (2014) Digital droplet PCR (ddPCR) for the precise quantification of human T-lymphotropic virus 1 proviral loads in peripheral blood and cerebrospinal fluid of HAM/TSP patients and identification of viral mutations. J Neurovirol 20(4):341–351
51. Baker M (2012) Digital PCR hits its stride. Nat Methods 9(6):541–544
52. Duewer DL, Kline MC, Romsos EL, Toman B (2018) Evaluating droplet digital PCR for the quantification of human genomic DNA: converting copies per nanoliter to nanograms nuclear DNA per microliter. Anal Bioanal Chem 410(12):2879–2887
53. Suo T, Liu X, Feng J, Guo M, Hu W, Guo D, Ullah H, Yang Y, Zhang Q, Wang X (2020) ddPCR: a more accurate tool for SARS-CoV-2 detection in low viral load specimens. Emerg Microbes Infect 9(1):1259–1268
54. Falzone L, Musso N, Gattuso G, Bongiorno D, Palermo CI, Scalia G, Libra M, Stefani S (2020) Sensitivity assessment of droplet digital PCR for SARS-CoV-2 detection. Int J Mol Med 46(3):957–964
55. Lv J, Yang J, Xue J, Zhu P, Liu L, Li S (2020) Detection of SARS-CoV-2 RNA residue on object surfaces in nucleic acid testing laboratory using droplet digital PCR. Sci Total Environ 742:
56. Alteri C, Cento V, Antonello M, Colagrossi L, Merli M, Ughi N, Renica S, Matarazzo E, Di Ruscio F, Tartaglione L, Colombo J, Grimaldi C, Carta S, Nava A, Costabile V, Baiguera C, Campisi D, Fanti D, Vismara C, Fumagalli R, Scaglione F, Epis OM, Puoti M, Perno CF (2020) Detection and quantification of SARS-CoV-2 by droplet digital PCR in real-time PCR negative nasopharyngeal swabs from suspected COVID-19 patients. PLoS ONE 15(9):
57. Rekha V, Rana R, Arun TR, Aswathi PB, Kalluvila J, John DG, Sadanandan GV, Jacob A (2014) Loop mediated isothermal amplification (LAMP) test–a novel nucleic acid based assay for disease diagnosis. Adv Anim Vet Sci 2:344–350
58. Hong TC, Mai QL, Cuong DV, Parida M, Minekawa H, Notomi T, Hasebe F, Morita K (2004) Development and evaluation of a novel loop-mediated isothermal amplification method for rapid detection of severe acute respiratory syndrome coronavirus. J Clin Microbiol 42(5):1956–1961
59. Wong YP, Othman S, Lau YL, Radu S, Chee HY (2018) Loop-mediated isothermal amplification (LAMP): a versatile technique for detection of micro-organisms. J Appl Microbiol 124(3):626–643

60. Abad-Valle P, Fernández-Abedul MT, Costa-García A (2005) Genosensor on gold films with enzymatic electrochemical detection of a SARS virus sequence. Biosens Bioelectron 20(11):2251–2260

61. Uwiringiyeyezu T, El Khalfi B, Belhachmi J, Soukri A (2019) Loop-mediated Isothermal amplification LAMP, simple alternative technique of molecular diagnosis process in medicals analysis: a review. Annu Res Rev Biol 1–12

62. Naveen K, Bhat A (2020) Development of reverse transcription loop-mediated isothermal amplification (RT-LAMP) and reverse transcription recombinase polymerase amplification (RT-RPA) assays for the detection of two novel viruses infecting ginger. J Virol Methods 282:

63. Esbin MN, Whitney ON, Chong S, Maurer A, Darzacq X, Tjian R (2020) Overcoming the bottleneck to widespread testing: a rapid review of nucleic acid testing approaches for COVID-19 detection. RNA 26(7):771–783

64. Bonnet EH (2019) The development and validation of a reverse transcription recombinase polymerase amplification assay for detection of flaviviruses. University of the Free State

65. Daunay A, Duval A, Baudrin LG, Buhard O, Renault V, Deleuze J-F, How-Kit A (2019) Low temperature isothermal amplification of microsatellites drastically reduces stutter arti-fact formation and improves microsatellite instability detection in cancer. Nucl Acids Res 47(21):e141–e141

66. Thi VLD, Herbst K, Boerner K, Meurer M, Kremer LP, Kirrmaier D, Freistaedter A, Papagiannidis D, Galmozzi C, Stanifer ML (2020) A colorimetric RT-LAMP assay and LAMP-sequencing for detecting SARS-CoV-2 RNA in clinical samples. Sci Transl Med 12(556)

67. Huang WE, Lim B, Hsu C-C, Xiong D, Wu W, Yu Y, Jia H, Wang Y, Zeng Y, Ji M, Chang H, Zhang X, Wang H, Cui Z (2020) RT-LAMP for rapid diagnosis of coronavirus SARS-CoV-2. Microb Biotechnol 13(4):950–961

68. Park G-S, Ku K, Baek S-H, Kim S-J, Kim SI, Kim B-T, Maeng J-S (2020) Development of reverse transcription loop-mediated isothermal amplification assays targeting severe acute respiratory syndrome coronavirus 2 (SARS-CoV-2). J Mol Diagn 22(6):729–735

69. Behrmann O, Bachmann I, Spiegel M, Schramm M, Abd El Wahed A, Dobler G, Dame G, Hufert FT (2020) Rapid detection of SARS-CoV-2 by low volume real-time single tube reverse transcription recombinase polymerase amplification using an exo probe with an internally linked quencher (exo-IQ). Clin Chem 66(8):1047–1054

70. El-Tholoth M, Bau HH, Song J (2020) A single and two-stage, closed-tube, molecular test for the 2019 novel coronavirus (COVID-19) at home, clinic, and points of entry. ChemRxiv

71. Mali P, Yang L, Esvelt KM, Aach J, Guell M, DiCarlo JE, Norville JE, Church GM (2013) RNA-guided human genome engineering via Cas9. Science 339(6121):823–826

72. Katalani C, Booneh HA, Hajizade A, Sijercic A, Ahmadian G (2020) CRISPR-based diagnosis of infectious and noninfectious diseases. Biol Proced Online 22(1):1–14

73. Myhrvold C, Freije CA, Gootenberg JS, Abudayyeh OO, Metsky HC, Durbin AF, Kellner MJ, Tan AL, Paul LM, Parham LA, Garcia KF, Barnes KG, Chak B, Mondini A, Nogueira ML, Isern S, Michael SF, Lorenzana I, Yozwiak NL, MacInnis BL, Bosch I, Gehrke L, Zhang F, Sabeti PC (2018) Field-deployable viral diagnostics using CRISPR-Cas13. Science 360(6387):444

74. Gootenberg JS, Abudayyeh OO, Lee JW, Essletzbichler P, Dy AJ, Joung J, Verdine V, Donghia N, Daringer NM, Freije CA, Myhrvold C, Bhattacharyya RP, Livny J, Regev A, Koonin EV, Hung DT, Sabeti PC, Collins JJ, Zhang F (2017) Nucleic acid detection with CRISPR-Cas13a/C2c2. Science 356(6336):438

75. Wang X, Zhong M, Liu Y, Ma P, Dang L, Meng Q, Wan W, Ma X, Liu J, Yang G, Yang Z, Huang X, Liu M (2020) Rapid and sensitive detection of COVID-19 using CRISPR/Cas12a-based detection with naked eye readout, CRISPR/Cas12a-NER. Sci Bull 65(17):1436–1439

76. Brandsma E, Verhagen HJMP, van de Laar TJW, Claas ECJ, Cornelissen M, van den Akker E (2020) Rapid, sensitive, and specific severe acute respiratory syndrome coronavirus 2 detection: a multicenter comparison between standard quantitative reverse-transcriptase polymerase chain reaction and CRISPR-based DETECTR. J Infect Dis

77. Joung J, Ladha A, Saito M, Segel M, Bruneau R, Huang M-lW, Kim N-G, Yu X, Li J, Walker BD (2020) Point-of-care testing for COVID-19 using SHERLOCK diagnostics. MedRxiv

Chapter 7
COVID-19 Diagnosis: A Comprehensive Review of Current Testing Platforms; Part B

Fataneh Fatemi, Zahra Hassani Nejad, Seyed Ehsan Ranaei Siadat, Sareh Arjmand, Behrad Ghiasi, and Samin Haghighi Poodeh

7.1 Introduction

In early December 2019, the coronavirus disease 2019 (COVID-19) was discovered in China when a cluster of patients with pneumonia symptoms of unknown pathogen were identified. Subsequent investigation and genomic analysis of the pathogen led to the discovery of the novel coronavirus (2019-nCoV) as the culprit viral species [1]. The virus is now named severe acute respiratory syndrome coronavirus 2 (SARS-CoV-2), one of the seven coronaviruses that can infect human and cause mild to lethal respiratory tract infections [2]. The SARS-CoV-2 is more transmittable and pathogenic than the previously identified SARS-CoV and Middle East respiratory syndrome coronavirus (MERS-CoV). Currently, COVID-19 has disseminated to more than 200 countries/territories causing a high rate of morbidity and mortality. Age and comorbid illness increase the risk of death among the infected [3, 4].

One of the issues in controlling the spread of COVID-19 stems from the fact that its transmission can be through pre-symptomatic (SARS-CoV-2 is detectable before symptoms development) or asymptomatic (SARS-CoV-2 is detectable without symptoms) individuals [5–8], making the mitigation and containment of the disease more problematic. The basic reproductive number (R_0) of SARS-CoV-2 is estimated to be

F. Fatemi (✉) · S. Arjmand (✉) · B. Ghiasi · S. Haghighi Poodeh
Protein Research Center, Shahid Beheshti University, Tehran, Iran
e-mail: f_fatemi@sbu.ac.ir

S. Arjmand
e-mail: s_arjmand@sbu.ac.ir

Z. Hassani Nejad
Institute of Biochemistry and Biophysics, University of Tehran, Tehran, Iran

S. E. R. Siadat
Sobhan Recombinant Protein, No. 22, 2nd Noavari St, Pardis Technology Park, 20th Km of Damavand Road, Tehran, Iran

around 5.7, meaning each COVID-19 patient can infect an additional 5.7 individuals on average [9]. The primary mode of transmission is by human to human contact, through respiratory droplets, aerosols, and fomites [4, 10, 11].

The clinical features of COVID-19 in patients range from mild to severe illness. Fever, sore throat, shortness of breath, chest pain, cough, headache, and the sudden appearance of olfaction and taste disturbances (OTDs) are among common symptoms seen in COVID-19 patients. Less common symptoms are muscle pain, sputum production, hemoptysis, and diarrhea [3, 4, 12]. The median incubation period for SARS-CoV-2 is approximately 5.1 days with 97.5% of patients exhibiting symptoms after 11.5 days [13].

The catastrophic economic impact the COVID-19 pandemic has caused is pushing governments to reopen their economies and lift restrictions that were initially implemented, and thereby giving rise to new COVID-19 cases. Given the lack of specific medication against COVID-19, the only currently available way to mitigate the spread of the virus is through vaccination, early and reliable diagnosis, social distancing, and isolation of those who are infected and contagious.

There are two main classes of diagnostic tests for COVID-19: nucleic acid amplification tests (NAAT), also known as molecular tests, and serological/antibody-based tests. NAAT are the most common diagnostic tests used for detecting viral RNA. NAAT has been largely discussed in part A of this review.

Alternative diagnostic platforms are based on the detection of viral antigens and antibodies produced in response to SARS-CoV-2. Antibody-based methods detect seroconversion in plasma or serum based on platforms like enzyme-linked immunosorbent assay (ELISA), colloidal gold-based immunochromatographic assay, lateral flow immunoassay (LFIA), and chemiluminescent immunoassay (CLIA). However, since seroconversion is time-dependent, these tests might fail to detect antibodies in patients who are already infected with SARS-CoV-2 and are more applicable for detecting previous infections.

In some cases, imaging is used for diagnosing patients. Some imaging modalities used for COVID-19 patients are computed tomography (CT), lung ultrasound (LUS), and chest radiograph (CXR). However, there are some drawbacks related to these methods. For instance, most of these tests are limited to healthcare settings, have little to no mobility, and require well-trained radiologists to examine the images. In addition, the use of CT for all patients seems impractical in terms of cost and time and the radiation exposure of patients.

The electrochemical biosensors have the potential to become reliable and timely diagnostic tools for COVID-19 diagnosis. Most of these biosensors target the nucleocapsid antigen and provide fast results. Currently, there are a number of research groups that have developed various biosensors for SARS-CoV-2 detection. Further evaluation of biosensor's clinical performance can significantly improve COVID-19 diagnostic testing.

The COVID-19 pandemic is not expected to get eradicated in the near future, and the growing number of COVID-19 infections warrant novel, accessible, and accurate testing strategies to improve diagnostic capacity. Here, we aim to summarize and provide a comprehensive overview of current diagnostic serological/antibody-based tests and available biosensor methods developed for detecting COVID-19.

7.2 Serological (Antibody-Based) Tests

Compared to molecular tests, antibody tests are faster, less expensive, and easier to implement. Serological tests can be useful in determining the real number of infections which is essential for determining the fatality rate of COVID-19 and making decisions on duration of social lockdowns. Serological tests are also essential for finding donors for convalescent plasma therapy. Furthermore, serological assays enable us to study the immune responses in qualitative and quantitative manner.

In contrast to molecular-based methods that detect viral nucleic acids, serological assays detect patients' antibody responses against SARS-CoV-2. Antibodies, also known as immunoglobins (Ig), are the main part of adaptive immune systems that take effect after adaptive immune responses are initiated and provide long-lasting protection. After detecting pathogens, the B lymphocyte cells are activated through a sequence of very specific events. Once a naïve B cell is activated, it begins to clonally expand, divide, and become specialized; as a result, it differentiates into an Ig secreting plasma cell or B memory cells [14, 15].

The antibody has a large "Y" shaped structure composed of four chains: two identical light chains (LC) and two identical heavy chains (HC) that are held together by a combination of noncovalent and covalent (disulfide) bonds. The heavy and light chains cooperate to form the two identical antigen-binding sites at the chains' N-terminal regions. This region is variable between antibody molecules that are of individual B cell descent, and determines their specificity [16]. It has been discovered that mild digestion of antibodies with papain produces three fragments; two identical ones that contain the antigen-binding sites (Fab fragments), and a crystallizable fragment (Fc), which interacts with the effector molecules and cells [17] (Fig. 7.1). There are five main classes (isotypes) of antibodies, defined by the Fc portion of the heavy chains; IgA, IgD, IgE, IgG, and IgM. Each isotype has an independent function in the adaptive immune system [18].

IgM is the most released antibody during viral infection, found mainly in blood and lymph fluids, provides short-term protection. The high-affinity IgG antibodies are secreted after IgM and are mainly responsible for long-term immunity after an infection or vaccination [19, 20].

The antibody tests detect IgA, IgM, IgG, or total antibodies against the virus. During the SARS epidemic in 2003, it was found that viral specific antibodies IgM and IgG indicate viral infection [21]. After the first outbreak of SARS, the studies have shown that antiviral IgM and IgG are detectable at week one of the onset of the symptoms, and IgM disappeared earlier than IgG [22, 23]. Since SARS-CoV-2 belongs to the same family of viruses, the detection of these antibodies could aid in the diagnosis and treatment of COVID-19.

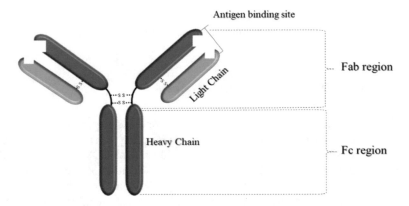

Fig. 7.1 Structure of antibodies. Fc domain of heavy chains determines antibody isotype

The antibody response against SARS-CoV-2 is not well understood. The seroconversion rates in COVID-19 patient has been reported to be 50% on day 7 and 100% on day 14 post symptom onset [24]. IgM and IgG are reported to be detectable in blood around day 3 and day 5 of infection, respectively [25]. Another study reports that both IgM and IgG antibodies against SARS-CoV-2 are detected as early as the 4th day post symptoms onset, and the seropositive rate of IgG has been reported to decrease around the 28th day after symptoms onset [26]. Another study on humoral immune response has detected IgM and IgA antibodies as early as the 5th day after infection which is earlier than IgG detection [27]. A study has reported earlier seroconversion (2 days after onset of symptoms) and higher levels of IgA antibody detection compared to IgM antibody in both severe and non-severe patients [28]. In addition, the levels of IgA and IgG were reported to be relatively higher in severe patients compared to non-severe patients, while the IgM levels were not much different between the two cases. Therefore, IgA might be a more reliable marker for early diagnosis of SARS-CoV-2, and its incorporation in serological tests can increase the sensitivity of the test [29].

The asymptomatic and symptomatic patients exhibit different kinetics of IgG/IgM responses to SARS-CoV-2. Asymptomatic patients might seroconvert later in the course of infection or may not seroconvert at all. IgG seroconversion is commonly seen in both asymptomatic and symptomatic patients. However, IgM seroconversion is less commonly seen in asymptomatic than in symptomatic patients. Furthermore, asymptomatic patients have lower IgG/IgM titers and plasma neutralization capacity at the virus clearance time compared to symptomatic patients [30]. Generally in SARS-CoV-2 infection, IgM declines more rapidly than IgG, and its disappearance may be an indicator of virus clearance [31]. Figure 7.2 shows the kinetic of serological response to SARS-CoV-2 in humans.

The serological tests have the advantage of being fast and simple to be performed in a typical laboratory or at a point of care. The serum and plasma specimen are

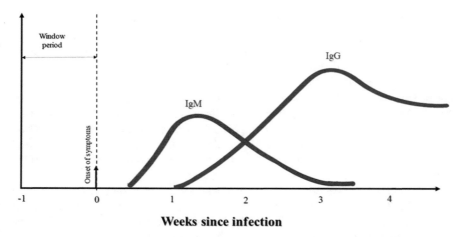

Fig. 7.2 Kinetics of serological (IgM, IgG) response, after SARS-CoV-2 infection

usually used to detect the antibodies in COVID-19 patients, but other biological fluids such as saliva or sputum could be used as well [32, 33].

The most commonly used viral antigens in antibody tests for coronavirus infections are the N and the S protein, including subunits S1 and S2 as well as receptor-binding domain RBD [1]. Unlike PCR tests that possesses high specificity, cross-reactivity is common in serological tests since there are six other coronaviruses that can infect humans. Among these coronaviruses, SARS-CoV-2 is genetically related to SARS-CoV sharing about 80% sequence similarity [1]. Significant cross-reactivity has been reported when N protein of either virus is used. However, the S1 or RBD region from the S protein provides better specificity [34].

There are three frequently used serological tests that enable fast detection of SARS-CoV-2 specific antibodies including lateral flow immunoassay (LFIA), chemiluminescence immunoassay (CLIA), and enzyme-linked immunosorbent assay (ELISA). These assays possess different sensitivities and detection time. ELISA technology is the most commonly used serological test format, used in the diagnostic laboratory, with an average time-to-answer of 2–5 h. Typically, the microtiter plates are coated with SARS-CoV-2 antigens and incubated with the patient's samples to bind and detect the corresponding antibodies. A secondary antibody (anti-human IgG or IgM) conjugated to a reporter molecule is used to identify the bound antigen–antibody complexes using colorimetric or fluorescent detection. The ELISA assays could be designed in direct, indirect, competitive, or sandwich formats. CLIA follows a similar concept to ELISA, but has more sensitivity, and is useful for measuring low antibody concentrations. In CLIA assay, the chemiluminescent indicators are used to label antigen or antibodies directly, which improves the analytical sensitivity of immunoassay. However, the relatively short duration of light emission limits its application to robot assistants [35]. RDT is an immunochromatography test based on lateral flow immunoassay (LFIA) technology, commonly used in pregnancy test kits. RDT uses the same principle of ELISA and is useful for the point of care or self-test application. The antigens are fixed to a nitro-cellulose strip, embedded in a cassette, and a drop of blood is used to detect antibodies' presence. The colored lines that typically appeared within 10–30 min indicate positive or negative results [36].

ELISA assays have been reported to be more accurate than LFIA in detection and quantification of SARS-CoV-2 IgM and IgG antibodies. To enhance the detection efficiency of antibody-based tests, simultaneous detection of IgM and IgG antibodies are recommended. Some studies have compared simultaneous detection of antibodies, and compared results with single antibody detection. The tests detecting IgM and IgG exhibited significantly higher sensitivity when compared to IgM or IgG single detection [37, 38].

Although serological tests have many advantages, they cannot be used for early diagnosis, and have limitations, such as cross-reactivity with other coronaviruses and false-positive results. However, these tests can complement molecular-based assays and provide a more accurate estimation of the disease stage.

7.3 Biosensors

Despite the high diagnostic sensitivity of real-time PCR for SARS-CoV-2 (and other viral and bacterial pathogens), the requirement for isolation of the viral genome, long turnaround time, and considerable risk of eliciting false-negative and false-positive results [39, 40] limit their application. In addition, serological tests are not suitable for screening early and asymptomatic cases [41]. After the emergence of the SARS-CoV-2 pandemic, scientists have commenced to develop novel and innovative designs for rapid and sensitive biosensing of the new coronavirus markers in clinical samples. Biosensors are suitable candidates for viral RNA and antigen detection since they require no amplification, and can offer fast results with minimum sample requirement. Several biosensor-based technologies for the diagnosis of COVID-19 have been established as novel devices to detect SARS-CoV-2 in a fast and efficient way.

Biosensors are powerful tools that have been tested for practical assessment of the presence and clinical progress of COVID-19, and could provide alertness on the severity or clinical trends of the infection.

The crucial development of biosensors began with the introduction of the glucose oxidase biosensor in 1962 [42]. Currently, biosensor systems technology is an extremely broad field that greatly impacts many industrial sectors, including pharmaceutical, diagnostics, and food industries as well as environmental monitoring. A biosensor is defined as a measuring system composed of a biological recognition element in close contact with a chemical or physical transducer (electrochemical, mass, optical, and thermal) that converts a biorecognition event into a detectable signal [43]. A typical biosensor consists of three parts; (a) biological recognition component that interacts with the analyte in the sample, (b) transducer, a sensor element that converts the signal of analyte/bioreceptor interaction into a measurable signal, and (c) an electronic system that amplifies and processes the acquiesced signal and displays the outcome in a user-friendly way (usually waveform or digital value) [44]. A scheme of a typical biosensor is shown in Fig. 7.3.

Due to their exceptional performance capabilities, including high sensitivity and specificity, rapid response, low cost, miniaturized size, and portable platform, biosensors have become a tool of choice in clinical, food, and environmental examination [45].

Biosensors can be classified based on the type of biological elements, biorecognition principle, or mode of physicochemical transduction [46]. On the basis of

Fig. 7.3 Schematic representation of a biosensor. A typical biosensor includes three parts: **a** a bioreceptor that interacts with the analyte, **b** a transducer that converts biorecognition signals into measurable ones, and **c** an electronic system that displays output

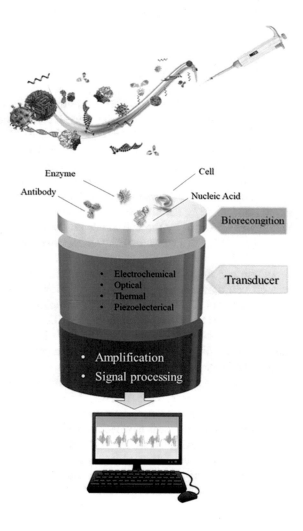

biological elements that are used, biosensors can be divided into nucleic acid/DNA-based, enzymatic-based, antibody/antigen-based, and whole cells-based. The biological components usually exhibit a bio affinity (bound specifically and selectively to the analyte) or biocatalytic (convert the analyte during the chemical reaction) role [47]. Depending on the transducer, the biosensor can be classified as electrochemical, optical, thermal, and piezoelectrical [48]. Electrochemical biosensors are the most commonly used platforms for biomedical analysis and early detection of diseases' biomarkers [49, 50]. The electrochemical technique can be further divided into four main groups: amperometric, poten-tiometric, cyclic voltametric, and impedimetric [46, 51].

Biosensors have the potential to become alternative analytical tools for pandemic outbreaks such as COVID-19. The previous experiences of the viral epidemics (e.g., SARS in 2002, avian flu in 2004, swine flu in 2009, MERS in 2012, and Ebola in 2014) encouraged scientists to do more research on developing innovative low-cost early detection systems based on biosensors.

So far, different components in SARS-associated viruses have been used as the subject to be detected. In one of the first attempts to design a biosensor for SARS-associated coronavirus, Zuo et al. developed a piezoelectric immunosensor to detect SARS-CoV whole virus in sputum in the gas phase [52]. The piezoelectric biosensors contain a piezoelectric crystal that is a mass device, and any mass change reflects on its oscillation frequency. The antibody or antigen is used as a biorecognition immobilized on this type of biosensor's surface. In this study, the horse polyclonal antibody, induced by SARS-CoV, was immobilized using a protein A layer on the piezoelectric crystal surface [52]. An ultra-sonicator generated the SARS-CoV antigen aerosols, which was adsorbed by the antibodies on the crystal surface, and the mass change on the crystal led to a frequency shift that linearly related to the concentration of the antigen [52].

In another study, a 30-mer unique sequence in the genome of the SARS-CoV was chosen as the target, and a hybridization-based genosensor was designed to detect the target sequence. The target's complementary strand (probe) was labeled with a thiol group and immobilized on a 100 nm sputtered gold film. The target sequence was conjugated to biotin and hybridized with the thiolated probe. The addition of alkaline phosphatase-labeled streptavidin allows enzymatic detection via the product's electrochemical signal [53].

Ishikawa and coworkers in 2009 successfully designed the first innovative electrochemical biosensor for SARS-CoV [54]. They described a field-effect transistor (FET)-based immunosensor that used antibody mimic proteins (AMPs) as the biorecognition element to selectively detect nucleocapsid protein as the SARS-CoV analyte. The AMPs are a class of small-sized and stable affinity binding agents produced in large quantities and at relatively low expenses. These polypeptides have the potential to surpass antibodies and nucleotide aptamers. They showed that the developed biosensor had the potential to detect the nucleocapsid at sub-nanomolar concentration and in a concentration-dependent manner [54].

The spike (S) protein in the SARS-related viruses is the most promising biomarker to be used as a target for detection. This membrane glycoprotein is composed of two subunits, S1 and S2. The S1 subunit is the receptor-binding domain responsible for receptor association and mediates viral entry. It has been reported that S1 subunit detection is more specific than other SARS-related virus proteins such as nucleocapsid, envelope, membrane, and either the whole S antigen [55].

Layqah and Eissa in 2019 described an electrochemical immunosensor for the detection of MERS-CoV, using the S1 subunit as the biomarker [56]. The biosensor's principle was based on an indirect competition between the recombinant S protein, immobilized on a carbon array electrode nanostructured with gold nanoparticle, and the free virus in the sample at a fixed antibody concentration. The voltametric response was detected by monitoring the peak current change upon adding different MERS-CoV antigen concentrations. The designed biosensor detected the MERS-CoV in nasal samples after 20 min with a detection limit of 1.0 pg/ml [56].

Seo et al. have fabricated a graphene-based FET biosensor in which the graphene sheets of the FET are coated with specific antibodies against SARS-CoV-2 spike protein. This FET biosensor was able to detect SARS-CoV-2 S protein in phosphate-buffer saline (PBS) and clinical transport medium. Furthermore, the FET biosensor was able to successfully detect SARS-CoV-2 in self-cultured medium and nasopharyngeal swab samples with a considerable low detection limit of 1.6×10^1 plaque-forming units (pfu)/mL and 2.42×10^2 copies/mL without any cross-reactivity with the MERS-CoV antigen [55]. The fabricated biosensor has been suggested as a highly sensitive rapid diagnostic method for COVID-19 with no need for sample preparation or labeling.

Mavrikoa and coworkers devised a novel biosensor for the ultra-rapid detection of the SARS-CoV-2 S protein. They performed membrane engineering by electro-inserting the human chimeric S1 antibody (as the target molecule) into the Vero cells (kidney epithelial cells) membrane. The attachment of S1 protein to the complementary antibodies resulted in the considerable and selective change in the engineered membrane bioelectric properties, measured by means of a bioelectric recognition assay. The results showed that this cell biosensor setup detected the analyte ultra-rapidly (3 min) with a low detection limit of 1 fg/ml [57].

In a recent study, Qiu and coworkers designed a dual functional plasmonic biosensor by integrating the plasmonic photothermal (PPT) effect and localized surface plasmon resonance (LSPR) sensing transduction on a single chip [58]. The DNA–RNA hybridization based on nucleic acid strand melting has been considered as the biorecognition principle. The single-stranded RNA genome of SARS-CoV-2 was the analyte and detected through the hybridization reaction. Three complementary DNA oligonucleotides have been synthesized with a thiol group as targets; two specific sequences from SARS-CoV-2 (the RdRp and the ORF1ab) and an oligonucleotide sequence from the coronaviral envelope protein gene. The complementary strands hybridize at a temperature slightly lower than the melting temperature, while a mismatch decreases the melting temperature significantly. The LSRP sensing chip was modified with a two-dimensional distribution of gold nano islands (AuNIs) functionalized with thiol-cDNA ligands. The plasmonic chip was able to generate the local PPT heat and transduce the in situ hybridization for highly sensitive and accurate SARS-CoV-2 detection [58].

Thus far, the aforementioned biosensors are the only developed biosensors designated for COVID-19 detection on lab scale; however, some authors have described innovative scenarios for future focuses. For instance, Xi et al., introduced G-quadruplex based biosensor as the potential promising tool for SAR-CoV-2 detection [59]. G-quadruplexes are stable non-canonical tetragonal secondary structure of nucleic acids, formed in some guanine rich sequences via Hoogsteen hydrogen bonds. This structure has been found in different organism's genomes and has biological significances such as transcription and translation inhibition. G-quadruplex has various unique properties in ligands binding. Compared to the standard biosensor, G-quadruplex probes can detect multiple substances such as metal ions, small molecules, proteins, and nucleic acids with an improved affinity. Recently, 25 putative G-quadruplex-forming sequences have been identified in different parts of the SARS-CoV-2 genome that, besides serving as the antiviral target, could be considered for being applied in the biosensors for viral detection.

Also, in other innovative studies, aptamers are used to detect SARS-CoV-2. Aptamers are frequently used for the detection of pathogens using electrochemical biosensors [60–62]. Aptamers are small ssDNA or RNA oligonucleotides that can selectively bind to various molecules with high affinity. Sometimes, they are referred to as "artificial antibodies" that can detect nucleic and non-nucleic acid molecules. Aptamers are synthesized through a selection process called systematic evolution of ligands by exponential enrichment (SELEX) [63]. The first DNA aptamers targeted toward the NP of SARS-CoV-2 were developed by Zhang and colleagues [64]. These aptamers can bind to Np successively in a sandwich type interaction. To validate these aptamers, ELISA was performed by using a combination of N protein aptamers and antibodies. In addition, the aptamers were fabricated into a gold nanoparticle immunochromatographic strip (GIS) to detect spiked N proteins in human serum and urine. The limit of detection (LOD) of this sensor for testing the N protein was better than that of the reported FET-based biosensor [55], and the process can be done in 15 min. These studies indicate that aptamers are a potent molecular tool with great potential in the fabrication of biosensors for COVID-19 detection.

Furthermore, a portable, multiplexed, wireless electrochemical platform based on mass-producible laser-engraved graphene named RapidPlex has been developed by Torrente-Rodriquez and coworkers. This electrochemical platform can be used for the detection of SARS-CoV-2's NP, specific antibodies IgM and IgG antibodies against the S1 protein, as well as inflammatory biomarker C-reactive protein (CRP), which is associated with the severity of COVID-19 infection90. Therefore, the RapidPlex can provide information on three important aspects of COVID-19 disease: severity of the disease (CRP), presence of viral infection (NP), and immune response (IgG and IgM antibodies). In addition, the applicability of this RapidPlex platform was tested with COVID-19 positive and negative blood and saliva samples. Thus, the RapidPlex has potential use in clinical analysis and point of care tests [65].

COVID-19 still remains a global crisis with many unresolved issues. Although no SARS-related biosensor has been applied for commercial application so far, there is no doubt that in the future, this technology will be widely employed for diagnosis. Biosensors' development for fast and efficient SARS-CoV-2 detection could offer an exciting alternative to traditional diagnostic techniques with the potential to revolutionize patient care quality. Therefore, further research on the specific and accessible biosensors for viral detection is of great interest. The list of biosensor-based techniques described for detecting SARS-related virus, divided by analyte type, is shown in Table 7.1.

7.4 Other Approaches

7.4.1 Antigen Detecting Tests

Antigen testing is based on the detection of viral proteins (antigens) or fragments of proteins from the virus. Among viral proteins of SARS-CoV-2, the N protein is an attractive target for SARS-CoV-2 detection. This viral protein can be detected up to 1 day before the onset of symptoms [100, 101]. Diao et al. developed N protein antigen-based fluorescence immunochromatographic (FIC) assay. Using RT-PCR assay as reference standard, this assay exhibits high specificity and sensitivity in the diagnosis of COVID-19 [101].

Although these tests are fast and easy to implement, they cannot replace molecular tests. Their accuracy is limited since antigen tests fail to detect all active infections, and have a high chance of giving rise to false-negative results. Low viral load or variability of viral loads in patients might be a reason. Antigen tests have the greatest utility with symptomatic patients with sufficient viral load. The combination of both molecular and antigen tests can substantially enhance diagnostic capability.

7.4.2 Imaging Modalities

7.4.2.1 Chest CT

The diagnosis of COVID-19 patients is mostly based on detection of viral RNA through molecular tests. However, in order to save time and reagents, radiological tests can be done. Chest CT is commonly used for pneumonia diagnosis. It involves taking many X-ray measurements at various angles across the chest, giving rise to cross-sectional images [102]. The CT features of COVID-19 depend on the state of the infection and improves approximately 6 to 12 days after symptoms onset [103]. The CT features of COVID-19 patients are nonspecific. The most common features are peripheral and bilateral ground glass opacities and lung consolidations. Even though CT is a potent tool in COVID-19 diagnosis, its low specificity (25%) limits its usage. There have been cases where the imaging features overlap with other viral pneumonia [104]. However, there are several discriminating features for

Table 7.1 List of biosensor-based techniques for detecting SARS-related virus

Analyte	Virus	The transduction method	Limits of detection (LOD)	References
RNA	SARS-CoV-2	Electrochemical	200 copies/mL	[66]
Monoclonal antiSARS-CoV-2 antibody	SARS-CoV-2	Electrochemical	–	[67]
S protein	SARS-CoV-2	Electrochemical	5.5×10^5 PFU/mL	[68]
ORF1ab; E; N regions of genome	SARS-CoV-2	Fluorescence	1,000 TU ml^{-1}	[69]
N protein	SARS-CoV-2	Electrochemical	0.8 pg/mL	[70]
S protein subunits	SARS-CoV-2	Optical	–	[71]
RNA	SARS-CoV-2	Optical	0.96 pM	[72]
H120 RNA	Infectious bronchitis virus (IBV) H120	Electrochemical	$2.96e^{-10}$ µM	[73]
N protein	SARS-CoV-2	Surface Plasmon resonance	1.02 pM	[74]
N and S protein	SARS-CoV-2	Electrochemical	1 copy/mL	[75]
Synthetic complementary DNA of SARS-CoV-2	SARS-CoV-2	Optomagetinc	0.4 fM	[76]
S protein	SARS-CoV-2	Optical	0–301.67 nM	[77]
RBD of S protein	SARS-CoV-2	Bioluminescent	15 pM	[78]
Special IgM and IgG of COVID-19	SARS-CoV-2	Fluorescent	–	[79]
N protein	SARS-CoV-2	Optical	10^{-18} M	[80]
IgG	SARS-CoV-2	Optical (LFA)	–	[81]
S protein	SARS-CoV-2	Electrochemical	1 fg/mL	[57]
IgG-IgM	SARS-CoV-2	Immunochromatographic	–	[82]
S protein	SARS-CoV-2	Electrochemical	90 fM	[83]
N protein	SARS-CoV-2	Optical	0.18 ng/µL	[84]
S protein	SARS-CoV-2	Field emitted transistor	2.42×10^2 copies/mL	[85]
RdRp and the ORF1ab	SARS-CoV-2	Plasmonic photothermal (PPT) effect and localized surface plasmon resonance (LSPR)	0.22 pM	[58]
IgG-IgM	SARS-CoV-2	Optical (Lateral flow assay)	–	[38]

(continued)

Table 7.1   (continued)

Analyte	Virus	The transduction method	Limits of detection (LOD)	References
Anti-SARS-CoV-2 antibodies	SARS-CoV-2	Surface plasmon resonance (SPR)/ Optical	–	[86]
IgG	SARS-CoV-2	Lateral flow immunoassay (LFIA)	–	[87]
Different fragments of the SARS-CoV-2 RNA	SARS-CoV-2	Color/luminescence/ optical	100 copies of RNA	[88]
Recombinant spike protein S1	MERS-CoV	Electrochemical	$1.0 \ pg.mL^{-1}$	[56]
IBV spiked serum	Infectious bronchitis virus (IBV)	Optical	4.6×10^2 EID50 per mL	[89]
Antibodies	IBV	Optical	79.15 EID/50 µL	[90]
Complementary DNA strands	MERS-CoV/ Mycobacterium tuberculosis (MTB)/ human papillomavirus (HPV)	Colorimetric	1.53 (MERS-CoV), 1.27 (MTB), and 1.03 nM (HPV)	[91]
Viral RNA	Influenza (IFN) type-A and B, respiratory syncytial virus (RSV) type-A and B, human coronavirus types OC43 and 229E	Optical	–	[92]

(continued)

COVID-19 pneumonia. Studies comparing COVID-19 CT results with that of non-COVID-19 pneumonia patients show that COVID-19 pneumonia results are more likely to have peripheral distribution, rounded opacities, fine reticular opacity, and vascular thickening, and less likely to have central and peripheral distribution, pleural effusion, and lymphadenopathy compared to non-COVID-19 pneumonia [105, 106].

The sensitivity of CT for COVID-19 is between 88 and 97%, whereas the sensitivity of PCR using nasopharyngeal swabs is known to be approximately 72%. Several studies have reported that chest CT scan can show abnormalities consistent with COVID-19 in patients whose RT-PCR results were negative at initial presentation. Therefore, Chest CT can be used as a complement to RT-PCR to increase the efficacy of the diagnosis and increase sensitivity [104].

Table 7.1 (continued)

Analyte	Virus	The transduction method	Limits of detection (LOD)	References
DNA	Including influenza A and influenza B, H1N1, respiratory syncytial virus (RSV), parainfluenza virus 1–3 (PIV1, 2, 3), adenovirus (ADV), and severe acute respiratory syndrome coronavirus (SARS)	SPR	Influ A:5 nM Influ B: 1 nM PIV$_1$:1 nM PIV$_2$: 2.5 nM PIV$_3$: 3.5 nM RSV: 3 nM ADV: 0.5 nM SARA: 2 nM H1N1:3 nM	[93]
N protein	SARS-CoV	Optical	0.003 nM	[94]
N protein	SARS-CoV	Optical	0.1 pg mL^{-1}	[62]
Surface antigen	SARS-CoV	SPR/optical	200 ng mL(-1)	[95]
Viral genome	SARS-CoV	Electrochemical	2.5 pM	[54]
N protein	SARS-CoV	FET	–	[54]
N protein	SARS-CoV	Localized surface plasmon coupled fluorescence (LSPCF)	~1 pg/mL	[96]
N protein	SARS-CoV	Chemiluminescence	–	[97]
Oligonucleotide microarray	Bacterial and viral upper respiratory infections (URI)	Electrochemical	–	[98]
Virus sequence	SARS-CoV	Electrochemical	–	[99]
SARS antigen	SARS-CoV	Piezoelectric	–	[52]

7.4.2.2 Lung Ultrasound (LUS)

The LUS is a fast, mobile, and noninvasive method that can be done by the patient's bedside reducing potential exposure of health workers. The LUS has been rapidly used for the evaluation of COVID-19 pneumonia lesions due to its widespread availability and low cost. Similar to chest CT, the accuracy of LUS depends on the phase of the infection. The LUS features in COVID-19 patients are thickened pleural lines, consolidations, single or confluent interstitial artifactual signs, and small hyperechoic lung regions [107]. LUS is known to be more sensitive than CXR in the diagnosis of consolidation and effusions [108]. In addition, LUS can detect lung injury even in the presence of normal CT results [109]. The studies have shown that the sensitivity of LUS for COVID-19 diagnosis is close to 100% with a specificity of approximately 78.6% [110]. However, there are some limitations to using LUS for COVID-19 diagnosis. For instance, LUS cannot differentiate early COVID-19 from other viral pneumonias. Further research is required to clarify the usage of LUS in diagnostic and prognostic COVID-19. Therefore, LUS should be implemented in conjunction with other diagnostic modalities [111].

7.4.2.3 Chest Radiograph (CXR)

An alternative to chest CT and lung ultrasound is chest radiograph (CXR). The common CXR features in COVID-19 patients are similar to that of CT with both demonstrating ground glass opacities and/or bilateral, peripheral consolidation. CXR signs improve approximately 10 to 12 days from symptom onset. Compared to CT and LUS, CXR has lower sensitivity (69%) and may miss up to 40% of confirmed COVID-19 cases [112, 113]. This is due to the lodging of virus particles in terminal alveoli near the pleural interface which are well visualized with CT and LUS, but are more difficult to see on plain imaging [114]. Therefore, CXR cannot always determine COVID-19 infection.

7.5 Conclusion

The occurrence of COVID-19 pandemic accelerates many types of research in the field of therapy, vaccine, and rapid diagnosis. Despite the lack of promising results in the therapeutic field, rapid vaccine development and the emergence of innovative detection methods would greatly help crisis management. The fight against the COVID-19 pandemic has directed a lot of attention to diagnostic tests. Thus far, various diagnostic platforms have been developed for rapid detection of SARS-CoV-2. Each testing method has a set of advantages accompanied with some inevitable drawbacks since there is no such thing as the "perfect test." Therefore, performing more than one detection method is highly advised instead of relying on a single detection method. Among COVID-19 diagnostic platforms, RT-PCR is a well-established method and is commonly used for COVID-19 diagnosis. Since some developing countries and low resource settings lack the means for accurate and timely COVID-19 RT-PCR tests to control the ongoing outbreak of COVID-19, the development of accurate, cost-effective, and easy-to-implement testing platform is necessary to identify patients' past and present infections. Electrochemical biosensors can be an ideal candidate for this purpose, and further clinical evaluation of biosensors can greatly optimize the COVID-19 testing capacity. Serological tests also can complement molecular-based assays and provide a more accurate estimation of the disease stage.

In this commentary, we categorized and reviewed some innovative methods that can be the next technologies for large-scale SARS-CoV-2 diagnosis. Nowadays, limited methods have been approved by regulatory agencies for commercial detection of SARS-CoV-2. However, there is no doubt that amongst the very diverse technologies that are being tested in laboratories worldwide, some of their best will soon enter the market and dramatically affect the virus detection cost and time.

References

1. Zhou P, Yang X-L, Wang X-G, Hu B, Zhang L, Zhang W, Si H-R, Zhu Y, Li B, Huang C-L, Chen H-D, Chen J, Luo Y, Guo H, Jiang R-D, Liu M-Q, Chen Y, Shen X-R, Wang X, Zheng X-S, Zhao K, Chen Q-J, Deng F, Liu L-L, Yan B, Zhan F-X, Wang Y-Y, Xiao G-F, Shi Z-L (2020) A pneumonia outbreak associated with a new coronavirus of probable bat origin. Nature 579:270–273. https://doi.org/10.1038/s41586-020-2012-7
2. Gorbalenya AE, Baker SC, Baric RS, de Groot RJ, Drosten C, Gulyaeva AA, Haagmans BL, Lauber C, Leontovich AM, Neuman BW, Penzar D, Perlman S, Poon LLM, Samborskiy DV, Sidorov IA, Sola I, Ziebuhr J, CSG of the IC on T of Viruses (2020) The species Severe acute respiratory syndrome-related coronavirus: classifying 2019-nCoV and naming it SARS-CoV-2. Nat Microbiol 5, 536–544. https://doi.org/10.1038/s41564-020-0695-z
3. Chen N, Zhou M, Dong X, Qu J, Gong F, Han Y, Qiu Y, Wang J, Liu Y, Wei Y, Xia J, Yu T, Zhang X, Zhang L (2019) Epidemiological and clinical characteristics of 99 cases of novel coronavirus pneumonia in Wuhan, China: a descriptive study. Lancet (London, England) 395(2020):507–513. https://doi.org/10.1016/S0140-6736(20)30211-7
4. Huang C, Wang Y, Li X, Ren L, Zhao J, Hu Y, Zhang L, Fan G, Xu J, Gu X, Cheng Z, Yu T, Xia J, Wei Y, Wu W, Xie X, Yin W, Li H, Liu M, Xiao Y, Gao H, Guo L, Xie J, Wang G, Jiang R, Gao Z, Jin Q, Wang J, Cao B (2019) Clinical features of patients infected with novel coronavirus in Wuhan. China, Lancet 395(2020):497–506. https://doi.org/10.1016/S0140-6736(20)30183-5
5. Qian G, Yang N, Ma AHY, Wang L, Li G, Chen X, Chen X (2020) COVID-19 transmission within a family cluster by Presymptomatic carriers in China. Clin Infect Dis 71:861–862. https://doi.org/10.1093/cid/ciaa316
6. Ye F, Xu S, Rong Z, Xu R, Liu X, Deng P (2020) Delivery of infection from asymptomatic carriers of COVID-19 in a familial cluster. Int J Infect Dis 94:133–138. https://doi.org/10.1016/j.ijid.2020.03.042
7. Bai Y, Yao L, Wei T, Tian F, Jin D-Y, Chen L, Wang M (2020) Presumed asymptomatic carrier transmission of COVID-19. JAMA 323:1406–1407. https://doi.org/10.1001/jama.2020.2565
8. Tang A, Tong Z-D, Wang H-L, Dai Y-X, Li K-F, Liu J-N, Wu W-J, Yuan C, Yu M-L, Li P, Yan J-B (2020) Detection of novel coronavirus by RT-PCR in stool specimen from asymptomatic child, China. Emerg Infect Dis 26:1337–1339. https://doi.org/10.3201/eid2606.200301
9. Sanche S, Lin YT, Xu C, Romero-Severson E, Hengartner N, Ke R (2020) High contagiousness and rapid spread of severe acute respiratory syndrome coronavirus 2. Emerg Infect Dis 26:1470–1477. https://doi.org/10.3201/eid2607.200282
10. Burke RM, Midgley CM, Dratch A, Fenstersheib M, Haupt T, Holshue M, Ghinai I, Jarashow MC, Lo J, McPherson TD, Rudman S, Scott S, Hall AJ, Fry AM, Rolfes MA (2020) Active monitoring of persons exposed to patients with confirmed COVID-19—United States, January–February 2020, MMWR Morb Mortal Wkly Rep 69, 245–246. https://doi.org/10.15585/mmwr.mm6909e1

11. Wang J, Du G (2020) COVID-19 may transmit through aerosol. Ir J Med Sci 189:1143–1144. https://doi.org/10.1007/s11845-020-02218-2
12. Villarreal IM, Morato M, Martínez-RuizCoello M, Navarro A, Garcia-Chillerón R, Ruiz Á, de Almeida IV, Mazón L, Plaza G (2020) Olfactory and taste disorders in healthcare workers with COVID-19 infection. Eur Arch Oto-Rhino-Laryngology. https://doi.org/10.1007/s00405-020-06237-8
13. Lauer SA, Grantz KH, Bi Q, Jones FK, Zheng Q, Meredith HR, Azman AS, Reich NG, Lessler J (2019) The incubation period of coronavirus disease, (CoVID-19) from publicly reported confirmed cases: estimation and application. Ann Intern Med 172(2020):577–582. https://doi.org/10.7326/M20-0504
14. Calame KL (2001) Plasma cells: finding new light at the end of B cell development. Nat Immunol 2:1103–1108. https://doi.org/10.1038/ni1201-1103
15. Lanzavecchia A, Sallusto F (2009) Human B cell memory. Curr Opin Immunol 21:298–304. https://doi.org/10.1016/j.coi.2009.05.019
16. Hoffman W, Lakkis FG, Chalasani G, Cells B (2016) Antibodies, and more. Clin J Am Soc Nephrol 11:137–154. https://doi.org/10.2215/CJN.09430915
17. Chiu CY, Miller SA (2019) Clinical metagenomics. Nat Rev Genet 20:341–355. https://doi.org/10.1038/s41576-019-0113-7
18. Mix E, Goertsches R, Zett UK (2006) Immunoglobulins—basic considerations. J Neurol 253:v9–v17. https://doi.org/10.1007/s00415-006-5002-2
19. Angenvoort J, Brault AC, Bowen RA, Groschup MH (2013) West Nile viral infection of equids. Vet Microbiol 167:168–180. https://doi.org/10.1016/j.vetmic.2013.08.013
20. Vidarsson G, Dekkers G, Rispens T (2014) IgG subclasses and allotypes: from structure to effector functions. Front Immunol 5:520. https://doi.org/10.3389/fimmu.2014.00520
21. Louie JK, Hacker JK, Mark J, Gavali SS, Yagi S, Espinosa A, Schnurr DP, Cossen CK, Isaacson ER, Glaser CA, Fischer M, Reingold AL, Vugia DJ, UD and CIW Group (2004) SARS and common viral infections. Emerg Infect Dis 10, 1143–1146. https://doi.org/10.3201/eid1006.030863
22. Li G, Chen X, Xu A (2003) Profile of specific antibodies to the SARS-associated coronavirus. N Engl J Med 349:508–509. https://doi.org/10.1056/NEJM200307313490520
23. Shi Y, Wan Z, Li L, Li P, Li C, Ma Q, Cao C (2004) Antibody responses against SARS-coronavirus and its nucleocaspid in SARS patients. J Clin Virol 31:66–68. https://doi.org/10.1016/j.jcv.2004.05.006
24. Wölfel R, Corman VM, Guggemos W, Seilmaier M, Zange S, Müller MA, Niemeyer D, Jones TC, Vollmar P, Rothe C, Hoelscher M, Bleicker T, Brünink S, Schneider J, Ehmann R, Zwirglmaier K, Drosten C, Wendtner C (2020) Virological assessment of hospitalized patients with COVID-2019. Nature 581:465–469. https://doi.org/10.1038/s41586-020-2196-x
25. Haveri A, Smura T, Kuivanen S, Österlund P, Hepojoki J, Ikonen N, Pitkäpaasi M, Blomqvist S, Rönkkö E, Kantele A, Strandin T, Kallio-Kokko H, Mannonen L, Lappalainen M, Broas M, Jiang M, Siira L, Salminen M, Puumalainen T, Sane J, Melin M, Vapalahti O, Savolainen-Kopra C (2020) Serological and molecular findings during SARS-CoV-2 infection: the first case study in Finland, January to February 2020. Euro Surveill 25:2000266. https://doi.org/10.2807/1560-7917.ES.2020.25.11.2000266
26. Xiang F, Wang X, He X, Peng Z, Yang B, Zhang J, Zhou Q, Ye H, Ma Y, Li H, Wei X, Cai P, Ma WL (2020) Antibody detection and dynamic characteristics in patients with COVID-19. Clin Infect Dis 1–5. https://doi.org/10.1093/cid/ciaa461
27. Ma H, Zeng W, He H, Zhao D, Jiang D, Zhou P, Cheng L, Li Y, Ma X, Jin T (2020) Serum IgA, IgM, and IgG responses in COVID-19. Cell Mol Immunol 17:773–775. https://doi.org/10.1038/s41423-020-0474-z
28. Yu H-Q, Sun B-Q, Fang Z-F, Zhao J-C, Liu X-Y, Li Y-M, Sun X-Z, Liang H-F, Zhong B, Huang Z-F, Zheng P-Y, Tian L-F, Qu H-Q, Liu D-C, Wang E-Y, Xiao X-J, Li S-Y, Ye F, Guan L, Hu D-S, Hakonarson H, Liu Z-G, Zhong N-S (2020) Distinct features of SARS-CoV-2-specific IgA response in COVID-19 patients. Eur Respir J 56:2001526. https://doi.org/10.1183/13993003.01526-2020

29. Padoan A, Sciacovelli L, Basso D, Negrini D, Zuin S, Cosma C, Faggian D, Matricardi P, Plebani M (2020) IgA-Ab response to spike glycoprotein of SARS-CoV-2 in patients with COVID-19: a longitudinal study. Clin Chim Acta 507:164–166. https://doi.org/10.1016/j.cca.2020.04.026

30. Jiang C, Wang Y, Hu M, Wen L, Wen C, Wang Y, Zhu W, Tai S, Jiang Z, Xiao K, Faria NR, De Clercq E, Xu J, Li G (2020) Antibody seroconversion in asymptomatic and symptomatic patients infected with severe acute respiratory syndrome coronavirus 2 (SARS-CoV-2). Clin Transl Immunol 9, e1182. https://doi.org/10.1002/cti2.1182

31. Zhou W, Xu X, Chang Z, Wang H, Zhong X, Tong X, Liu T, Li Y (2021) The dynamic changes of serum IgM and IgG against SARS-CoV-2 in patients with COVID-19. J Med Virol 93:924–933. https://doi.org/10.1002/jmv.26353

32. Dhamad AE, Abdal Rhida MA (2020) COVID-19: molecular and serological detection methods. Peer J 8, e10180–e10180. https://doi.org/10.7717/peerj.10180

33. To KK-W, Tsang OT-Y, Leung W-S, Tam AR, Wu T-C, Lung DC, Yip CC-Y, Cai J-P, Chan JM-C, Chik TS-H, Lau DP-L, Choi CY-C, Chen L-L, Chan W-M, Chan K-H, Ip JD, Ng AC-K, Poon RW-S, Luo C-T, Cheng VC-C, Chan JF-W, Hung IF-N, Chen Z, Chen H, Yuen K-Y (2020) Temporal profiles of viral load in posterior oropharyngeal saliva samples and serum antibody responses during infection by SARS-CoV-2: an observational cohort study. Lancet Infect Dis 20:565–574. https://doi.org/10.1016/S1473-3099(20)30196-1

34. Chia WN, Tan CW, Foo R, Kang AEZ, Peng Y, Sivalingam V, Tiu C, Ong XM, Zhu F, Young BE, Chen MI-C, Tan Y-J, Lye DC, Anderson DE, Wang L-F (2020) Serological differentiation between COVID-19 and SARS infections. Emerg Microbes Infect 9:1497–1505. https://doi.org/10.1080/22221751.2020.1780951

35. Zhang Q-Y, Chen H, Lin Z, Lin J-M (2012) Comparison of chemiluminescence enzyme immunoassay based on magnetic microparticles with traditional colorimetric ELISA for the detection of serum α-fetoprotein. J Pharm Anal 2:130–135. https://doi.org/10.1016/j.jpha.2011.10.001

36. Ghaffari A, Meurant R, Ardakani A (2020) COVID-19 serological tests: how well do they actually perform? Diagnostics (Basel, Switzerland) 10:453. https://doi.org/10.3390/diagnostics10070453

37. Pan Y, Li X, Yang G, Fan J, Tang Y, Zhao J (2020) Serological immunochromatographic approach in diagnosis with SARS-CoV-2 infected COVID-19 patients 81, e28–e32

38. Li Z, Yi Y, Luo X, Xiong N, Liu Y, Li S, Sun R, Wang Y, Hu B, Chen W, Zhang Y, Wang J, Huang B, Lin Y, Yang J, Cai W, Wang X, Cheng J, Chen Z, Sun K, Pan W, Zhan Z, Chen L, Ye F (2020) Development and clinical application of a rapid IgM-IgG combined antibody test for SARS-CoV-2 infection diagnosis. J Med Virol 92:1518–1524. https://doi.org/10.1002/jmv.25727

39. Caygill RL, Blair GE, Millner PA (2010) A review on viral biosensors to detect human pathogens. Anal Chim Acta 681:8–15. https://doi.org/10.1016/j.aca.2010.09.038

40. Wang Y, Kang H, Liu X, Tong Z (2020) Combination of RT-qPCR testing and clinical features for diagnosis of COVID-19 facilitates management of SARS-CoV-2 outbreak. J Med Virol 92:538–539. https://doi.org/10.1002/jmv.25721

41. Cui F, Zhou HS (2020) Diagnostic methods and potential portable biosensors for coronavirus disease 2019. Biosens Bioelectron 165, 112349. https://doi.org/10.1016/j.bios.2020.112349

42. Clark LC Jr, Lyons C (1962) Electrode systems for continuous monitoring in cardiovascular surgery. Ann N Y Acad Sci 102:29–45. https://doi.org/10.1111/j.1749-6632.1962.tb13623.x

43. Mello LD, Kubota LT (2002) Review of the use of biosensors as analytical tools in the food and drink industries. Food Chem 77:237–256. https://doi.org/10.1016/S0308-8146(02)00104-8

44. Cagnin S, Caraballo M, Guiducci C, Martini P, Ross M, Santaana M, Danley D, West T, Lanfranchi G (2009) Overview of electrochemical DNA biosensors: new approaches to detect the expression of life. Sensors (Basel) 9:3122–3148. https://doi.org/10.3390/s90403122

45. Zarei M (2017) Portable biosensing devices for point-of-care diagnostics: recent developments and applications. TrAC Trends Anal Chem 91:26–41. https://doi.org/10.1016/j.trac.2017.04.001

46. Sawant SN (2017) 13—development of biosensors from biopolymer composites. In: Sada-sivuni KK, Ponnamma D, Kim J, Cabibihan J-J, MABT-BC in AlMaadeed E (eds). Elsevier, pp 353–383. https://doi.org/10.1016/B978-0-12-809261-3.00013-9.

47. Kozitsina AN, Svalova TS, Malysheva NN, Okhokhonin AV, Vidrevich MB, Brainina KZ (2018) Sensors based on bio and biomimetic receptors in medical diagnostic. Environ, Food Anal, Biosens 8:35. https://doi.org/10.3390/bios8020035

48. Thévenot DR, Toth K, Durst RA, Wilson GS (2001) Electrochemical biosensors: recom-mended definitions and classification1 international union of pure and applied chemistry: physical chemistry division, commission I.7 (Biophysical Chemistry); analytical chemistry division, commission V.5. Electroanal, Biosens Bioelectron 16, 121–131. https://doi.org/10. 1016/S0956-5663(01)00115-4

49. Frías IAM, Avelino KYPS, Silva RR, Andrade CAS, Oliveira MDL (2015) Trends in biosen-sors for HPV: identification and diagnosis. J Sens 913640. https://doi.org/10.1155/2015/913640

50. Nezami A, Dehghani S, Nosrati R, Eskandari N, Taghdisi SM, Karimi G (2018) Nanomaterial-based biosensors and immunosensors for quantitative determination of cardiac troponins. J Pharm Biomed Anal 159:425–436. https://doi.org/10.1016/j.jpba.2018.07.031

51. Ronkainen NJ, Halsall HB, Heineman WR (2010) Electrochemical biosensors. Chem Soc Rev 39:1747–1763. https://doi.org/10.1039/B714449K

52. Zuo B, Li S, Guo Z, Zhang J, Chen C (2004) Piezoelectric immunosensor for SARS-associated coronavirus in sputum. Anal Chem 76:3536–3540. https://doi.org/10.1021/ac035367b

53. Abad-Valle P, Fernández-Abedul MT, Costa-García A (2007) DNA single-base mismatch study with an electrochemical enzymatic genosensor. Biosens Bioelectron 22:1642–1650. https://doi.org/10.1016/j.bios.2006.07.015

54. Ishikawa FN, Chang H-K, Curreli M, Liao H-I, Olson CA, Chen P-C, Zhang R, Roberts RW, Sun R, Cote RJ, Thompson ME, Zhou C (2009) Label-free, electrical detection of the SARS virus N-protein with nanowire biosensors utilizing antibody mimics as capture probes. ACS Nano 3:1219–1224. https://doi.org/10.1021/nn900086c

55. Seo G, Lee G, Kim MJ, Baek SH, Choi M, Ku KB, Lee CS, Jun S, Park D, Kim HG, Kim SJ, Lee JO, Kim BT, Park EC, Il Kim S (2020) Rapid detection of COVID-19 causative virus (SARS-CoV-2) in human nasopharyngeal swab specimens using field-effect transistor-based biosensor. ACS Nano 14, 5135–5142. https://doi.org/10.1021/acsnano.0c02823

56. Layqah LA, Eissa S (2019) An electrochemical immunosensor for the corona virus associated with the middle east respiratory syndrome using an array of gold nanoparticle-modified carbon electrodes. Mikrochim Acta 186:224. https://doi.org/10.1007/s00604-019-3345-5

57. Mavrikou S, Moschopoulou G, Tsekouras V, Kintzios S (2020) Development of a portable, ultra-rapid and ultra-sensitive cell-based biosensor for the direct detection of the SARS-CoV-2 S1 spike protein antigen. Sens (Basel) 20:3121. https://doi.org/10.3390/s20113121

58. Qiu G, Gai Z, Tao Y, Schmitt J, Kullak-Ublick GA, Wang J (2020) Dual-functional plasmonic photothermal biosensors for highly accurate severe acute respiratory syndrome coronavirus 2 detection. ACS Nano 14:5268–5277. https://doi.org/10.1021/acsnano.0c02439

59. Xi H, Juhas M, Zhang Y (2020) G-quadruplex based biosensor: a potential tool for SARS-CoV-2 detection. Biosens Bioelectron 167, 112494. https://doi.org/10.1016/j.bios.2020.112494

60. Xi Z, Gong Q, Wang C, Zheng B (2018) Highly sensitive chemiluminescent aptasensor for detecting HBV infection based on rapid magnetic separation and double-functionalized gold nanoparticles. Sci Rep 8:1–7. https://doi.org/10.1038/s41598-018-27792-5

61. Ghanbari K, Roushani M, Azadbakht A (2017) Ultra-sensitive aptasensor based on a GQD nanocomposite for detection of hepatitis C virus core antigen. Anal Biochem 534:64–69. https://doi.org/10.1016/j.ab.2017.07.016

62. Roh C, Jo SK (2011) Quantitative and sensitive detection of SARS coronavirus nucleocapsid protein using quantum dots-conjugated RNA aptamer on chip. J Chem Technol Biotechnol 86:1475–1479. https://doi.org/10.1002/jctb.2721

63. Stoltenburg R, Reinemann C, Strehlitz B (2007) SELEX—A (r)evolutionary method to generate high-affinity nucleic acid ligands. Biomol Eng 24:381–403. https://doi.org/10.1016/j.bioeng.2007.06.001

64. Zhang L, Fang X, Liu X, Ou H, Zhang H, Wang J, Li Q, Cheng H, Zhang W, Luo Z (2020) Discovery of sandwich type COVID-19 nucleocapsid protein DNA aptamers. Chem Commun 3:1–4. https://doi.org/10.1039/d0cc03993d

65. Torrente-Rodríguez RM, Lukas H, Tu J, Min J, Yang Y, Xu C, Rossiter HB, Gao W (2020) SARS-CoV-2 RapidPlex: a graphene-based multiplexed telemedicine platform for rapid and low-cost COVID-19 diagnosis and monitoring. Matter. https://doi.org/10.1016/j.matt.2020.09.027.10.1016/j.matt.2020.09.027

66. Zhao H, Liu F, Xie W, Zhou T-C, OuYang J, Jin L, Li H, Zhao C-Y, Zhang L, Wei J, Zhang Y-P, Li C-P (2021) Ultrasensitive supersandwich-type electrochemical sensor for SARS-CoV-2 from the infected COVID-19 patients using a smartphone. Sens Actuators B Chem 327, 128899. https://doi.org/10.1016/j.snb.2020.128899

67. Rashed MZ, Kopechek JA, Priddy MC, Hamorsky KT, Palmer KE, Mittal N, Valdez J, Flynn J, Williams SJ (2021) Rapid detection of SARS-CoV-2 antibodies using electrochemical impedance-based detector. Biosens Bioelectron 171, 112709. https://doi.org/10.1016/j.bios.2020.112709

68. Mojsoska B, Larsen S, Olsen DA, Madsen JS, Brandslund I, Alatraktchi FA (2021) Rapid SARS-CoV-2 detection using electrochemical immunosensor. Sens (Basel) 21:390. https://doi.org/10.3390/s21020390

69. Wang D, He S, Wang X, Yan Y, Liu J, Wu S, Liu S, Lei Y, Chen M, Li L, Zhang J, Zhang L, Hu X, Zheng X, Bai J, Zhang Y, Zhang Y, Song M, Tang Y (2020) Rapid lateral flow immunoassay for the fluorescence detection of SARS-CoV-2 RNA. Nat Biomed Eng 4:1150–1158. https://doi.org/10.1038/s41551-020-00655-z

70. Eissa S, Zourob M (2021) Development of a low-cost cotton-tipped electrochemical immunosensor for the detection of SARS-CoV-2. Anal Chem 93:1826–1833. https://doi.org/10.1021/acs.analchem.0c04719

71. Al Ahmad M, Mustafa F, Panicker N, Rizvi TA (2020) Development of an optical assay to detect SARS-CoV-2 spike protein binding interactions with ACE2 and disruption of these interactions using electric current. MedRxiv. https://doi.org/10.1101/2020.11.24.20237628

72. Jiao J, Duan C, Xue L, Liu Y, Sun W, Xiang Y (2020) DNA nanoscaffold-based SARS-CoV-2 detection for COVID-19 diagnosis. Biosens Bioelectron 167. https://doi.org/10.1016/j.bios.2020.112479

73. Yang Y, Yang D, Shao Y, Li Y, Chen X, Xu Y, Miao J (2020) A label-free electrochemical assay for coronavirus IBV H120 strain quantification based on equivalent substitution effect and AuNPs-assisted signal amplification. Microchim Acta 187:624. https://doi.org/10.1007/s00604-020-04582-3

74. Bong J-H, Kim T-H, Jung J, Lee SJ, Sung JS, Lee CK, Kang M-J, Kim HO, Pyun J-C (2020) Pig sera-derived anti-SARS-CoV-2 antibodies in surface plasmon resonance biosensors. BioChip J 14:358–368. https://doi.org/10.1007/s13206-020-4404-z

75. Chaibun T, Puenpa J, Ngamdee T, Boonapatcharoen N, Athamanolap P, O'Mullane AP, Vongpunsawad S, Poovorawan Y, Lee SY, Lertanantawong B (2021) Rapid electrochemical detection of coronavirus SARS-CoV-2. Nat Commun 12:802. https://doi.org/10.1038/s41467-021-21121-7

76. Tian B, Gao F, Fock J, Dufva M, Hansen MF (2020) Homogeneous circle-to-circle amplification for real-time optomagnetic detection of SARS-CoV-2 RdRp coding sequence. Biosens Bioelectron 165. https://doi.org/10.1016/j.bios.2020.112356

77. Song J, Qu J, Peng X, Zhou Y, Nie K, Zhou F, Yuan Y (2020) Promising near-infrared plasmonic biosensor employed for specific detection of SARS-CoV-2 and its spike glycoprotein. New J Phys 22. https://doi.org/10.1088/1367-2630/abbe53

78. Quijano-Rubio A, Yeh HW, Park J, Lee H, Langan RA, Boyken SE, Lajoie MJ, Cao L, Chow CM, Miranda MC, Wi J, Hong HJ, Stewart L, Oh B-H, Baker D (2020) De novo design of modular and tunable allosteric biosensors. BioRxiv Prepr Serv Biol. https://doi.org/10.1101/2020.07.18.206946

79. Feng M, Chen J, Xun J, Dai R, Zhao W, Lu H, Xu J, Chen L, Sui G, Cheng X (2020) Development of a sensitive immunochromatographic method using lanthanide fluorescent

microsphere for rapid serodiagnosis of COVID-19. ACS Sens 5:2331–2337. https://doi.org/10.1021/acssensors.0c00927

80. Murugan D, Bhatia H, Sai VVR, Satija J (2020) P-FAB: a fiber-optic biosensor device for rapid detection of COVID-19. Trans Indian Natl Acad Eng 5:211–215. https://doi.org/10.1007/s41403-020-00122-w

81. Wen T, Huang C, Shi F-J, Zeng X-Y, Lu T, Ding S-N, Jiao Y-J (2020) Development of a lateral flow immunoassay strip for rapid detection of IgG antibody against SARS-CoV-2 virus. Analyst 145:5345–5352. https://doi.org/10.1039/D0AN00629G

82. Zeng L, Li Y, Liu J, Guo L, Wang Z, Xu X, Song S, Hao C, Liu L, Xin M, Xu C (2020) Rapid, ultrasensitive and highly specific biosensor for the diagnosis of SARS-CoV-2 in clinical blood samples. Mater Chem Front 4:2000–2005. https://doi.org/10.1039/D0QM00294A

83. Mahari S, Roberts A, Shahdeo D, Gandhi S (2020) eCovSens-ultrasensitive novel in-house built printed circuit board based electrochemical device for rapid detection of nCovid-19. BioRxiv 1–20. https://doi.org/10.1101/2020.04.24.059204

84. Moitra P, Alafeef M, Dighe K, Frieman MB, Pan D (2020) Selective naked-eye detection of SARS-CoV-2 mediated by N gene targeted antisense oligonucleotide capped plasmonic nanoparticles. ACS Nano 14:7617–7627. https://doi.org/10.1021/acsnano.0c03822

85. Seo G, Lee G, Kim MJ, Baek SH, Choi M, Ku KB, Lee CS, Jun S, Park D, Kim HG, Kim SJ, Lee J-O, Kim BT, Park EC, Il Kim S (2020) Rapid detection of COVID-19 causative virus (SARS-CoV-2) in human nasopharyngeal swab specimens using field-effect transistor-based biosensor. ACS Nano 14, 5135–5142. https://doi.org/10.1021/acsnano.0c02823

86. Djaileb A, Charron B, Jodaylami MH, et al (2020) A rapid and quantitative serum test for SARS-CoV-2 antibodies with portable surface plasmon resonance sensing. ChemRxiv. https://doi.org/10.26434/chemrxiv.12118914.v1

87. Chen Z, Zhang Z, Zhai X, Li Y, Lin L, Zhao H, Bian L, Li P, Yu L, Wu Y, Lin G (2020) Rapid and sensitive detection of anti-SARS-CoV-2 IgG, using lanthanide-doped nanoparticles-based lateral flow immunoassay. Anal Chem 92:7226–7231. https://doi.org/10.1021/acs.analchem.0c00784

88. Chakravarthy A, Anirudh KN, George G, Ranganathan S, Shettigar N, Suchitta U, Palakodeti D, Gulyani A, Ramesh A (2021) Ultrasensitive RNA biosensors for SARS-CoV-2 detection in a simple color and luminescence assay. MedRxiv. https://doi.org/10.1101/2021.01.08.21249426

89. Weng X, Neethirajan S (2018) Immunosensor based on antibody-functionalized MoS(2) for rapid detection of avian coronavirus on cotton thread. IEEE Sens J 18:4358–4363. https://doi.org/10.1109/JSEN.2018.2829084

90. Ahmed SR, Kang SW, Oh S, Lee J, Neethirajan S (2018) Chiral zirconium quantum dots: a new class of nanocrystals for optical detection of coronavirus. Heliyon 4. https://doi.org/10.1016/j.heliyon.2018.e00766

91. Teengam P, Siangproh W, Tuantranont A, Vilaivan T, Chailapakul O, Henry CS (2017) Multiplex paper-based colorimetric DNA sensor using pyrrolidinyl peptide nucleic acid-induced AgNPs aggregation for detecting MERS-CoV, MTB, and HPV oligonucleotides. Anal Chem 89:5428–5435. https://doi.org/10.1021/acs.analchem.7b00255

92. Koo B, Jin CE, Lee TY, Lee JH, Park MK, Sung H, Park SY, Lee HJ, Kim SM, Kim JY, Kim S-H, Shin Y (2017) An isothermal, label-free, and rapid one-step RNA amplification/detection assay for diagnosis of respiratory viral infections. Biosens Bioelectron 90:187–194. https://doi.org/10.1016/j.bios.2016.11.051

93. Shi L (2015) Development of SPR biosensor for simultaneous detection of multiplex respiratory viruses. Biomed Mater Eng S2207–S2216. https://doi.org/10.3233/BME-151526

94. Hsu Y-R, Lee G-Y, Chyi J-I, Chang C-K, Huang C-C, Hsu C-P, Huang T-H, Ren F, Wang Y-L (2013) Detection of severe acute respiratory syndrome (SARS) coronavirus nucleocapsid protein using AlGaN/GaN high electron mobility transistors. ECS Trans 50:239–243. https://doi.org/10.1149/05006.0239ecst

95. Park TJ, Hyun MS, Lee HJ, Lee SY, Ko S (2009) A self-assembled fusion protein-based surface plasmon resonance biosensor for rapid diagnosis of severe acute respiratory syndrome. Talanta 79:295–301. https://doi.org/10.1016/j.talanta.2009.03.051

96. Huang JC, Chang Y-F, Chen K-H, Su L-C, Lee C-W, Chen C-C, Chen Y-MA, Chou C (2009) Detection of severe acute respiratory syndrome (SARS) coronavirus nucleocapsid protein in human serum using a localized surface plasmon coupled fluorescence fiber-optic biosensor. Biosens Bioelectron 25:320–325. https://doi.org/10.1016/j.bios.2009.07.012

97. Ahn D-G, Jeon I-J, Kim JD, Song M-S, Han S-R, Lee S-W, Jung H, Oh J-W (2009) RNA aptamer-based sensitive detection of SARS coronavirus nucleocapsid protein. Analyst 134:1896–1901. https://doi.org/10.1039/B906788D

98. Lodes MJ, Suciu D, Wilmoth JL, Ross M, Munro S, Dix K, Bernards K, Stöver AG, Quintana M, Iihoshi N, Lyon WJ, Danley DL, McShea A (2007) Identification of upper respiratory tract pathogens using electrochemical detection on an oligonucleotide microarray. PLoS ONE 2:e924–e924. https://doi.org/10.1371/journal.pone.0000924

99. Abad-Valle P, Fernández-Abedul MT, Costa-García A (2005) Genosensor on gold films with enzymatic electrochemical detection of a SARS virus sequence. Biosens Bioelectron 20:2251–2260. https://doi.org/10.1016/j.bios.2004.10.019

100. Che X-Y, Hao W, Wang Y, Di B, Yin K, Xu Y-C, Feng C-S, Wan Z-Y, Cheng VCC, Yuen K-Y (2004) Nucleocapsid protein as early diagnostic marker for SARS. Emerg Infect Dis 10:1947–1949. https://doi.org/10.3201/eid1011.040516

101. Diao B, Wen K, Chen J, Liu Y, Yuan Z, Han C, Chen J, Pan Y, Chen L, Dan Y, Wang J, Chen Y, Deng G, Zhou H, Wu Y (2020) Diagnosis of acute respiratory syndrome coronavirus 2 infection by detection of nucleocapsid protein. MedRxiv Prepr Serv Heal Sci. https://doi.org/10.1101/2020.03.07.20032524

102. Whiting P, Singatullina N, Rosser JH (2015) Computed tomography of the chest: I basic principles. BJA Educ 15:299–304. https://doi.org/10.1093/bjaceaccp/mku063

103. Bernheim A, Mei X, Huang M, Yang Y, Fayad ZA, Zhang N, Diao K, Lin B, Zhu X, Li K, Li S, Shan H, Jacobi A, Chung M, Chest CT (2019) Findings in coronavirus disease, (COVID-19): relationship to duration of infection. Radiology 295(2020):685–691. https://doi.org/10.1148/radiol.2020200463

104. Ai T, Yang Z, Hou H, Zhan C, Chen C, Lv W, Tao Q, Sun Z, Xia L (2019) Correlation of chest CT and RT-PCR testing for coronavirus disease (COVID-19) in China: a report of 1014 cases. Radiology 296(2020):E32–E40. https://doi.org/10.1148/radiol.2020200642

105. Liu M, Zeng W, Wen Y, Zheng Y, Lv F, Xiao K (2020) COVID-19 pneumonia: CT findings of 122 patients and differentiation from influenza pneumonia. Eur Radiol. https://doi.org/10.1007/s00330-020-06928-0

106. Bai HX, Hsieh B, Xiong Z, Halsey K, Choi W, My T, Tran L, Pan I, Shi L-B, Hu P-F, Agarwal S, Xie F, Li S (2020) Performance of radiologists in differentiating COVID-19 from viral pneumonia on chest CT. Radiology 296:E46–E54. https://doi.org/10.1148/radiol.2020200823

107. Colombi D, Petrini M, Maffi G, Villani GD, Bodini FC, Morelli N, Milanese G, Silva M, Sverzellati N, Michieletti E (2020) Comparison of admission chest computed tomography and lung ultrasound performance for diagnosis of COVID-19 pneumonia in populations with different disease prevalence. Eur J Radiol 109344. https://doi.org/10.1016/j.ejrad.2020.109344

108. Tierney DM, Huelster JS, Overgaard JD, Plunkett MB, Boland LL, St Hill CA, Agboto VK, Smith CS, Mikel BF, Weise BE, Madigan KE, Doshi AP, Melamed RR (2020) Comparative performance of pulmonary ultrasound, chest radiograph, and CT among patients with acute respiratory failure. Crit Care Med 48

109. Lopes AJ, Mafort TT, da Costa CH, Rufino R, de Cássia Firmida M, Kirk KM, Cobo CG, da Costa HD, da Cruz CM, Mogami R (2020) Comparison between lung ultrasound and computed tomographic findings in patients with COVID-19 pneumonia. J Ultrasound Med n/a. https://doi.org/10.1002/jum.15521

110. Tung-Chen Y, Martí de Gracia M, Díez-Tascón A, Alonso-González R, Agudo-Fernández S, Parra-Gordo ML, Ossaba-Vélez S, Rodríguez-Fuertes P, Llamas-Fuentes R (2019) Correlation between chest computed tomography and lung ultrasonography in patients with coronavirus disease, COVID-19. Ultrasound Med Biol 46(2020):2918–2926. https://doi.org/10.1016/j.ultrasmedbio.2020.07.003

111. McDermott C, Daly J, Carley S (2020) Combatting COVID-19: is ultrasound an important piece in the diagnostic puzzle? Emerg Med J 37:644–649. https://doi.org/10.1136/emermed-2020-209721
112. Wong HY, Lam HY, Fong AH, Leung ST, Chin TW, Lo CS, Lui MM, Lee JC, Chiu KW, Chung TW, Lee EY, Wan EYF, Hung FNI, Lam TPW, Kuo M, Ng M-Y (2020) Frequency and distribution of chest radiographic findings in COVID-19 positive patients authors. Radiology 296, E72–E78. https://doi.org/10.1148/radiol.2020201160
113. Qian-Yi P, Xiao-Ting W, Li-Na Z, CCCUSG. (CCUSG) (2020) Findings of lung ultrasonography of novel corona virus pneumonia during the 2019–2020 epidemic. Intensiv Care Med 46 849–850. https://doi.org/10.1007/s00134-020-05996-6
114. Huang Y, Wang S, Liu Y, Zhang Y, Zheng C, Zheng Y, Zhang C, Min W, Yu M, Hu M (2020) A Preliminary study on the ultrasonic manifestations of peripulmonary lesions of non-critical novel coronavirus pneumonia (COVID-19). Res Sq. https://doi.org/10.21203/rs.2.24369/v1

Chapter 8
COVID-19 and Development: Effects and Consequences

Mahmood Sariolghalam

8.1 Introduction

This article discusses the subject in two sections. First, the state of development and its political and geopolitical aspects prior to the COVID-10 pandemic. Second, an analysis of the possible trends in development during the 2020–21 period.

Development has an intrinsic relationship with globalization. Even before the COVID-19 pandemic, globalization both theoretically and practically faced immense challenges. The economic instability resulting from globalization led to the rise of right-wing governments in Europe with authoritarian governments securing their foothold in Central America, the Arab World and Asia. Considering the centrality of the concept of globalization, the pre-pandemic state of development in the international system can perhaps be summarized in the following trends [1]:

- The measured confrontation between the U.S. and China,
- The powerful financial and economic emergence of China in the international system,
- The gradual economic emergence of India,
- The decline in the level of multilateralism in international relations,
- The intensification of environmental challenges,
- The prominence of information technology (IT) and artificial intelligence (AI),
- The increase in geopolitical tensions between the United States and Russia,
- The widening of the social inequality in most countries of the world,
- The doubts concerning the advantages of inclusive globalization.

A comparative review of these nine variables reveals that the emergence of Chinese economic power during the past three decades is perhaps the most pivotal development in international relations both politically and economically. The transfer of

M. Sariolghalam (✉)
School of Economics and Political Science, Shahid Beheshti University, Tehran, Iran
e-mail: m-sariolghalam@sbu.ac.ir

© The Author(s), under exclusive license to Springer Nature Singapore Pte Ltd. 2021
M. Rahmandoust and S.-O. Ranaei-Siadat (eds.), *COVID-19*,
https://doi.org/10.1007/978-981-16-3108-5_8

production and wealth from the West to East Asia and the reduction of GDP share of G7 from two-thirds of the global economy to less than a half exposed the new vulnerability of Western economies. To a considerable degree and based on its diversified economy and foreign policy, Germany, being fourth in the world, is still a viable and sustained economy. Ten years ago, no Chinese bank ranked among the top international banks, but now half of the top global banks are Chinese. Centralized finance and production in China incrementally prepared the ground for a reduced distribution of wealth and job opportunities in the West expanding intense competition and thereafter causing gradual socio-economic crises in the West. Without granting much mutual opportunities and advantages, China began a process of measured confrontation characterized like a chess game particularly with the U.S. both in economics and politics. This was the beginning of a new cold war between the top two economic powers in the world. All countries in the world conduct somewhere between 20 and 80% of their trade with China. The Chinese who have an insatiable demand for raw materials have established extensive and long-term relations with almost all African countries, and hundreds of development projects are managed by half a million Chines engineers in return for raw materials and metals. The Chinese established two large financial and political organizations (the Shanghai Cooperation Organization and the Asian Infrastructure Investment Bank) as two financial arms to further globalize their national economic power. The emergence of China has influenced international relations to the extent that no global issue can be raised and resolved unless Beijing plays a role in the discussion and decision-making process.

Multilateralism, environmental challenges, management of most international organizations and global economy are utterly inconceivable without Chinese participation [2]. The silent crawling political movement of the Chinese that has its origins in Confucius teachings did not alarm the world and the international community did not notice the gradual accumulation of Chinese economic and political power. Power in international relations from 1945 to 1990 was overshadowed by the U.S. and Soviet relations. However, power relations since 1990 have been played out in a U.S.-Russia-China Triangle. It is now believed that the U.S. has never in its history faced a rival as powerful as China. During the past three decades, expanding relations with China was perceived to be so cherished that Americans tolerated the conditions mostly shaped by the Chinese. The Chinese benefited from the transfer of technology and vast expansion of joint ventures between the two countries. Moreover, the United States hoped for a 'reset' of its relations with Russia. Successive U.S. administrations failed to fathom the entirety of the long-term behaviors of Moscow and Beijing. Perhaps complacency prevailed due to the assertive American global role. Compared to the US, decision-making in Russia and China enjoys greater continuity. Through the end of the Obama administration, Americans generally welcomed trade and financial relations with China [3], while the Chinese intensified their mergers and acquisitions drew on the rich scientific-technological communities of the U.S. including sending some 369,000 Chinese students to American universities [4]. The Chinese policy of the Trump administration emerged as a bipartisan approach. Even if Hillary Clinton had won the presidency, she would have pursued a similar approach to Beijing. The new dynamics in the U.S.-Russia-China Triangle resulted in an implicit disagreement and divergence of policy options between the United States and main partners in the European Union [3]. It appears that intense competition, measured confrontation and renewed assertiveness in Russia have resulted in two essential consequences: enhancement of economic nationalism, right-wing policies among medium-level powers and increased military expenditures by countries.

The cold war produced high levels of predictability in the international system, but the contemporary new triangle of international competition has both increased uncertainty and made international cooperation far more problematic. Interestingly, expanded political and economic cooperation between Europe and Asia has created fissures between the US and particularly Germany and France. The U.S. developed the mechanism of Better Utilization of Investments to Leverage Development (BUILD) to challenge Chinese economic and financial initiatives. It has increased its economic and security activities in Central and Southeastern Europe on the one hand and in the Eastern Mediterranean on the other. In Asia also, the U.S. set up the International Development Finance Corporation against the Chinese coalitions to provide countries with financial and industrial resources. At the same time, the United States has expanded its military relations with India, Vietnam, the Philippines and Taiwan [3]. In other words, with the rise of China's financial and economic power, the issue of development has indirectly regained an ideological nature, shifted political alignments and renewed military foundations that it had during the cold war. One fundamental consequence is a noticeable level of unpredictability in the emerging international system. The pre-COVID-19 global structure in development already led to many tensions among countries. Trump's trade and security policies towards Asia and Europe exacerbated these tensions for understandable reasons. America no longer wished to allow other countries to have a free ride. While China and many other rising powers did not question the liberal foundations of the international system and played by the rules and procedures of the American-built international system yet they collectively aspired to expand their political and economic share in the system. Although the Chinese did not make much concessions to Washington after several years of trade negotiations with the U.S. and adopted the policy of protraction, the Trump administration however succeeded in changing many provisions of the NAFTA agreement in its favor. With the advent of China, the issue of development in international relations has been altered in such proportions where no country can disregard China in its international and trade relations. The emerging international system in the last few years has provided countries such as Brazil, Argentina, Indonesia and India with a rare opportunity to diversify their foreign economic and political relations. Of interest is the sophistication of diplomacy of these middle-level countries where they reach out to all great powers to advance their political and economic interests. Arab countries that have normally pursued strategic relations with Europe and the U.S. are now slowly expanding their interactions with Moscow and Beijing. Weak developing countries in Africa and Asia are now dependent on international financial organizations such as the World Bank on the one hand and on potential barter projects with China on the other.

During a period of half a century since 1970, the world population has grown from 3.5 billion to 7.7 billion, whereas 220 years ago, in 1800, it was just one billion

[5]. Urbanization, improvements in health and increased production prepared the ground for population growth. Population growth exerts considerable pressure on the resources required for production and on fuel and at the same time, causes environmental erosion. During the forty years since 1980, half of the Earth's oil reserves have been used to achieve the considerable economic growth in the West but also in East Asia and the BRICS nations. Globalization has meant increased production of goods and services, wealth, financial markets, urbanization, consumption, middle-class empowerment, and economic interdependence. Furthermore, globalization did not only become a barrier to the gradual growth of armaments production and exports but it also became an economic and wealth-producing issue for the West and Russia. However, because of the interdependence of great powers, globalization reduced the probability of war and direct military conflict between them while proxy wars have characterized new confrontations. In 2019, development was no longer monopolized by the West but was realized in its financial, productive and economic sense in Asia, Latin America and the Arab states of the Persian Gulf.

The Chinese and Asian model of development substantially refuted the fundamental assumption of economic liberalism introduced by Milton Friedman in the 1960s that economic growth, industrialization, and economic privatization will lead to democratic political systems [6]. State-led privatization in Asia consequently produced substantial wealth without a visible democratic imprint. Trends since 2000 may be summarized as follows:

1. The rise of China as the world's second-largest economy,
2. Production of about 40% of the global GDP by China and the United States,
3. The expansion of about four hundred million people of middle-class citizens in Asia,
4. American and Chinese monopoly on 5G,
5. The daily consumption of about one hundred million barrels of oil in the world,
6. The zero economic growth in Japan,
7. The reduced competitiveness of many industrial European countries,
8. The growing importance of the global south in production and supply networks due to low labor costs,
9. The considerable expansion of tourism, travel and luxury goods consumption,
10. The rising regional geopolitical tensions in the Far East and the Middle East,
11. The surge in US national debt to $23 trillion,
12. The increase in the production and export of armaments in the world.

8.2 The COVID-19 Pandemic and the Gradual Disruption in Growth and Development

Before Covid-19, other communicable diseases occurred in different parts of the world: SARS in 2003, H1N1 in 2009, MERS in 2012, Ebola in 2014 and Zika in 2015 [7]. Although health experts recommended preventive measures against these contagious diseases, planning for the future was not taken very seriously because the spread of these diseases was relatively controlled and their nature was not as dangerous as that of Covid-19 [7]. The global trade, financial and economic issues were so critical that from December 2019, major industrialized governments chose to downplay the importance of the COVID-19 pandemic and consider it controllable. The U.S. and China were and are so dependent on each other's markets that they were reluctant to make public the worries concerning the new pandemic. Mutual dependence is not just limited to consumer goods. For example, 156 important life-saving medicines are exported from China and India to the U.S. [7]. Concerns about disruptions in what everyone had become used to, planned, and depended on may have psychologically caused the delay in responding to Covid-19 and to its extensive consequences. One of the outcomes of globalization on China is the astronomical expansion of China's tourism. About 14.5 million Chinese tourists travelled to 28 EU countries in 2018 and, in the same year, about 2.5 million Schengen visas were issued to the Chinese [8]. The share of tourism in Greek and Italian GDP is about 30% and 13%, respectively [9]. The estimated trade and therefore the mutual dependence between the U.S. and China was about $600 billion in 2019 [10]. Economic growth became so important during a few decades that about one billion people travelled each year, and thousands of manufacturing companies, especially in Asia, Latin America and Africa, were connected to the $142 trillion supply chain producing the global GDP [11].

However, these exceptional conditions in the history of economic growth and development changed rapidly when Covid-19 became a pandemic in late December 2019 and early January 2020. The global economy is predicted to shrink by about 5.2% in 2020 on average which is the most severe downturn in the last 80 years. Countries with more international markets and more dependent on tourism and supply chains will suffer more. Along with major industrialized countries, emerging markets and developing economies (EMDEs) will hit harder [12]. During the months that followed February 2020, nearly 4 billion people did not go to the workplace; schools and universities were closed; most flights around the world were cancelled, and companies were shut down. Table 8.1 demonstrates a forecast of the decline in GDP growth rates in some countries:

The decline in economic activities for millions of enterprises will mean that they can neither pay their employees nor manage their debts. Economic bankruptcy can gradually be accompanied by financial crises in many large and small economies. Investments in all industries around the world have declined more sharply due to corporate financial crises. The tourism industry which grew by 6.5% annually in the last decade due to the growth of the middle class in Asia came to a near-complete halt with a very steep slope in the first quarter of 2020. Among the six regions of the world, Latin America will experience the largest reduction in economic growth rate. Since 1870, the international economy has experienced fourteen recessions with the 2020 recession the first experience as a result of a pandemic. The previous recessions were caused by war, financial crises and rising oil prices [12]. Arrears also have faced many banks with liquidity problems and fewer investment opportunities. In the economic cycle, the service and labor sectors, especially in countries that used foreign labor, experienced a considerable decline in the first half of 2020. The

Table 8.1 GDP growth decline rates

	GDP growth (annual percentage)	
	2019	2020
China	6.1	1
U.S	2.3	−6.1
EU	1.2	−9.1
Brazil	1.1	−8
Turkey	−0.9	−3.8
Saudi Arabia	−0.3	−3.8
Iran	−8.2	−5.3
Russia	1.3	−6
India	4.2	−3.2
High-income economies	1.7	−6.8
Low-income economies	5.1	1

immediate consequence of this decline in activity was clearly observed in the energy sector also so that oil demand fell by about 10 percent [12]. Demand for metals also dropped by up to 16 percent in the post-COVID-19 period. If the demand for metals increases in the second half of 2020, and gradually in 2021, it will be dependent on China's demand and GDP growth since it accounts for about 50% of the global demand for metals [12].

The majority of developing countries will be affected by this considerable contraction in economic activities more than the Western countries and the powerful Asian economies such as Japan, South Korea and China that can withstand declining production and demand due to their financial strength and savings rates. Production in the global south is dependent on consumption in Western and East Asian countries. For example, the production chains in a country like Bangladesh which plays a critical role in the garment industry depend on consumption in Europe and North America. Now that 42 million people are unemployed in the U.S. and there is a significant reduction in demand for garments in Europe, many of the garment factories in Bangladesh will be closed and will start production if the demand for garments rises in the Western markets in 2021. Due to the low savings rate and the presence of poor classes in this type of Asian, African and Latin American countries, it can be predicted that the level of poverty and degree of socio-economic instability in developing countries will increase at least during 2020. As the rate of decline in GDP is about three times that of the 2009 recession, and in contrast to that year when the global crisis happened only in the financial markets, in 2020 the recession occurred in all sectors of the economy and in all globalized economies. Hence, a return to pre-COVID-19 conditions will also be subject to the global variables of production, supply and demand.

The issue of growth and development will be overshadowed by the global change in the trends and behavioral patterns of consumption and employment caused by this pandemic. This change will have serious effects on the entire demand in 2021 and beyond. It is not clear exactly what the stable responses of demand and financial markets will be to the post-COVID-19 conditions. It seems that both households and enterprises will be inclined to increase their savings rates which will naturally have repercussions for investment and consumption. The sudden emergence of Covid-19 with very high mortality rates in Brazil, Chile and Mexico can have serious consequences for consumption and production of agricultural products and even for food security in the second half of 2020 and possibly into 2021 [12]. Naturally, a return to the pre-COVID-19 economic conditions will be at least slower if health controls and physical distancing continue. European concerns caused by immigration from Africa and the Middle East, the slow penetration of Asian companies into European economy, the possible slow U.S. disengagement from geopolitics of Europe and European defense, and a serious decline in European competitiveness in the global economy will face the EU [12]. The Chinese who enjoyed at least 5% annual economic growth rate for decades will experience a 1–2 percent growth in 2020–21. In the totality of the Asian region, especially in East Asia, the economic growth rate will decline to a half percent. The reduction in consumption, production, investment and trade in the region will equal those of the 1990 crisis. Most of these countries have adopted the policy of granting credit and loans to companies to gradually encourage a return to economic activities and increased demand. In the process of implementing this policy, the Council of Europe proposed 750 billion euros for the EU. The Chinese hope to create the required conditions for demand and investment by reducing taxes, making direct payments to vulnerable households, and activating credit for the provinces. Malaysia's stimulus package is estimated at about 17% of GDP [13].

The Middle East, which suffers from structural failures, the refugee crisis, war and insecurity, experienced a far more unfavorable position with the advent of the Covid-19. Egypt, Jordan, the Palestinian territories and Tunisia that had somewhat reduced their budget deficit by trading with the Arab nations of OPEC and receiving financial aid from them are now suffering intensely from falling oil prices and oil exports. If tourism fell by 4% in the 2009 financial crisis, it contracted by 60–80% in 2020 and countries such as Jordan, the UAE and Turkey lost a considerable part of their tourism revenues. Investments in the Middle East will face a serious crisis due to the declining oil revenues that will hurt hundreds of local, European and Asian companies [14]. In Pakistan and India where the economy is generally based on small and medium-sized enterprises, the sharp decline in production and demand has resulted in the bankruptcy of many enterprises leading to chronic unemployment and considerable social anomalies. Governments have coped with social and political instability by providing assistance to the poor, making direct payments, increasing investments in health, and supporting employment and economic activity. The Indian population in the Arab States of the Persian Gulf, numbering about 9 million, send $20 billion in remittances which is expected to decline by one-fifth in 2020 [12]. Covid-19 has caused capital flight from Africa, almost shut down mining activities and stopped investments in mining projects resulting in widespread economic and social crises. Therefore, the current vulnerabilities in Africa will deepen because of the widespread poverty. Health protocols are very difficult for two-thirds of the Africans living in very densely populated areas and have raised many concerns regarding the uncontrollable spread of the pandemic. Many African countries have short- and medium-term loans that can only be paid back by activities in the mining and oil and gas sectors. It is feared that the debt crisis on the one hand and food insecurity for many people in African countries (72 million people in 35 countries) on the other will deepen [12]. Covid-19 is the biggest economic crisis of the last century and, in general, no region in the international system has been immune from its negative economic consequences. COVID-19 is the worst economic crisis during the periods of peace in the past 100 years. In all, no region in the international system has escaped from the adverse economic implications of COVID-19.

8.3 Conclusion: COVID-19 and the Future Developmental Trends

The United States, China and the EU collectively represent about half of the global economy. If US economic growth declines by one percent, it will have a negative impact of 0.7–0.8 percent on the economies of the rest of the world. If all these three economies together experience a one percent reduction in growth, then the rest of the global economy will face a decline of about 1.3%. This assessment indicates a deep interdependence in the global economy [12]. Due to this interdependence, the developing countries have far fewer mechanisms of economic stability than they had during the 2009 financial crisis. Many third-world countries need to import up to 50% of the raw materials used in their production lines. The slowdown in the activities of the supply chain of raw materials, goods and services has had a lasting effect on the unemployment rates in these countries. On average, borrowing costs in most of these countries have risen by 11% over the past decade so that borrowing accounts for up to 55% of GDP. Economic contraction in the spectrum of rich and poor countries leads to a decline in capital accumulation as the source of national power and in savings for development and national security projects. If the consequences of COVID-19 are scrutinized from this perspective, it is not very clear how long it will take for the global economy to recover [12]. In other words, the growth and development horizon have suffered a major stroke, at least in the medium term. Even if some sectors of the economy such as the food industry, internet providers, online retail firms have not only remained in business but also expanded, the return of other industries including the automotive, tourism, housing, and the consumer goods industries have no bright future. For example, the cruise ship industry which has a turnover of about $150 billion a year has lost up to 82% of its activity, and it is predicted that this industry will not have the opportunity to return to pre-COVID-19 conditions because older people constitute most of its passengers [12].

These developments are the natural results of a global pandemic. Since the rise of China as an economic power, the international system had become accustomed to a framework of the political economy of privatization based on the active and direct role of the government. Covid-19 disrupted these trends prevailing set of rules. Production disruption has created a set of negative consequences for all national and international growth and development indicators. In addition, labor and capital markets have also been severely disrupted. When income levels fall, they have immediate negative impacts on health, nutrition and education. The formation of eight-kilometer queues of vehicles in some of the states in the U.S. to receive free food packages suggests that even in advanced economies the pandemic can have serious and immediate effects on employment and economic activities [12]. Nearly 50 million people in the United States applied for unemployment benefits during four months [15]. Such developments in the field of employment and economic activity have also consequences for innovation. Enterprises refrain from investing in technological innovations at least before semi-normal conditions return [16]. The environmental catastrophes or epidemics in the past such as SARS had limited economic and geographical consequences, but the importance of this pandemic is that it has a global dimension and a return to normalcy is subject to extensive international cooperation which is not visible. Given the overall strategy of the Biden administration of measured confrontation with the alleged economic and geopolitical excesses of China and Russia, it is not clear whether global cooperation on the pandemic can be expected. In addition, physical distancing will at least slow down many activities such as tourism. New employment in manufacturing, reopening of schools and universities and educational environments in general as well as sports competition will be disrupted into the foreseeable future. If healthcare measures are not effective and reliable vaccination is not provided for the wider public, the serious global consequences of Covid-19 can raise severe concerns, also called super-hysteresis, regarding the future of production and economic growth. Damages to production chains, global trade, capital markets and global interactions can be more sustained [17]. The loss of about 300 million jobs by mid-2020 has dealt a considerable economic shock to the globalization processes [12]. The International Labor Organization estimates that nearly 80% of the world's approximately two billion workers have in some way been financially affected by Covid-19. In the last four months also, about $11 trillion have been allocated by governments in the form of financial and monetary packages to cope with COVID-19. Many European countries, the U.S., Brazil, Japan and Mexico are forced to delay the reopening of economic activities in order to prevent a re-emergence of COVID-19. This will postpone the realization of efficiency, production and economic growth to the future. The global budget deficit is projected to increase to 101% of GDP during 2020–2021 (a growth of 19% compared to 2019), with the international average rising from 4 to 14%. Government revenues will be severely reduced due to the decline in consumption, production and household income and corporate revenues [12].

Along with the changes in economic growth and development, there have also been changes in the political economy at the global level. In the post-COVID-19 era, the role of the state has been considerably reinvigorated. The emergency caused by the pandemic caused a disruption in the state-business and state-society relations leading to a powerful assertive role of the state [18]. Opinion polls show that this new role of the government does not reflect its credibility or the public trust but it reflects an implication under emergency conditions [18]. Interestingly, in countries such as Demark and Sweden where the level of confidence in the political system has been traditionally high, there has been greater acceptance of the increase in government power whereas in Italy, France and Spain, the authority of the political system has been under question [19]. COVID-19 has augmented the inclination of the public, at least at the EU level, to accept cooperation between both societies and countries, and at the same time has sharply damaged America's reputation as the EU's most important partner. Renewed nationalistic predispositions in the EU have given way to a new idea dubbed as 'strategic sovereignty'. Overall, the interest in maintaining and expanding European convergence with other countries has amplified because of COVID-19. The joint French-German reconstruction program for dealing with the pandemic has enhanced this ground for intensified coordination so that some have called it the United States of Europe. The new outlook on sovereignty has also echoed assertions that Europe needs to free itself from the "dominance" of Facebook and Huawei. Covid-19 has created the idea that if the Europeans do not cooperate with each other they will suffer in the trade war and also by the geopolitics of China and the U.S. Populism and opposition to the EU are expected to decline and the pandemic has created a new political climate and the political economy of growth and development among the EU members [19]. The ASEAN region also exhibits this aspect of cooperation and collective commitment to combat the negative consequences of the pandemic because its economic convergence will suffer severely in the absence of cooperation between the member countries [19].

In the post-COVID-19 era, countries will pay more attention to their priorities. Production chains will not function merely based on calculations of profit. The free movement of millions of travelers for purposes of growth and development will not be the same as before. The social-class gaps, poverty, debt and budget deficit will deepen. Collective cooperation among countries in new pandemics and environmental issues will become more energetic. The importance of information technology and artificial intelligence will intensify. Countries will be more in need of unions, regionalism and exchanges. The sharp increase in the purchase of weapons before the pandemic and the greater militarization of politics in countries will slow down. Economics will take precedence over politics [20]. It appears what the Freedom House has reported as the decline in the levels of freedom and as the steady rise of authoritarianism over the past 14 years will be even more deepened in the post-COVID-19 era [21]. The unimaginable drop in oil prices and the decline in oil revenues will cause serious fluctuations in the position of the GCC as its member countries spent 5.8% of their GDP on purchasing weapons. This is at a time when the Middle East region is facing the highest youth unemployment rate (35%) and the need for economic reconstruction after Covid-19. In a world full of political and economic competition after COVID-19, it is feared that the number of failed states in the Middle East and Africa will increase due to budget deficits, low oil prices and rising poverty [22].

Some consequences of COVID-19 will be permanent. For example, 2.2 million (10% of the total number of) restaurants in the world will be closed forever [23]. Vaccine production, efficacy and reliability and its availability to the public will be the key issues in shaping the characteristics of the global economy and political economy in the post-COVID-19 era [24]. Vaccination will determine the duration of the pandemic and physical distancing. It has been argued that physical distancing may be a permanent feature of social life. People who are vulnerable due to their age may even reduce their level of social activity despite the availability of the vaccine. Physical distancing will have a direct relationship to consumption. Consumption and demand will be the key economic variables in the return of growth and development in 2021. Along with demand, the ability of millions of unemployed people to find new jobs or return to their previous work will be the next deciding variable. There is no guarantee that people who lost their jobs will start working in the same enterprises. In addition, industries and companies will change the form and content of their work and activities during the pandemic. The inclination to automate and reduce the cost of observing health protocols among the workforce will reduce the rate of return to work in those industries where automation is possible. If demand in the consumer industry does not grow as it did in the past, many crises resulting from unemployment, low incomes, illness and also government subsidies will continue and improvement in economic affairs will be delayed even beyond 2021.

As for COVID-19 itself, cooperation between countries, the assistance provided by WHO, the readiness of governments and international organizations to help low-income third world countries in health matters, and especially the role played by China and the U.S. on the international political scene in preventing this disease from spreading again will be substantially effective [25]. Once the vaccine is made available to the wider public, it is the role of the governments in market regulation and economic policymaking that can stimulate demand, consumption and the economic cycle, thus managing higher rates of political and political stability. Policy packages of governments to facilitate domestic and foreign investment, especially in developing policies and taking actions in both public investment and digital infrastructure, will be imperative in returning to the pre-COVID-19 conditions. One of the serious responsibilities of governments in the post-COVID-19 world will be to make new technological investments in the field of public health. Third world governments that are more concerned with security in their national priorities will pay less attention to such responsibilities. If the political economy of growth in the global south does not correlate with the industrialized and developed countries and an imbalance is generated in the international economy, trends in global demand and consumption will not easily improve [18]. Even among developed countries, there is no guarantee that budgets, credit and expertise will generate the necessary organization and lead to the needed efforts in preventing the spread of this pandemic or other possible pandemics in the future. In the United States, health experts have recommended that federal and state officials should take the Strategic National Stockpile for Biodefense seriously [18]. It has even been suggested that a Public Health Treaty Organization should be formed with the participation of all countries to coordinate future health programs [7].

In the discourses of developmental studies, the state and government were usually regarded as a regulator and balancer. This role is well-demonstrated in countries such as Norway, Denmark and Germany. However, in the post-pandemic world, the role of the state in garnering rule of law, national consensus and social contract in order to balance growth and development will be even far more critical. The successful implementation of coronavirus management in countries such as South Korea raise the salience of government efficiency and structures of decision making. The post-COVID-19 international system will be largely subject to the level of efficacy of governmental processes in decision making [7]. What is concerning in 2020 and beyond is the level of adoption of local and national procedures in dealing with Covid-19 and learning mechanisms among countries. Even coordination and cooperation with the World Health Organization has been somewhat limited. Evidence vividly suggests that almost all countries were shocked when this pandemic occurred and were least prepared to deal with its consequences. Even it took the G-20 member states three months to initiate dialogue and coordination to adopt collective measures to deal with the pandemic [26].

It appears that a large number of countries today are suffering from a state of complacency. The days when there were robust attitudes of serving the public and being 'public servants' are perhaps over. Two dominant groups act far above the political process in a large number of countries: the corporate and the intelligence class [27]. The post-pandemic world may be a historic opportunity to reexamine not so much the merits or the perils of capitalism or globalization but rather how they are enforced in many countries and the processes of decision making. Can the state make decisions in the public interest? Germany, the Netherlands, New Zealand, Australia, Canada, and the Scandinavian region appear to do well and hold public interest above all other institutions and stakeholders. High levels of egalitarian policies and politics in these states display the reality that it is possible to practice economic privatization and be fully engaged in the globalization process and yet implement policies to serve the public. New Zealand stands out as an exemplary where national, public and private interests converge [28].

In the post-COVID 19 international system, the role of the state and government will become more prominent. But tending towards what kind of features and responsibilities? Carrying what type of structure, organization, prioritization, orientation? No author of political economy from Adam Smith to Thomas Piketty has denied the role of government in regulating economic growth and development [29]. Perhaps the new emerging discourse will involve the nature of the state in the post-pandemic world. Will governments play a more assertive role in protecting the environment and the unlimited use of natural resources, or will corporations and the demand for goods and services guide policy making? The interrelationships between the members of the triad "the government, the private sector and the public good" have once again returned to the focus of developmental theories. Covid-19 caused many to pause and think about the questions and the challenges of economic development. The year 2021 will be a crucial one in reorienting the economic growth and development discourse. However, one political question on the position and substance of development in the post-COVID-19 era will be the decisive one: Will the centers of power like China, the United States, Russia, and the EU be able to converge on a systematic approach and roadmap to achieve an international system characterized by an eco-friendly environment, wider social equity and arms control?

References

1. Innovation and National Security. Council on Foreign Relations. https://www.cfr.org/report/keeping-our-edge/. Accessed 04 Mar 2021
2. The international system and the middle east's new geopolitics [translation]. Sariolghalam, Mahmoud. https://www.sariolghalam/2017/08/27. Accessed 04 Mar 2021
3. American, China, Russia, and the return of great-power politics. Foreign Affairs. https://www.foreignaffairs.com/articles/2019-12-10/age-great-power-competition. Accessed 04 Mar 2021
4. Number of Chinese students in the U.S. 2019. Statista. https://www.statista.com/statistics/372900/number-of-chinese-students-that-study-in-the-us/. Accessed 04 Mar 2021
5. Population explosion. HuffPost. https://www.huffpost.com/entry/population-explosion_b_9837278. Accessed 04 Mar 2021
6. Capitalism and Freedom, Friedman, Appelbaum. https://press.uchicago.edu/ucp/books/book/chicago/C/bo68666099.html. Accessed 06 Mar 2021
7. Coronavirus: chronicle of a pandemic foretold. https://www.foreignaffairs.com/articles/united-states/2020-05-21/coronavirus-chronicle-pandemic-foretold. Accessed 06 Mar 2021
8. Number of tourist arrivals from China in the EU 2012–2018. Statista. https://www.statista.com/statistics/901146/number-of-arrivals-from-china-in-tourist-accommodations-in-the-eu/. Accessed 06 Mar 2021
9. Tourism now generates over one quarter of Greece's GDP. https://greekreporter.com/2019/05/30/tourism-now-generates-over-one-quarter-of-greeces-gdp. Accessed 06 Mar 2021
10. Official source us export and import statistics—foreign trade—US Census Bureau. https://www.census.gov/foreign-trade/index.html. Accessed 06 Mar 2021
11. Highlights from Chapter 1 Pandemic, recession: the global economy in crisis
12. (No Title). http://pubdocs.worldbank.org/en/267761588788282656/Global-Economic-Prospects-June-2020-Highlights-Chapter-1.pdf. Accessed 06 Mar 2021
13. Surging U.S. virus cases raise fear that progress is slipping. https://www.modernhealthcare.com/clinical/surging-us-virus-cases-raise-fear-progress-slipping. Accessed 06 Mar 2021
14. Five ways that the coronavirus should transform the EU—European Council on Foreign Relations. https://ecfr.eu/article/commentary_five_ways_that_the_coronavirus_should_transform_the_eu/. Accessed 06 Mar 2021
15. Cruise industry in coronavirus aftermath: what's in store? CNN Travel. https://edition.cnn.com/travel/article/cruise-industry-coronavirus-aftermath/index.html. Accessed 06 Mar 2021
16. Coronavirus US: thousands of cars line up at LA food bank. Daily Mail Online. https://www.dailymail.co.uk/news/article-8218267/Thousands-cars-line-Los-Angeles-food-bank.html. Accessed 06 Mar 2021
17. 1.48m more Americans file for unemployment as pandemic takes toll. US unemployment and employment data. The Guardian. https://www.theguardian.com/business/2020/jun/25/us-unemployment-figures-latest-coronavirus-pandemic-economy. Accessed 06 Mar 2021
18. World Economic Outlook update, June 2020: a crisis like no other, an uncertain recovery. https://www.imf.org/en/Publications/WEO/Issues/2020/06/24/WEOUpdateJune2020. Accessed 06 Mar 2021
19. Europe's pandemic politics: how the virus has changed the public's worldview—European Council on Foreign Relations. https://ecfr.eu/publication/europes_pandemic_politics_how_the_virus_has_changed_the_publics_worldview/. Accessed 06 Mar 2021
20. ASEAN solidarity and response in the face of COVID-19—Opinion. The Jakarta Post. https://www.thejakartapost.com/academia/2020/06/02/asean-solidarity-and-response-in-the-face-of-covid-19.html. Accessed 06 Mar 2021
21. The Future of U.S. Foreign Policy. Carnegie Connects—YouTube. https://www.youtube.com/watch?v=_YCW_HiYwek. Accessed 06 Mar 2021
22. Freedom in the World 2020: a leaderless struggle for democracy. https://freedomhouse.org/report/freedom-world/2020/leaderless-struggle-democracy. Accessed 06 Mar 2021
23. Beirut Institute Summit e-Policy Circle 7—YouTube. https://www.youtube.com/watch?v=TFdpomyjjyY. Accessed 06 Mar 2021

24. Bloomberg. https://www.bloomberg.com/news/audio/2020-06-24/bloomberg-daybreak-june-24-2020-hour-2-radio. Accessed 06 Mar 2021
25. Conversations on COVID-19 and development: Lawrence summers. Center for Global Development. https://www.cgdev.org/event/conversations-covid-19-and-development-lawrence-summers. Accessed 06 Mar 2021
26. How to protect American democracy from pandemics. https://www.foreignaffairs.com/articles/united-states/2020-06-01/more-resilient-union. Accessed 06 Mar 2021
27. Why international institutions failed to contain the coronavirus pandemic. https://www.foreignaffairs.com/articles/world/2020-06-09/when-system-fails. Accessed 06 Mar 2021
28. Is it an end to the intellectual world? [translation]. Sariolghalam, Mahmoud. https://sariolghalam.com/2020/02/24/1889/. Accessed 06 Mar 2021
29. Sariolghalam, Mahmoud. Dec 2019. https://sariolghalam.com/2019/12/10/. Accessed 06 Mar 2021

Chapter 9
Covid-19 and Cooperation/Conflict in International Relations

Heidarali Masoudi

9.1 Introduction

Based on a TRIP survey among the International Relations (IR) scholars in the United States, about 65% of the respondents believed that international cooperation in the fight against the COVID-19 crisis has not been sufficiently efficient in coping with the problem [1]. In other words, about two-thirds of the IR elites in the United States think that COVID-19 has had a destructive effect on the deep layers of cooperation among the international actors. The paper thus seeks to answer how and to what extent COVID-19 crisis will affect international cooperation/conflict. The author's main thesis is that the COVID-19 crisis will increase the divergent and contentious trends in the short term and enhance the cooperative trends on the international scene in the longer term.

To answer the main question of this article, two issues must be addressed: the nature and history of pandemic in the world, and the nature of conflict and cooperation in the international area prior to the current pandemic. In other words, we can study the effect of COVID-19 pandemic as an independent variable on the possibility of international cooperation and conflict as a dependent variable.

Throughout history, human being has usually faced several pandemics such as plague, cholera, and many other communicable diseases which have killed millions of people worldwide. One of the factors that caused armed conflicts between the city-states of Athens and Sparta was the spread of such communicable and fatal diseases [2]. In the recent century, there has been some pandemic such as the Spanish flu, which happened after World War I, and HIV, SARS, and Ebola in the recent decades. The Spanish flu spread to the world during WWI, and as a result, millions of people died, especially because of the inability of the countries involved in the war to cope with it effectively and to mobilize the necessary resources to eradicate it [3, 4].

H. Masoudi (✉)
Faculty of Economics and Political Science, Shahid Beheshti University, Tehran, Iran
e-mail: h_masoudi@sbu.ac.ir

However, in recent times, we witness that combating HIV and Ebola has become a starting point for international cooperation. HIV has been one of the most dangerous communicable diseases in the twentieth century, especially in the poor countries. The spread of HIV hit many countries in the world, especially poor African countries. However, many inflicted countries used all their resources to fight with and prevent this disease by producing or receiving vaccines to provide health care and protection for their populations. The World Health Organization (WHO) utilized all its expertise to play the global role in fighting HIV and succeeded in this undertaking. Of course, there were some obstacles in eradicating HIV, such as dysfunctionality of production and distribution of drugs and commercial attitude of pharmaceutical companies. However, in general, it is fair to say that the international cooperation in fighting and preventing HIV has been improved during the last decades [5]. Concerning cooperation and conflict between countries in fighting COVID-19, however, we need to first analyze the context of cooperation and conflicts between countries before the COVID-19 pandemic, and then add the COVID-19 as the new factor to carefully study its consequences and effects.

9.2   International Cooperation/Conflict Prior to the COVID-19 Crisis

The COVID-19 crisis happened in an era of revival of nationalism and decline of international cooperation. Trumpism in the US, the UK's exit from the EU, and the expansion of far-right ideology in many European countries have been the most important signs of this turning point in international affairs. Trump won the election in 2016 with the slogans of "America First" and "Make America Great Again" and did its best to destroy the foundations of liberal international order which have been designed by the US after WWII. The United States' withdrawal from the Trans-Pacific Partnership (TPP), the Human Rights Council, UNESCO, the Paris Climate Agreement, and the Iran nuclear deal are examples of the U.S. anti-internationalism in Trump era. Europe also faced a powerful wave of right-wing extremism. The consolidation of the social and political status of far-right parties in many European countries, including Germany, Austria, France, and Italy, indicates the widening gap that already existed in the EU as the most important and largest project of cooperation and convergence between countries.

The wave of right-wing extremism in the world promotes the idea that the global trends, especially in the economic realm, have led to increased socioeconomic vulnerabilities in many countries. For these vulnerabilities to be prevented, countries have to regulate economic globalization in a way that nation-states can exercise an effective control on the movement of capital, goods, and labor within national borders. As a result, nation-states can plan the society and economy in accordance with the local demands, needs, and necessities rather than global ones. From the nationalistic viewpoint, radical commitment to globalization caused many significant economic and social anomalies for many countries around the world. For instance, it resulted in decreasing demographic and human solidarity and homogeneity in the societies, caused by continuous waves of immigration and high level of mobility of the workforce beyond the national borders [6]. Therefore, the identity of the local communities in European countries has completely changed, and their homogeneous social structure has been frighteningly shaken by the emergence of xenophobic thoughts [7]. Beyond social solidarity, the inappropriate dependence on the global supply chain can have negative long-term socioeconomic impacts and also political effects on the societies. In order for the countries to be less vulnerable during global crises, they have to be able to produce their vital commodities and goods based on the internal production mechanism rather than complete reliance on global supply chains [8, 9].

The COVID-19 crisis showed that imbalanced and asymmetric reliance on the unrestrained globalization trend can result in the destruction of the indigenous economic structures, and consequently, under emergent situations may minimize the possibility of having access to urgent goods and commodities. Therefore, the globalized economies suffer from inadaptability to critical urgent conditions, and this can be a warning sign for many countries connected to the global economy. Although the global divergences on political, economic, and cultural issues were being strengthened even before the COVID-19 crisis, travel ban, quarantine, and trade problems weakened the global economy at least in the short term [10]. In this context, many countries affected by the COVID-19 crisis, including Iran, are thinking of reconstructing the endogenous structures of their national economy and producing the necessary commodities internally [11].

Therefore, the main question is whether or not the impact of COVID-19 would be limited only to deteriorating the gaps that already existed in the globe or it can have deep and long-enduring consequences for international affairs. A theoretical framework is required to answer such questions. The mainstream IR theories can play an important role in our understanding of the COVID-19 crisis and its consequences for international cooperation. In the following, firstly, a realist view will be discussed with its emphasis on nation-state power, economic isolationism, the rise of nationalism, and anti-globalization viewpoints. Secondly, a liberal view will be introduced emphasizing the importance of global and international cooperation in response to newly emerging unconventional threats in borderless world. Finally, the constructivist view will be discussed with an emphasis on the emergence of new transnational identities, redefinition of the fundamental concepts of IR, and the importance of narratives and representations on the international scene.

9.2.1 Realist Perspective

Realism emphasizes the centrality of nation-state, national interests, and state power in international relations. Based on this assumption, it is expected that states will assume extraordinary powers to check and control their people in an emergency situation like COVID-19 crisis. Consequently, state power will increase whereas that of the civil society will decrease [12, 13]. Moreover, considering the vulnerability of countries in supplying and distributing strategic goods and commodities during crises, the global supply chain will weaken more than before, especially in health commodities and treatment items. Therefore, countries will strengthen the structure of national production which my lead to the rise of economic isolationism in international society. Some experts have even stated that reverse protectionism in global trade will increase; that is, states will prevent export of necessary and strategic commodities to other countries [14]. According to this pessimistic scenario, anti-globalization trend will be strengthened [15], and disagreements and conflicts between countries will be increased over production of health goods and medicines including COVID-19 vaccines. Consequently, it leads to the ever-increasing intervention of states in the economy to promote appropriate policies, prevent bankruptcy of small and medium-size companies, and more importantly, guarantee production and supply of strategic goods such as pharmaceuticals and medical equipment.

Realists also point to the dysfunctionality and incapability of international institutions, such as UN Security Council and the WHO, in fighting the COVID-19 crisis as a confirmation of their pessimism. Ambiguous and sometimes contradictory recommendations, lack of provision of accurate and timely information and of sufficient budget, politicization and support of some governments, and lack of effective operational tools are among the most important factors intensifying the pessimism regarding UN institutions and agencies, especially the WHO [16]. Moreover, abuse of these institutions by some countries as a political tool, including President Trump's halting U.S. funding for WHO, can paralyze health projects in various countries. In addition, it is expected that COVID-19 crisis will reduce the potentiality of international aid agencies, suspend humanitarian aid, and stop normal vaccination programs, thus, worsening the lives of millions of people around the world and increasing the number of hungry people in the world from 135 to 250 million in near future [17, 18].

The growth in right-wing extremist ideas like xenophobia and anti-immigration attitudes can result in a new wave of violence against migrants and minority groups, thus, weakening the democratic processes in various countries [19]. Based on the statistics published by Freedom House, democratic principles and standards are in decline in many democratic countries. One of the factors intensifying this trend is the increased control of governments on people's social relationships in order to prevent the spread of COVID-19 [20].

In sum, from the realist perspective, the COVID-19 crisis can result in increased conflict between countries in the form of isolationism and economic protectionism, intensified government control over the society, fragility of global supply chain and weakening global institutions, growth in right-wing extremism and democratic deficits in various countries, and intensification of the international conflict in the short term. In contrast to this pessimism, in the following, I will investigate whether or not it is possible to be more optimistic regarding the possibility of international cooperation in long-term.

9.2.2   Liberal Perspective

From an optimistic point of view, the COVID-19 crisis can lead to two important developments in the international relations. The first one is the strengthening of the idea of a borderless world, and the second is the emergence of new institutions and actors. Understanding a borderless world, as interpreted by Richard Haass, means that the internal affairs of any country around the globe strongly affect those of other countries. In fact, we see that domestic and international affairs inter-linked and inter-twined [21]. Therefore, physical borders will not be able to prevent the spread of threats to other countries and regions, and any domestic threat in any country can rapidly turn into a threat to all the countries around the world. In other words, the security of any country is intertwined with that of all other countries, and essentially, all affairs must be managed at a global level. Put differently, as stated by Joseph Nye, even the most powerful countries such as the U.S. and China will not be able to maintain their security alone, and substantial achievements at the international level will not be accomplished by emphasizing slogans such as "America First" [22].

The second important point is the emergence of new international actors and institutions. The consequences of the COVID-19 crisis showed that global health governance requires major improvements and revisions. Despite all its endeavors, the WHO was not very successful in fighting COVID-19. Consequently, countries need to increase their cooperation in strengthening functional institutions. It is also necessary to make better use of scientific achievements in making foreign policies; i.e., the community of experts and even philanthropic institutions and individuals, such as Bill Gates, will have more opportunity to be involved in developing long-term policies for many countries and global institutions.

In sum, liberal internationalists believe that the global COVID-19 crisis crossed all physical borders, and revealed this reality to human beings that future threats will be mainly global in its nature, requiring the unwavering determination of global actors to cooperate with each other and guarantee global access to public goods and also to network commodities. If the world is faced with new threats, it is necessary to use the capacities of the new actors in the framework of new cooperation structures on the international scene. In the next section, I will examine constructivism with its emphasis on ideational and semantic aspects of COVID-19 crisis.

9.2.3   Constructivist Perspective

Constructivists emphasize the importance of identity, narratives, norms, and representations in shaping international realities. It seems that the COVID-19 crisis and its consequences have led to the formation of new international identities and norms which can facilitate or impede international cooperation. Hence, the formation of new transnational identities [23]. Based on the common feeling of hopelessness facing common threats, the necessity of having integrated global rules for health, the possibility of developing new procedures in trade and tourism including strict public health laws in shipping goods overseas and traveling to countries [24], and increases in the share of virtual communications in domestic and foreign affairs are among the most important signs of new identities and practices on the international realm that can lead to novel forms of international cooperation between countries.

Another important aspect of constructivist explanation is the redefinition of IR fundamental concepts such as security, balance of power, and the self/other construct. The self/other construct under the conditions of COVID-19 has undergone a conceptual change. For instance, the other is no longer an adversarial threatening actor, easily can be identified and contained. Rather, the other would be the complex mysterious and unknown virus with a significant destructive power for all countries. Such newly created common threat would be far more dangerous than any military operation for the safety and security of human beings on the globe. So it is safe to say that the COVID-19 crisis is the most important international crisis since WWII, or put differently, it is the greatest twenty-first century crisis [25].

Moreover, the concept of security has become increasingly intertwined with human security. In the post-COVID-19 world, security does not simply and necessarily mean to maintain the physical continuity of societies. As happened to the American aircraft carrier USS America which was infected by COVID-19 [26], it is likely that a lethal virus can paralyze the most modern military equipment and its personnel. Security in the post-COVID-19 world will be strongly influenced by the resilience levels of countries in coping with unpredicted human crises. In addition, the concept of power distribution will undergo semantic change; i.e., countries with a high status in the hierarchy of international balance of power may exhibit a far weaker response compared to those with lower status in this hierarchy. Therefore, distribution of cognitive and ideational power will gain more importance than hard power. In this context, having more power requires intelligent and smart software preparedness in governance for coping with unpredicted threats.

Another aspect of the constructivist approach is the role of narratives and representations in creating ideational constructs that influence the behavior of the actors. Since the preliminary stage of COVID-19 spread, there has been a war of image between different media in different countries to stabilize their own desired narratives and representations on the origin of COVID-19, especially between the two rival powers, i.e., the U.S. and China [26]. In the primary stage of this war, the narrative of the main media in the Western countries was that the ancient Chinese diet can be regarded as the main cause of this pandemic. This anti-Chinese narrative produced a feeling of insecurity from Chinese rising power in the world.

In contrast, the second stage showed a more balance narrative on the origin of COVID-19. According to this narrative, some Iranian, Russian, and Chinese media and news agencies addressed the possibility of the role played by the military and security organizations of the U.S. in building and testing COVID-19 as a biological weapon. In this narrative, the concept of biosecurity, which means coping with the intentional spread of destructive biological agents, is highlighted instead of biosafety, which means tackling the unintentional spread of these agents.

9.3 Conclusion

The main question in this article was that how and to what extent COVID-19 can affect international cooperation/conflict. First, the history of pandemic in the world (for example, during the ancient Greek era) and also in the twentieth and twenty-first centuries shows that they can be contradictory on international affairs, i.e., they can lead to both increased conflict between countries in the near term, and also, prepare the ground for cooperation in the long-term. Then, based on the realist approach, it was shown that the consequences of the COVID-19 crisis increase isolationism, intensified government control over the society, weakened global supply chain and multilateral international institutions, growth of right-wing extremist ideas, and finally, increased conflicts between international actors in the short term. Based on the liberal perspective, it was said that the COVID-19 crisis can strengthen the understanding of a borderless world, common new-emerging threats and global network commodities, and the necessity for efficient and science-based confrontation with international crises among the international actors. This can give us the hope that cooperation between all international actors including governments and philanthropic individuals and specialist institutions for coping with common threats will increase in the medium-term. Finally, based on the constructivist approach, the formation of new international identities and norms in trade and international interactions and also the changes in the IR core concepts, such as security and balance of power and self/other, were emphasized. The COVID-19 crisis indicated that narratives and representations enjoy an important role in making international realities, and with the rising trend in virtual and on-line communications and interactions between communities, it is expected that the role played by these narratives and representations will become substantially greater. Consequently, from the constructivist perspective, we can claim that the consequences of the COVID-19 crisis have led to a redefinition of the ideational contexts of cooperation/conflict between international actors.

References

1. Snap poll: what foreign-policy experts make of Trump's coronavirus response and U.S. standing post-pandemic. https://foreignpolicy.com/2020/05/08/snap-poll-what-foreign-policy-experts-think-trump-coronavirus-response-election/. Accessed 02 Mar 2021
2. The Great Plague of Athens has Eerie parallels to today—The Atlantic. https://www.theatlantic.com/ideas/archive/2020/03/great-plague-athens-has-eerie-parallels-today/608545/. Accessed 02 Mar 2021
3. How did the Spanish flu pandemic end and what lessons can we learn from a century ago? Euronews. https://www.euronews.com/2020/06/03/how-did-the-spanish-flu-pandemic-end-and-what-lessons-can-we-learn-from-a-century-ago. Accessed 02 Mar 2021
4. Harvard expert compares 1918 flu, COVID-19—Harvard Gazette. https://news.harvard.edu/gazette/story/2020/05/harvard-expert-compares-1918-flu-covid-19/. Accessed 02 Mar 2021
5. Global AIDS monitoring 2019. https://digitallibrary.un.org/record/3801751?ln=en. Accessed 02 Mar 2021

6. Far-right politics in Europe—Jean-Yves Camus, Nicolas Lebourg. Harvard University Press. https://www.hup.harvard.edu/catalog.php?isbn=9780674971530. Accessed 02 Mar 2021

7. Ariely G (2012) Globalization, immigration and national identity: how the level of globalization affects the relations between nationalism, constructive patriotism and attitudes toward immigrants? Gr Process Intergr Relat 15(4):539–557. https://doi.org/10.1177/136843021143 0518

8. Supply chains and the coronavirus—the Atlantic. https://www.theatlantic.com/ideas/archive/2020/03/supply-chains-and-coronavirus/608329/. Accessed 02 Mar 2021

9. Covid-19 pandemic exposes severe weaknesses in U.S. domestic manufacturing. Barron's. https://www.barrons.com/articles/domestic-manufacturing-coronavirus-supply-chain-emerge ncy-supplies-masks-51586549335. Accessed 02 Mar 2021

10. Isabelle F (2020) OECD interim economic assessment coronavirus: the world economy at risk

11. Financial Tribune. Financial Tribune. https://financialtribune.com/?__cf_chl_jschl_tk__= 45c620c239df81b5a7b947c65a84ad7790de139f-1614696861-0-AYdXiCbwoqIo1x0Dqp NsWLfXsoXJLU9rTddmPLnWEu0QDUu7qzFKCsQmi10jB3FND--Boq8sEpBqlo3qfPyS3 9SHjsY6nH-bAjlgynfPeLRtJ19wnPdOPzRM9ymEbDWrpUJsUnChDetl8nWtAcgh4hw. Accessed 02 Mar 2021

12. The Coronavirus Bill: extraordinary legislation for extraordinary times. The Institute for Government. https://www.instituteforgovernment.org.uk/blog/coronavirus-bill. Accessed 02 Mar 2021

13. State of emergency: how different countries are invoking extra powers to stop the coronavirus. https://theconversation.com/state-of-emergency-how-different-countries-are-inv oking-extra-powers-to-stop-the-coronavirus-134495. Accessed 02 Mar 2021

14. Hufbauer GC, Jung E (2020) What's new in economic sanctions? Eur Econ Rev 130:103572. https://doi.org/10.1016/j.euroecorev.2020.103572

15. COVID-19 could spur automation and reverse globalisation—to some extent. VOX, CEPR Policy Portal. https://voxeu.org/article/covid-19-could-spur-automation-and-reverse-globalisa tion-some-extent. Accessed 02 Mar 2021

16. The WHO v coronavirus: why it can't handle the pandemic. World Health Organization. The Guardian. https://www.theguardian.com/news/2020/apr/10/world-health-organization-who-v-coronavirus-why-it-cant-handle-pandemic. Accessed 02 Mar 2021

17. Global hunger increasing even before pandemic, UN warns, putting Zero Hunger 2030 target in doubt. The Independent. The Independent. https://www.independent.co.uk/news/world/global-hunger-coronavirus-pandemic-united-nations-zero-hunger-2030-a9616186.html. Accessed 02 Mar 2021

18. Opinion. The post-coronavirus global hunger crisis—The New York Times. https://www.nyt imes.com/2020/06/12/opinion/coronavirus-global-hunger.html. Accessed 02 Mar 2021

19. U. Nations, COVID-19: UN counters pandemic-related hate and xenophobia. United Nations. [Online]. https://www.un.org/en/coronavirus/covid-19-un-counters-pandemic-related-hate-and-xenophobia. Accessed 02 Mar 2021

20. Spotlight on freedom: impact of coronavirus on basic freedoms. Freedom House. https://freedo mhouse.org/article/spotlight-freedom-impact-coronavirus-basic-freedoms. Accessed 02 Mar 2021.

21. The coronavirus pandemic will accelerate history rather than reshape it. Foreign Affairs. https://www.foreignaffairs.com/articles/united-states/2020-04-07/pandemic-will-acc elerate-history-rather-reshape-it. Accessed 02 Mar 2021

22. Leadership in China and US both 'failed' in coronavirus response. https://www.cnbc. com/video/2020/05/11/leadership-in-china-and-us-both-failed-in-coronavirus-response.html. Accessed 02 Mar 2021

23. Global citizenship: COVID-19 global pandemic, global identity, and literature—NCTE. https:// ncte.org/blog/2020/04/global-citizenship-covid-19/. Accessed 02 Mar 2021

24. Tourism policy responses to the coronavirus (COVID-19). http://www.oecd.org/corona virus/policy-responses/tourism-policy-responses-to-the-coronavirus-covid-19-6466aa20/. Accessed 04 Mar 2021

25. In conversation with Ambassador Nicholas burns on the Covid-19 crisis. Belfer Center for Science and International Affairs. https://www.belfercenter.org/publication/conversation-amb assador-nicholas-burns-covid-19-crisis. Accessed 02 Mar 2021
26. US Navy destroyer returning to port with coronavirus outbreak—The Diplomat. https:// thediplomat.com/2020/04/us-navy-destroyer-returning-to-port-with-coronavirus-outbreak/. Accessed 02 Mar 2021

Chapter 10
Iran's Foreign Diplomacy During the COVID-19 Pandemic

Amir Mohammad Haji-Yousefi

10.1 Introduction

Despite doubts about its ideological agenda, the Islamic Republic of Iran has adopted a realistic paradigm in its foreign policy in maintaining national security. In its traditional sense, national security mainly encompasses the persistence of the political system and geographical borders; however in its modern implications, as human security, it also includes population survival. In other words, the main goals of a country's foreign policy should be the survival of the state and society. In the geopolitical competition arena, which is the traditional agenda of any foreign policy, Iran has been facing serious security threats in West Asia. Perhaps the most important rivalry is between Iran and Saudi Arabia over regional supremacy, but Iran's regional threats can be traced back to its hostile relations with Saudi Arabia, on the one hand, and with the United States and Israel, on the other. The outbreak of COVID-19 and its global spread since the beginning of 2020 has forced Iran's security agenda to focus, in addition to traditional geopolitical rivalries, on a new security threat that has directly targeted the lives of its citizens. Accordingly, Iran's foreign policymakers inevitably had to take measures in order to achieve national security in the new context.

This study seeks to examine Iran's foreign diplomacy after the COVID-19 outbreak, particularly since March 2020, when the first symptoms of it appeared in Iran. The question is, what steps has Iran taken to achieve its national security, i.e., the survival of its people? In other words, what are Iran's diplomatic measures to protect its people against the Coronavirus and how are they evaluated?

This study's main claim is that Iran predominantly used "naming and shaming" diplomacy to show international sanctions, which were inhumane, more inhumane

A. M. Haji-Yousefi (✉)
Faculty of Economics and Political Science, Shahid Beheshti University, Tehran, Iran
e-mail: am-yousefi@sbu.ac.ir

© The Author(s), under exclusive license to Springer Nature Singapore Pte Ltd. 2021   253
M. Rahmandoust and S.-O. Ranaei-Siadat (eds.), *COVID-19*,
https://doi.org/10.1007/978-981-16-3108-5_10

and brutal due to the outbreak of the COVID-19 and attract the attention of other international actors, including states and international organizations to address the issue. At the same time, like other states, Iran adopted a domestic policy in confronting the COVID-19, and as the role of the government became more prominent, it tried to rely more and more on its domestic economic, scientific, and technological power to fight against the COVID-19. The increased maximum pressure by the United States strengthened this introspective approach and consequently, Iran's foreign policy and diplomacy were affected.

In order to answer our question and examine the claim raised in this study, the new regional and international context after the outbreak of the Corona must be studied and its features must be analyzed in order to better realize and evaluate Iran's diplomatic actions. On the one hand, in regional and international arenas, the Corona pandemic has increased solidarity and sympathy among countries. Accordingly, cooperation among countries, as well as the role of international organizations, was expected to increase. This could have led states and international organizations to clearly express their objection to the prolongation of international sanctions on Iran, particularly the sanction imposed by the US. In addition, it was expected that different state parties to the JCPOA (Iran nuclear deal) insist on the need to maintain it. However, in the international arena, traditional geopolitical disputes extended. The decline of the American leadership and the liberal international system, which was evident before the COVID-19 pandemic, became more tangible. The inefficiency of the US and European states in their initial response to the coronavirus confirmed such weakness. At the same time, China's role became even more important. Not only did China promptly manage COVID-19, but it pursued an active policy through "mask diplomacy" and intended to show that it could be a viable alternative to US withdrawal from the leadership of the international system led by Trump.

In the West Asian region, countries are still concerned about traditional geopolitical competition. The strategic competition between Iran and Saudi Arabia and the formation of various alliances by these two major actors in West Asia have led to an unstable bipolar system in this region in which disputing parties are constantly seeking to maintain and expand their sphere of influence from Libya to Afghanistan. The outbreak of the Corona pandemic made no changes in this competition. The war in Yemen and Syria continues, and there are significant conflicts and rivalry in other regions such as Iraq, Lebanon, Afghanistan, and the Persian Gulf.

10.2   International and Regional Systems Under the Shadow of the COVID-19

Many international relation scholars believe that the outbreak of the Corona pandemic has substantial and lasting effects on human life, including the international system [1]. As a result, three scenarios have been proposed for the future of the international system. The first scenario is based on the continuation of past trends. Perhaps the expression "Back to the Future" can be applied to this scenario. This scenario, which is more prevalent among adherents to the realist school in International Relations, argues that the continued conflict/rivalry in the international system, particularly between the United States and China, the continued European divergence, as well as increased regional conflicts, especially in West Asia, and the continued weakening multilateralism in the international organizations, will remain the most important features of the international relations in the Corona and post-Corona pandemic era. Kishore Mahbubani, the prominent Singaporean diplomat, believes that the COVID-19 epidemic will not change global economic direction fundamentally but will accelerate the move from US-centered globalization to China-centric globalization. John Ikenbery, the prominent international relations thinker, argues that given the economic damages and social collapse emerging as a result of COVID-19, there can be no future for the international system other than strengthened movement towards nationalism, competition among great powers, separation, and non-strategic cooperation of governments. Generally, almost a year after the outbreak of the coronavirus, there has been no serious change in the states' behavior beyond what realists claim, but Trump's defeat in the recent US presidential election could promise the US's return to a more responsible role for the management of the great powers.

Using the past experiences of infectious pandemics, the second scenario is about the positive impact of COVID-19 on international relations and systems as increased cooperation among different actors. This view, which is more prevalent among liberal theorists of international relations, by dividing goods into private, public, group, and network goods, argues that immunity against COVID-19 is neither a private good that a particular country and community, such as a given global superpower, can insure itself against (e.g., a nuclear weapon), nor a public good provided by some and used by others for free (e.g., security), nor a group good used by a specific group like rich countries (e.g., NATO), rather it is a network good that the more it is provided, the better for everyone, and it is owned by everyone or no one. In other words, immunity against infectious diseases is a network good that even the weakest links in the chain should benefit from; otherwise, no one can guarantee their immunity [2]. According to the liberals, the only possible policy to counter the COVID-19 pandemic is cooperation among states, and therefore COVID-19 pandemic can promise a participatory and cooperative order at the international and regional systems. Although the claim that health and safety are network goods is correct, it did not have the slightest effect on the US's behavior and policy of maximum pressure on Iran during the Corona pandemic.

The third scenario emphasizes the formation of a new international order and claims that with the withdrawal of the United States and the West from the leadership of world affairs, which shows weakness (America) and divergence (Europe) on the one hand and China's rise on the other, there will be a formation of a multipolar system with an Asian flavor as a new Cold War. The transition from the Western international system to an Asian international system with such powers as China, Russia, and India has precedence in the international relations literature, but the new claim is that COVID-19 will accelerate this process. Accordingly, the post-Corona world promises a new Cold War at the international level, marked by the divergence of Europe, the withdrawal of the United States from world leadership, and the rise of China. From the perspective of Iranian foreign policymakers, this scenario seems to be closer to the current reality of the international system. This new emphasis in Iran's foreign policy on strategic relations with China and Russia, as well as adopting a new version of the policy towards the East (Eurasian-ism), has become more evident.

From the author's point of view, the current reality of international relations and the system in the COVID-19 pandemic era can be depicted as follows. First, contrary to widespread expectations, intergovernmental cooperation could not become a major feature of international relations. On the one hand, the mistrust between China and Western societies was consolidated. According to The Economist, quoting Pew, Western countries' trust in China has significantly declined [3]. China was accused of hiding the outbreak of COVID-19 in Wuhan in 2020 and suppressing any revelations by domestic or foreign forces, which has exacerbated the disease and its worldwide spread. The Chinese government was not only forced to violate human rights in the fight against COVID-19 and as a result, implemented some repressive policies [4] but also imposed trade sanctions on some countries, such as Australia, which called for a global probe into the origins of COVID-19 [5]. This led some Western countries, particularly the United States and Trump himself, to try to call COVID-19 a Chinese virus and to insist on its Chinese nature all the time in interviews and meetings such as the G20 summit [6]. China, on the other hand, sought to both improve its global image and counter the policy of its main rivals through the mask diplomacy [7]. However, this did not weaken the international cooperation, and some countries openly acknowledged China for its assistance in confronting COVID-19.

There were also differences among regional states. For instance, COVID-19 made a gap among EU member states. The North European countries did not cooperate in assisting Italy and even France, Germany and the Czech Republic prevented the export of medical equipment to Italy. In the Persian Gulf region, however, countries such as Kuwait, Qatar, and the United Arab Emirates sent healthcare products to Iran. Trump's America, as a global superpower, had no interest in playing a leading role in resolving crises and advancing international affairs. In fact, compared to the post-World War II period until 2008, the US's international role in resolving most of the crises since the war in Georgia in 2008 has been significantly reduced [2]. At the same time, strengthening multilateralism and international organizations for global governance was not on Trump's agenda, and China sought to fill the vacancy produced by the US withdrawal from the global leadership [7]. It can be said that the move towards a multipolar international system is quite obvious, in which the US-Russia-China triangle will play a key role. In the West Asian region and Iran's immediate environment, there will be no serious change in the regional geopolitical game in the short term, but in the long run, the United States, Russia, and China are in favor of stability in the region, and Iran will have no choice but to take this path and adapt its foreign policy to the requirements of this new setting.

10.3 Iran's Status in the International System During Corona Pandemic

Fifty days after the corona was announced epidemic in China, Iran was the second country to announce on February 18, 2020, that two people had died from the Coronavirus in Qom [8]. Iran, which was under the harshest sanctions, particularly after Trump withdrew the U.S. from the JCPOA on May 8, 2018, faced another tragedy: The Coronavirus. Both Iran's economy and the lives of its people were under the most severe threats. As a result of the US's maximum pressure policy, which was adopted as the main US policy towards Iran since the withdrawal of the United States from JCPOA, Iran's crude oil exports declined rapidly and reached below 10% of exports before Trump's withdrawal, and consequently, Iran's foreign exchange income decreased. Additionally, the value of Iran's national currency fell by more than two-thirds, followed by rampant inflation (41% increase according to 2019 prices) in the Iranian economy. According to the estimates, Iran's economy during the onset of the corona pandemic was in one of its worst conditions after the Islamic Revolution, even compared to Iran's economic situation after the end of the Iran-Iraq war [9]. The new sanctions, driven by a policy of maximum pressure on the one hand and a corona pandemic (impact on the economy by reducing production and unemployment of millions of people) on the other, brought Iran's economy to the verge of collapse.

This made the Rouhani administration seek the solution inside rather than outside Iran. Furthermore, achieving national interest through interaction with the West, which was a roadmap in Rouhani's diplomacy, was replaced by strengthening interactions with the East, particularly neighboring countries, including closer ties with China and Russia, strengthening Eurasianism (signing Iran's preferential trade agreement with the members of the Eurasian Economic Union (EAEU) on 27 October 2019), emphasis on trade with the East as a way out of Western trade restrictions, emphasis on trade and regional ties with neighboring countries, and efforts to sign preferential trade agreements with neighboring countries such as ECO [10].

10.4 Iran and "Naming and Shaming" Diplomacy

The Corona pandemic caused significant changes in conventional diplomatic processes and practices, including the strengthening of digital diplomacy compared to other methods and dimensions of diplomacy, and all countries, including Iran, were forced to use digital diplomacy to achieve their goals and interests. Diplomacy that is establishing peaceful contacts among official diplomats of different countries has been one of the most important ways of establishing cooperation and peace in international politics. The digital revolution has highlighted digital diplomacy, which means the use of social media to achieve foreign policy goals, and has changed the way information is acquired and managed, as well as the planning, decision-making, and management of foreign policy, including crisis management by diplomats. Although digital diplomacy has its advantages and is sometimes considered a revolution in diplomacy [11], it also had negative consequences, including reduced interpersonal relationships that, according to an Iranian diplomat, is like "blood in the course of its conventional diplomacy" (interview with Mr. Ghaebi, an Iranian diplomat in Austria, January 6, 2021). Accordingly, the possibility of Iranian diplomats participating in bilateral and multilateral talks and vital meetings to advance the country's goals, particularly in the face of escalating sanctions, was reduced and this made things more difficult for Iran (interview with Mr. Ali bakhshi, the Iranian diplomat in Madagascar). Iran is a country that is under the maximum pressure of a global superpower and therefore must adopt active diplomacy, therefore, face-to-face meetings, both bilaterally and multilaterally, could better facilitate the exchange of information in providing cooperation and peace through interpersonal communication and creating trust. As an Iranian diplomat states, "Even countries that do not have any problems, that is, they adopt a completely calm and conservative diplomacy, have encountered difficulties in performing their diplomacy, let alone Iran, which is in difficult circumstances as a result of sanctions" (interview with Mr. Alibakhshi, the Iranian ambassador in Madagascar, October 26, 2020).

In fact, Iranian officials and diplomats, like other diplomats, have become accustomed to the traditional way that is based on physical presence in negotiations, and therefore the Corona pandemic caused a shock in diplomacy and foreign policy of countries including Iran. The COVID-19 pandemic has made face-to-face meetings and discussions at various levels between Iranian officials, diplomats, and envoys with officials, diplomats, and representatives of other countries problematic and in some instances impossible. This added to the dual problems of sanctions and the COVID-19 pandemic for Iran. In fact, the triangle of sanctions, the COVID-19 pandemic, and the difficulty of face-to-face negotiations made Iran's critical situation even more apparent. In order to reduce or mitigate the pressures of sanctions, Iran had to have an active diplomacy. The most important thing in conducting diplomacy is to have sufficient skills by diplomats, but Iran has had serious problems in this regard. In other words, although Iran has had skilled and professional diplomats in different periods, it has lacked strong diplomacy in general, and in the Corona era, due to the need to change from face-to-face diplomacy to virtual diplomacy, it has certainly had newer obstacles.

Apart from the diplomats' skills, Iran should use different platforms and media to convey its message, and the lack of balanced and fair access to them added to this difficulty. In other words, the lack of professional diplomats who can use social media to try to change the positions of other countries to reduce the impact of sanctions on the one hand, and the lack of balanced and fair access to social media, mainly in the West, on the other, shows the difficulty of implementing the foreign policy goals of many countries, including Iran (interview with Ghaebi, an Iranian diplomat in Austria, January 6, 2021). Lack of proficiency with digital diplomacy, particularly lack of necessary skills to use various platforms such as Skype, Zoom, and Hangouts as well as the impossibility of using some of these platforms by countries such as Iran due to US sanctions, is among examples of complications of digital diplomacy. Moreover, the stability and security of such platforms are the concern of diplomats from different countries, such as Iranian diplomats. According to Mr. Alibakhshi, Iran's diplomat in Madagascar, "The issues of countries, including ours, is generally confidential and not issues that can be raised via telephone, the Internet, social networks, and cyber communications" (interview with Alibakhshi, the Iranian ambassador in Madagascar, October 26, 2020). In his view, countries cannot raise difficult and classified issues in public meetings, and therefore this makes it difficult to implement diplomatic relations and, consequently, the efforts towards a country's national goals become ineffective. One of the characteristics of diplomacy is its secrecy and confidentiality, and therefore, the COVID-19 has made diplomacy an issue. As Mr. Dehshiri, Iran's ambassador to Senegal, asserts, "The Corona era reduced face-to-face contact. Instead, the emphasis has been on cyberspace and webinars. This reduces the intimacy of the relationships. In any case, there is no solution such as face-to-face meetings and contacts because many classified issues are raised in face-to-face and private meetings. However, in cyberspace, due to the possibility of hacking, the confidential diplomacy has practically decreased and transparent and formal diplomacy is more visible" (interview with Dehshiri, Iran's ambassador to Senegal, October 25, 2020). However, according to Iranian diplomats, adopting digital diplomacy by Iranian diplomats and officials, particularly by the Foreign Minister, Mohammad Javad Zarif, has increased dramatically since the Corona pandemic.

The most important focus of Iranian diplomacy in the Corona era has been to achieve goals through a "naming and shaming" policy. Many psychologists and sociologists believe that shame can control our lives, and although it has unwritten rules, people (including state leaders) use shame as a guide to action. Thus, in order not to be ashamed, they do not commit immoral acts, such as human rights violations, or apologize if they did something wrong [12]. "Naming and shaming" is a political tool for punishing rivals and rewarding friends in international relations [13]. "Naming and shaming" is expected to lead political leaders and governments to refrain from actions that harm others. The "naming and shaming" policy is usually operated by countries and actors that do not have much international power to force their rivals or enemies to do what they want. In other words, "naming and shaming" is a soft policy that seeks to provoke a type of moral atmosphere in international relations, and the rival or enemy country will inevitably fall short due to its unwillingness to be recognized as a violator of legal and moral rules and regulations and give up its policy. However, this policy may not be effective for reasons beyond the scope of this paper.

In the past year and since the start of the Corona pandemic, Iranian officials have repeatedly tried to make US behavior, particularly that of President Trump, a disgrace to the United States and to use the COVID-19 conditions to call for lifting the sanctions. For instance, some of the most important speeches of Iranian diplomats in the past year, which fall under the "naming and shaming" policy, are listed in Table 10.1.

Indeed, there is no consensus on whether adopting such a policy, "naming and shaming" by Iran, is effective or not. However, it seems that Iran's diplomatic efforts

Table 10.1 Diplomatic quotations under "naming and shaming" policy

Official	Quotation	Date	Source
Dr. Zarif, Foreign Minister of Iran	It is essential to lift all illegal US sanctions in order to fight against Corona. Letter to the Secretary-General of the United Nations	12/03/2020	[14]
Dr. Zarif	In the midst of the Corona epidemic, the US regime is trying to destroy the remaining channels for paying for medicine and food	09/10/2020	[14]
Dr. Zarif	Collusion to starve a nation is a crime against humanity, and those who block our money will be brought to justice	09/10/2020	[14]
Dr. Takht-Ravanchi, Iran's Permanent Representative to the United Nations	Implementing "compulsory unilateral" measures by the United States has "affected our economy" and "severely exacerbated general health conditions during the Corona epidemic."	09/10/2020	[14]
Dr. Zarif	Corona revealed the truth of the oppressive sanctions to the world, and that the sanctions on the health of Iranians were from the very beginning at the core of economic terrorism and sanctions	18/03/2020	[15]
Dr. Baeidinejad the Iranian Ambassador to the United Kingdom	In a Twitter post: Since it is not possible to hold a meeting with the British media due to COVID-19, he wrote a letter asking them to fulfill their humanitarian duty, and while raising public awareness about the inhumane effects of US sanctions on limiting Iran's medical and economic/financial resources to counter Corona, calling for the lifting of such sanctions	18/03/2020	[16]
Dr. Ghasemi, the Iranian Ambassador to France	Iran, a large and historical country with a rich culture, has been subject to unilateral, illegal, inhuman, and immoral sanctions by the United States, particularly Trump, for many years, and this has stopped the exports of medical and healthcare goods and equipment and any financial transfer between Iran by many countries in the world	18/03/2020	[17]

(continued)

Table 10.1 (continued)

Official	Quotation	Date	Source
Dr. Khatibzadeh, Iran's Foreign Ministry Spokesman	The crimes committed by the Americans these days will never be forgotten by the Iranians	19/10/2020	[18]

have sometimes led officials in countries and international organizations on the one hand, and public opinion in countries including the United States on the other, to stand in favor of Iran and against sanctions.[1] In other words, Iran's digital diplomacy, particularly the messages of Iranian officials, was able to persuade global and domestic public opinion (including American public opinion), which was against Trump's specific policies and methods, to oppose his administration's policy of maximum pressure and harsh sanctions against Iran. The author believes that Iran's "naming and shaming" diplomacy, particularly the messages by the Foreign Minister Zarif, first of all, created an atmosphere of empathy and cooperation with Iran, which was manifested in the speeches by the leaders of countries, especially China and Russia, as well as many of Iran's neighbors on the one hand, and some protests in different countries against the US sanctions imposed on Iran on the other. Second, despite the maximum pressure from the United States and very severe sanctions, Iran was able to show its political, economic, and social resilience, and this enabled it in its relations with international organizations, particularly the World Health Organization, as well as bilateral relations with various countries to achieve some of its COVID-19 foreign policy goals, including purchasing and importing vaccines against COVID-19. The Security Council's opposition to the extension of Iran's arms embargo sanctions and the return of sanctions on Iran on August 15 and 25, 2020, respectively, demonstrated Iran's active and successful diplomacy. As the former Foreign Ministry spokesman Mr. Mousavi said, "Over the 75-year course of the United Nations history, the United States has never been so isolated. In spite of all the travels, pressures, and ramblings, the United States was able to attract only a small country to itself" [19].

10.5 Conclusion

As noted, Iran needed to engage in active diplomacy to counter international sanctions and Trump's policy of maximum pressure, but the outbreak of the COVID-19 and the need to rely on distant forms of diplomacy, including digital diplomacy, were a serious obstacle to Iranian diplomats' efforts. However, it seems that in spite of Iran's structural weaknesses in digital diplomacy, in many cases Iran was able to make its voice heard by the other parties, and through "naming and shaming" policy

---

[1] However, Mr. Amir Sherkat, an Iranian diplomat in Germany, disagrees, "The naming and shaming policy has been part of our approach, but due to the supremacy of nationalism and national interests, it has not affected ethics in international relations, particularly after the COVID-19 pandemic" (interview with Amir Sherkat, an Iranian diplomat in Germany, October 25, 2020).

and using the conditions caused by the COVID-19, to achieve some of its goals in changing the behavior of other countries and international organizations, which is a major goal of diplomacy, in favor of Iran.

References

1. The Coronavirus Pandemic Will Change the World Forever. https://foreignpolicy.com/2020/03/20/world-order-after-coroanvirus-pandemic/. Accessed 02 Mar 2021
2. Gardini GL (2020) The world before and after covid-19 intellectual reflections on politics, diplomacy and international relations
3. Unfavorable Views of China Reach Historic Highs in Many Countries I Pew Research Center. https://www.pewresearch.org/global/2020/10/06/unfavorable-views-of-china-reach-historic-highs-in-many-countries/. Accessed 02 Mar 2021
4. China's Covid Success Story is Also a Human Rights Tragedy I Human Rights Watch. https://www.hrw.org/news/2021/01/26/chinas-covid-success-story-also-human-rights-tragedy. Accessed 02 Mar 2021
5. Australia called for a COVID-19 probe. China responded with a trade war - ABC News. https://www.abc.net.au/news/2021-01-03/heres-what-happened-between-china-and-australia-in-2020/13019242. Accessed 02 Mar 2021
6. Donald Trump's 'Chinese virus': the politics of naming. https://theconversation.com/donald-trumps-chinese-virus-the-politics-of-naming-136796. Accessed 02 Mar 2021
7. Verma R (2020) China's 'mask diplomacy' to change the COVID-19 narrative in Europe. Asia Eur J 18(2):205–209. Accessed 01 June 2020 (Springer). https://doi.org/10.1007/s10308-020-00576-1
8. Rassouli M, Ashrafizadeh H, Shirinabadi Farahani A, Akbari ME (2020) COVID-19 management in iran as one of the most affected countries in the world: advantages and weaknesses. Front Public Heal 8, 510. https://doi.org/10.3389/fpubh.2020.00510
9. Iran: the double jeopardy of sanctions and COVID-19. https://www.brookings.edu/opinions/iran-the-double-jeopardy-of-sanctions-and-covid-19/. Accessed 02 Mar 2021
10. Iran determined to continue cooperation with Asian, Eurasian powers: Spox - IRNA English. https://en.irna.ir/news/83884248/Iran-determined-to-continue-cooperation-with-Asian-Eurasian. Accessed 06 Mar 2021
11. Digital diplomacy: theory and practice—1st edn—Corneliu Bjola. https://www.routledge.com/Digital-Diplomacy-Theory-and-Practice/Bjola-Holmes/p/book/9781138843820. Accessed 02 Mar 2021
12. Shame plays an important role in political life—or at least it used to. https://theconversation.com/shame-plays-an-important-role-in-political-life-or-at-least-it-used-to-124755. Accessed 02 Mar 2021
13. Koliev F (2020) Shaming and democracy: explaining inter-state shaming in international organizations. Int Polit Sci Rev 41(4):538–553. https://doi.org/10.1177/0192512119858660
14. "تخت روانچی: هدف آمریکا تغییر نظام ایران است". تخت-روانچی-هدف-آمریکا-تغییر-/https://www.asianews.ir/fa/newsagency/2697 Accessed نظام-ایران 06 Mar 2021
15. "پیام ویدیویی ظریف درباره تحریم های ظالمانه و مبارزه مردم ایران با کرونا/ کرونا، حقیقت تحریم‌ها را برجهان آشکار کرد/ تحریم سلامت ایرانیان و مرگ پراکنی، دستور کار تروریسم اقتصادی". پیام-ویدیویی-ظریف-درباره-تحریم-های-/https://fa.alalamtv.net/news/4803401 Accessed ظالمانه-و-مبارزه-مردم-ایران-با 06 Mar 2021

16. جهان در بحران شدید انسانی ناشی از کرونا/ بعیدی‌نژاد خواستار مشارکت رسانه‌های انگلیس علیه " تحریم‌های ضد ایرانی شد." https://fa.alalamtv.net/news/4803221/-بحران-در-جهانAccessed شدید-انسانی-ناشی-از-کرونا-بعیدی-نژاد-خواستار-مشارکت 06 Mar 2021

17. نامه سفیر ایران به مقامات فرانسه در باره لزوم شکستن تحریم های ضد بشری آمریکا در شرایط " کرونا اپیدمی." https://fa.alalamtv.net/news/4802276/-مقامات-به-ایران-سفیر-نامهAccessed ضد-های-تحریم-شکستن-لزوم-باره-در-فرانسه- 06 Mar 2021

18. نشستAccessed وزارت امور خارجه جمهوری اسلامی ایران- نشست خبری سخنگوی وزارت امور خارجه با رسانه‌های " خارجی و داخلی." https://www.mfa.gov.ir/portal/newsview/614544/- خارجی-و-داخلی-های-رسانه-با-خارجه-امور-وزارت-سخنگوی-خبری 06 Mar 2021

19. شورای امنیت قطعنامه پیشنهادی آمریکا برای تمدید تحریم تسلیحاتی ایران را رد کرد " Euronews. https://per.euronews.com/2020/08/15/un-rejection-of-a-us-resolution-to-extend-the-arms-embargo-in-iran. Accessed 02 Mar 2021

Printed in the United States
by Baker & Taylor Publisher Services